James Sully

The Human Mind

Vol. 2

James Sully

The Human Mind
Vol. 2

ISBN/EAN: 9783337365462

Printed in Europe, USA, Canada, Australia, Japan

Cover: Foto ©berggeist007 / pixelio.de

More available books at **www.hansebooks.com**

A TEXT-BOOK OF PSYCHOLOGY

BY

JAMES SULLY, M.A., LL.D.

EXAMINER IN MENTAL AND MORAL SCIENCE IN THE UNIVERSITY OF LONDON ;
AUTHOR OF " ILLUSIONS," ETC.

IN TWO VOLUMES

VOL. II.

LONDON

LONGMANS, GREEN & CO.

1892

THE HUMAN MIND.

ERRATA.

Vol. II.

Page   1, last line, *for* " are " *read* " is ".

  ,,   9, 3rd par., 10th line, *for* " fowl's eye " *read* " corn ".

  ,,   21, footnote 2, *for* " xer " *read* " xes ".

  ,, 138, 1st par., 2nd line, *for* " attitude " *read* " attribute ".

  ,, 139, 2nd par., 7th line, *for* " form " *read* " forms ".

  ,, 182, footnote 1. 9th line, *for* " Fleichsig " *read* " Flechsig ".

  ,, 188, 2nd par., *for* " Weissmann " *read* " Weismann ".

  ,, 211, last par., 2nd line, *for* " conscious " *read* " a conscious ".

  ,, 263, 2nd par., last line but one, *for* " as a " *read* " a ".

  ,, 278, 2nd par., 11th line, *for* " brought " *read* " are brought ".

  ,, 279, 5th line, *for* " religious " *read* " a religious ".

  ,, 284, 2nd par., 8th line, *for* " dominant " *read* " virtuous ".

  ,, 303, 3rd par., 6th line, *for* " account " *read* " amount ".

  ,, 306, 2nd par. and note, *for* " Weissmann " *read* " Weismann ".

  ,, 349, note 3, last line, *for* " *The development* " *read* " *The origin* ".

CONTENTS.

PART IV.

THE FEELINGS.

CHAPTER XIII.

FEELING: PLEASURE AND PAIN.

*

(c) *Complication of Activities.*

CHAPTER XIV.

FEELING : ITS VARIETIES AND DEVELOPMENT.

Varieties of Feeling.

(A) *Sense-Feelings.*

(B) *Emotions.*

Development of Emotion.

CHAPTER XV.

THE EMOTIONS: INSTINCTIVE EMOTIONS AND SYMPATHY.

(A) *General Feeling of Happiness and Misery.*

CHAPTER XVI.

THE EMOTIONS (*Continued*): ABSTRACT SENTIMENTS.

(C) *Representative Emotions:*

(2) *Abstract Sentiments.*

CONTENTS.

PART V.

CONATION OR VOLITION.

CHAPTER XVII.

VOLUNTARY MOVEMENT.

CHAPTER XVIII.

COMPLEX ACTION: CONDUCT.

CHAPTER XIX.

CONCRETE MENTAL DEVELOPMENT: INDIVIDUALITY, NORMAL AND ABNORMAL
PSYCHOSES.

xii   CONTENTS.

PART IV.

THE FEELINGS.

CHAPTER XIII.

FEELING: PLEASURE AND PAIN.

Having now reviewed the successive stages of the development of intellection or cognition, we may pass on to consider the development of the second of the three phases of mind, namely, the affective phase or feeling.

§ 1. *Feeling and its Importance.* As already pointed out, we include under the head of feeling all psychical states or phenomena so far as they have the element or aspect of the agreeable and disagreeable. How far such agreeable or disagreeable aspects extend, how they are related to the presentative element of our consciousness, and whether they cover everything which is properly called feeling, are questions to be considered presently.

This preliminary rough demarcation of the region of feeling may help us to see its peculiar significance as an aspect of our mental life.

To begin with, feeling marks off the *interesting* side of our experience. External objects only have a value for us when they touch our feelings. Mere cognition of an object may leave us cold, but the appreciation of its beauty, involving a wave of pleasure, warms and thrills us. It is evident that what we mean by happiness, and its opposite, unhappiness or misery, are made up of elements of feeling. We are happy so far as we are the subject of pleasure, unhappy so far as the subject of pain.   Our estimate of things and of human life as a whole will thus depend on the experiences which come under the head of feeling.

Again, feeling is subjective experience *par excellence*.   In all perception of, and all thought about, objects we are in the "objective attitude," that is, representing a world of common cognition.   Our actions, too, involve changes carried out in the external world, and so have an objective aspect.   But our feelings, save in their external manifestation, are all our own. To be affected by joy or by sorrow, to fear or to hope, is to have an experience which we detach from the object-world and refer to the inner world of self.   Feeling in all its higher and developed forms stands in close connexion with self-conscious- ness, though owing to the great fact of sociality and the interchanges of feeling in sympathy, this consciousness of self tends to merge in a larger consciousness of a common emotion.[1]

While feeling has thus a special intrinsic interest as subject of study it has a further extrinsic interest because of its bearing on the other aspects of our mental life.   The interactions of feeling on the one side with intellection and conation on the other will be more fully considered by-and-by.   Here it will be enough to say that the cultivation of the feelings stands in close organic connexion with that of intelligence.   To develop the powers of observation and of thought is to awaken interests, that is to say, to excite and raise to the position of strong incentives certain varieties of feeling.   The cultivation of feel- ing connects itself on another side in the closest way with the development of volition.   As we shall see by-and-by, the prompting forces in our voluntary action are feelings.   We exert ourselves for the sake of some future gratification of feeling, as pride or love.   Hence the consideration of the feelings connects itself closely with that of conation, and has, indeed, by some been altogether comprehended under this head.

---

[1] *Cf.* above, p. 65 f.   The objective reference in many of our emotional states, as when we say we fear, love, admire an object, is, as we shall see, bound up with the representative factor in emotion.

§ 2. *Definition of Feeling.* We may now seek to mark off the element of feeling more precisely by examining into its essential characteristics.

By common consent psychical states that are distinctly pleasurable, or the opposite, come under the head of feelings. Thus, to take the lower region of " bodily " feeling, it is generally agreed that the pain of a burn, or the pleasure of quenching thirst, is properly described as a feeling. So in the higher region of " mental " feeling or emotion, the joy of success, the pain of bereavement, are recognised by all as feelings.

In addition to such well-marked cases of pleasurable and painful consciousness we have to include under the head of feeling every psychical state so far as it has any agreeable or disagreeable aspect, however slight. Thus every consciousness of a difficulty or hitch in an operation, whether bodily or mental, is as such disagreeable, and so falls under the head of painful feeling. When then we say that feeling consists of all varieties of pleasurable and painful consciousness we must be taken to use these terms more widely than is done in popular discourse. A pleasure is any degree of agreeable consciousness which as such contents us, and is voluntarily held to ; a pain any degree of disagreeable consciousness which as such discontents us, and is voluntarily repelled.

It is evident that this comprehensive use of the terms pleasurable and painful enables us to say that most of our common experiences are coloured by some degree of feeling or affective ingredient. Thus a close introspective observation tells us that sensations are in the large majority of cases, if not universally, accompanied by an element of feeling.[1] The same is true of all modes of ideation. To imagine and to think are not processes altogether colourless as regards feeling, but on close inspection can be seen to have a shade of the agreeable or the disagreeable. Whether, however, the affective phase is absolutely universal is a difficult point that will be best considered presently in connexion with the discussion of the relation of feeling to the presentative element.

---

[1] James Mill is consequently wide of the mark when he says that probably the greater number of our sensations are indifferent. (*Analysis*, ed. by J. S. Mill, ii. p. 184.)

2a. *Feeling and Excitement.* One point in the demarcation of feeling is so important that it must be specially considered forthwith. It is said that pleasure and pain as defined above do not exhaust all that we mean by feeling, that in addition to these opposed modes there is a third mode, *viz.*, neutral feeling or bare, colourless excitement. Thus it is said that the primitive experience of shock, which later develops into the feeling of surprise or wonder, is an example of such a neutral or indifferent feeling.

The determination of this point is one of the *cruces* in psychology. Without attempting an exhaustive discussion of it, one may suggest that the term excitement here covers two distinct things. It is used, first of all, to denote mere intensity and range or amplitude of presentative or ideational element. Thus a loud sound merely as occasioning an unusually intense auditory sensation, together with secondary effects in the shape of a rush of ideas, may be called exciting without reference to the specific variety of experience known as feeling. But, in the second place, it may be said that no known mode of excitement is absolutely colourless, and that the meaning of the term as commonly employed has reference to the properly affective element which enters into it. Much of what we call excitement is distinctly pleasurable, as is seen in the use of the expression " love of excitement," and in people's eagerness in the pursuit of it. Other forms, *e.g.*, a sudden shock, are, as sudden and confusing, distinctly disagreeable. Lastly, it must not be forgotten that states of feeling are rarely pure pleasures or pure pains. The complexity of our mental life produces a constant intermingling of affective elements. Thus a surprise is commonly at once disagreeable and agreeable : disagreeable as producing momentary disturbance of the mental mechanism, agreeable as a new, unexpected experience which as such exhilarates us, stimulating us to more intense thought-activity. In many of our affective states, as we shall see, there is this meeting of the opposed elements of the agreeable and the disagreeable ; and the collision of the two, *e.g.*, in the love of the marvellous, in the pleasure of pathetic story and tragedy, constitutes a part, and a very important part, of what is known as excitement.

We appear justified then in saying that feeling proper is

nothing but the various shades of the agreeable and the disagreeable, apart and in their comminglings.

The assumption of neutral or indifferent feeling which may be found in an
implicit form in a number of psychological writings, *e.g.*, those of Beneke and
T. Reid, is most explicitly made by Prof. Bain. (*The Emotions and the Will*, 3rd ed.,
p. 13 ff.) He succeeds in showing that feeling as pleasure or pain varies greatly
in its degree, being in many cases inconspicuous, and further that states of excitement may have an intensity greatly in excess of their pleasurable or painful aspect.
This fact, however, does not, according to the above argument, establish the existence of any third variety of feeling. Prof. Wundt, in his account of the intensity-
scale of sensation, fixes a point of neutrality where a sensation, *e.g.*, of sound, that
was pleasurable passes by further increase of stimulus into an excessively, *i.e.*,
disagreeably, powerful one. Such a mathematical conception, however, though
useful does not prove the actual existence of any concrete sensation outside both
the agreeable and the disagreeable regions of the scale. And even were the existence of such a perfectly neutral intensity of sensation demonstrated it would still be
a question whether we ought to attribute to it any affective character at all.[1]

§ 3. *Relation of Pleasure and Pain.* All feeling thus seems
reducible to pleasure and pain. Here, then, we have two
ultimate or elementary varieties of affective state, and it
becomes necessary to define their precise relation one to
another.

That pleasure and pain are distinct and strongly opposed
experiences is a matter of simple subjective observation. The
distinction appears to be the deepest, most impressive, and the
earliest recognised among all the distinctions in our conscious
life. Observation shows us that the infant distinguishes its
pleasurable and painful states before it attains any clear
differentiation of the presentative elements of its experience;
and the same thing seems to be true in the evolution of consciousness in the zoological series.[2] The contrast of the
pleasurable and the painful bulks more than any other in our
thought about life and in our current forms of language.
Poetry is one long expression of the contrast. As we shall

[1] On the question of neutral feeling, see Hamilton's Edition of Reid's Works,
p. 311; Bain, *loc. cit.*; Ladd, *Elements of Physiological Psychology*, p. 509 ff.;
Volkmann, *Lehrbuch der Psychologie*, ii. § 128; and Fr. Bouillier, *Du Plaisir et de
la Douleur*, chap. viii. Consult further the full discussion of the subject in *Mind*,
xiii. p. 80 ff. and p. 248 ff., and xiv. p. 97 ff.

[2] It is this fact that leads some psychologists to regard feeling as more fundamental than the presentative or intellectual phase of consciousness. (See above,
p. 168 ff.)

see later on, the opposition between the two is a radical and determining factor in voluntary action. To act with conscious purpose is to seek the pleasurable and to shun the painful. The relation is further seen in the fact that pleasure and pain can balance, and to this extent counteract or neutralise one another. A suffering child may have its pain diminished, and in extreme cases "quenched" by the addition of simultaneous pleasure. Voluntary action proceeds on the pre-supposition that an excess of pleasure over pain is tantamount to a pure pleasure.[1]

While, however, pleasure and pain thus appear to our ordinary consciousness two fundamentally positive and co-ordinate modes of feeling, some psychologists have regarded them as having another relation. Thus, it has been said by certain ancient and modern writers that, while pain is a positive state of consciousness, pleasure is a negative state, that is, the mere absence of pain. That the removal of pain is of itself sufficient to bring a pleasure is a fact that will be illustrated by-and-by. But such removal, so far from being the essential condition of pleasure, is in many cases wanting altogether. Thus the pleasures of the higher senses, as those of art, are not in ordinary circumstances preceded by any pain. It must be added that not only does the removal of pain bring pleasure, but the removal of pleasure brings pain. There is consequently nothing in these transitions that proves pain to be fundamental, and we may abide by the common judgment that pleasure and pain are equally real, positive, and fundamental varieties of our experience.

The doctrine that pleasure is only negative, which has been advocated by Plato and the modern German pessimists, rests on the assumption that the primary and the predominant condition of our consciousness is one of painful unrest or craving, from which pleasure is but a momentary escape. We cannot consider this point fully till we come to deal with the causes of pleasure and pain and the nature of volition. It must suffice at this stage to point out that this view of pleasure appears to apply, as Aristotle, in his criticism of Plato's theory, remarked, only to the lower regions of bodily gratification. Here, undoubtedly, pleasure follows and depends upon a previous state of craving or appetite. In all the higher regions of affective experience, including that growing out of the activities of the special

[1] This is the common view, though the pessimist, as we shall see presently, denies this compensatory equivalence of the two modes of feeling.

senses, and of imagination and thought, the view seems in flagrant contradiction with the facts.[1]

§ 4. *Distinction of Feeling and Presentation.* As has been already implied, pleasure and pain do not occur as isolated experiences, but in close connexion with presentative elements, that is to say, sensations, and their derivatives, percepts, and ideas. Thus we commonly speak of a pleasure as one " of taste," " of colour," " of imagination," and so forth ; and similarly we speak of the pain " of a burn," and " of a contradiction". Before we can with profit investigate the causes of pleasure and pain we must seek to define more precisely the relation between the presentative and the affective side of our mental experience. And in order to simplify matters we shall confine ourselves in the main to the elementary stage of sensation.

The first thing to do here is to mark off as sharply as possible the presentative and the affective element. This is only very imperfectly done in common thought owing to the close connexion between the two and the difficulties of analysis.[2]

Let us take a simple case of pleasurable sensation, say that of a soft, caressing touch. How shall we discriminate between the presentative and the affective side of this experience ? The best way seems to be to say that the presentative element is marked off from the concomitant of feeling by a certain determinate quality and local complexion. Thus the touch as soft, as experienced at a particular region, or over a particular surface of the skin, is pure presentation.

Feeling as such has no quality (apart from the radical difference of the pleasant and the unpleasant) and no local attribute. We cannot speak of a rough or acrid *feeling;* nor, strictly speaking, can we give local reference to it. When we are said to localise our ' feelings,' what we really do is to localise the sensations of which the feelings are concomitants.

---

[1] On the question of the negativity of pleasure and of its relation to pain, see Hamilton, *Lectures on Metaphysics,* ii, xliii.; Léon Dumont, *Théorie de la Sensibilité,* pt. i. chap. i. (where there is a historical *résumé* of the doctrine) ; Fr. Bouillier, *Du Plaisir et de la Douleur,* chap. ix. ; also my volume, *Pessimism,* p. 218 ff. Further references may be seen in the historical *résumé* given below, Appendix I.

[2] The tendency to confuse the two in the region of tone-sensation is well illustrated by Stumpf in his *Tonpsychologie.* (See especially ii. p. 527.)

'On the other hand, feeling certainly has, like sensation, intensity, duration, and probably also extensity or volume, so far as this can be divested of all local or spatial implication. Thus the pain of a laceration may be more or less keen, *i.e.*, intense, more or less prolonged, and more or less massive. These aspects determine the quantity of the feeling, and also serve to invest it with a certain semblance of qualitative difference. Thus when we speak of a dull, of a pricking, or of a throbbing pain, we refer to the form which the feeling assumes by reason of its temporal aspect (*e.g.*, sudden and momentary, or prolonged), and of its changes (or absence of changes) in intensity from moment to moment. Such differences come under the general head of affective quality, complexion, or tone.

Höffding refers all apparent differences among pains, *e.g.*, "burning," "cutting," "piercing," etc., to the underlying presentative element or sensation. (*Outlines of Psychology*, p. 224.) No doubt this is true of properly qualitative difference, as that which characterises "burning". We here transfer to the concomitant feeling what strictly belongs to the sensation. On the other hand, Höffding seems to overlook the fact that in the case of the sense-feelings as in that of the higher feelings or emotions, the peculiar intensity, massiveness, temporal character (*e.g.*, momentary or intermittent) and course of change in intensity and volume (*e.g.*, gradually rising and falling), invest the feeling itself with a *quasi*-formal attribute. Just as it has been shown within the region of sensation proper that what is called musical quality or timbre consists in part of such peculiarities,[1] so we may say that these give the semblance of qualitative difference to our feelings. It is to be added that the distinct consciousness of any such formal peculiarity through a process of differentiation falls under the head, not of feeling, but of intellect. Even the *discriminative* consciousness of pleasure itself, *i.e.*, in its difference from pain, is an intellective phenomenon. The proposition that feeling as such has no quality (apart from the feeling-quality itself, agreeableness, disagreeableness) is held by most psychologists. The attempt to invest feelings with qualitative differences for ethical purposes, *e.g.*, by J. S. Mill, has not met with much support in psychology. The contrast between acute and massive seems to be applied to the element of sense-feeling by Prof. Bain, as when he writes of the pleasure of muscular exercise : "As to the degree of the pleasure, it is massive rather than acute".[2]

[1] See Stumpf's careful analysis of clang-effect, *op. cit.*, ii. p. 527 ff.

[2] *Compendium of Mental Science*, bk. i. chap. i. p. 21. Prof. Bain, by using the term 'quality' with reference to the pleasurable, painful, or indifferent side of the sensation, seems to miss the advantage of marking off sensation as presentative, that is, intellectual material, by a specific qualitative character. (See especially *loc. cit.*, p. 19.) On the way of demarcating the presentative and the affective side of sense-experience consult further Ward, article "Psychology" (*Encyclop. Britann.*), p. 40.

§ 5. *Connexion between Feeling and Presentative Element.* That in the region of our sense-experience the presentative and the affective element are closely bound up one with another has been already suggested. This is clearly shown in the common way of describing feeling by epithets borrowed from sensation, *e.g.*, a "burning," a "pricking" pain. We have now to look into this connexion somewhat more closely.

To a large extent it appears at first to be true that the presentative element and feeling are given together in strict simultaneity as different "elements in," or "aspects of," one experience. Thus in the region of touch the sensation produced by the pressure of a sharp point and the feeling of pain seem to be one individual experience. Indeed, as already hinted, owing to this close concomitance of the sensation and the attendant feeling in such cases, we are in the habit of calling a pleasurable or painful feeling of touch a sensation ; and, for the same reason, psychologists have frequently tended to confuse the two.[1]

A closer examination shows, however, that the relation of the element of feeling to that of sensation is far less simple or uniform than at first appears. Thus, in cases where the two elements are given in close concomitance, it has been found that they are in a measure independent one of another. Common experience tells us we may have the sensation of a blow before we feel its painfulness, and experiment has confirmed the observation. Thus Beau found that from one to two seconds may elapse between the sensation and the feeling of pain when a fowl's eye is struck. E. H. Weber observed that in plunging the hand into very hot or very cold water the thermal sensation precedes the pain. And the same lateness of the pain is illustrated by the facts of electrical stimulation.[2]

While feeling thus takes longer time for its excitation than sensation, it can be artificially separated from the latter. Pathological evidence shows that in the state called by Beau

---

[1] See, for example, Jas. Mill's account of "the pleasurable and painful sensations". (*Analysis*, ii. chap. xvii.)

[2] The same thing is still more plainly seen in certain diseases. Thus in *tabes dorsalis* there is a distinct and constant time-interval between the occurrence of a sensation of pressure, *e.g.*, that of a pin-point, and of the connected pain. (See Funk, Hermann's *Handbuch der Physiologie*, iii. ii. p. 297 f.)

and others analgesia pain may be done away with and sensation retained. Again, Schiff found that in certain stages of anæsthesia brought on by chloroform or ether pain is deadened, whereas tactual sensibility to stimuli of moderate strength is unimpaired.[1] Conversely, there is some reason to suppose that sensation may be impaired while feeling remains intact. Thus in certain hypnotic conditions the patient is said to feel the sadness of blue and the gladness of yellow though he perceives no object.[2]

While we thus find a certain independence of sensation and feeling in the region of touch, where they are commonly experienced together, we further note that among the several varieties of sensation the two elements, the presentative and the affective, are far from appearing together with equal degrees of strength or prominence. Some sensations, e.g., that of contact with an ordinary, moderately smooth surface, can hardly be said to have, for the adult consciousness at least, any appreciable concomitant of feeling. Certain varieties also of sensations of smell and taste are proximately at least indifferent, that is, feelingless. The same relation obtains, as we shall see, among our percepts and ideas : they may have but a faint and inappreciable concomitant of feeling. The percept or image of a stone is for most of us feelingless, as compared with that of a strikingly beautiful object. On the other hand, we see in the region of organic sensation a marked preponderance of the affective over the presentative side. The sensation of a laceration or of indigestion has, as we saw above, no well-defined specific quality like that of colour, and is, indeed, by some regarded as a case of pure feeling.[3] As we shall see later on, feeling preponderates over the presentative element in all those mental states which we call emotions. A similar preponderance of the affective

---

[1] Quoted by Wundt, op. cit., i. p. 114.

[2] For a careful account of the evidence, see Ladd, op. cit., p. 510 ff. ; Höffding, op. cit., p. 280 ff. ; and more fully Weber in Wagner's Wörterbuch der Physiol. iii. ii. p. 563 ff.

[3] E. H. Weber, in the place referred to above, deals with organic sensation or cœnæsthesis as a case of painful feeling. As J. Ward, however, has shown, even organic sensation has a certain presentative element in its vague local character. (See article " Psychology," Encyclop. Britan. p. 40, col. ii. ; cf. Wundt, op. cit., i. p. 539.)

over the presentative element is met with throughout the sub-conscious or semi-conscious region of our mental experience. We often 'feel,' that is, are affected by, though only dimly apprehending, the presence of an element in a sensation-complex, *e.g.*, a new feature in a lady's dress, as also of an idea, *e.g.*, a teasing recollection of some omitted duty, before we clearly differentiate and recognise this ingredient.

We may summarise the results as follows :—

(1) There is a general concomitance between the presentative element, that is, sensation or its ideal representative, and feeling—to the extent, at least, that there is no feeling which does not imply a minimum of presentative consciousness.

(2) The affective and the cognitive element do not appear with equal prominence in our sensational and ideational experience ; the higher degrees of definiteness of presentation tend to keep down feeling, and conversely the higher degrees of intensity of feeling tend to hinder the full development of the presentative element as a sharply discriminated quality.

§ 5*a*. *Nervous Conditions of Feeling and Presentation.* These facts point, first of all, to the conclusion that the nervous conditions of sensation (together with its ideational equivalent) and of feeling are, to some extent, common : the nervous process that issues in a sensation or an idea *tends* to produce further a concomitant of feeling. At the same time, they point to a certain distinctness in the nervous conditions of the two phenomena. As we shall see presently, feeling involves in most, if not all, cases a more extended central nerve-process than the presentative element.

But this is not all. The cognitive and the affective consciousness are two distinct modes, involving, it may be assumed, two unlike varieties of central excitation. The psycho-physical arrangements that issue in a finely discriminated presentation, say, that of a straight line, are, it would seem, incompatible with the diffusion of nerve-current which underlies all fully developed feeling. When we are closely attending to a sense-presentation, as a colour, so as to sharply discriminate and assimilate its particular shade of quality, we are in a peculiar attitude of intellectual strain. Our minds are occupied with a given sensuous material, not on its agreeable side, but on its intellectual side, that is to say, *in its relations* of difference, likeness, and associative concomitance (objective significance). When, on the other hand, we give ourselves up to enjoying the mere sensation of colour, this ingredient of intellectual strain is absent. It is, comparatively speaking, a state of repose, of relaxation of tension. No doubt, as pointed out above, there is an element of attention in such affective observation or contemplation ; but since the need of intellectual elaboration is done away with the attention becomes relatively easy and spontaneous. These differences point to a contrast in the whole psycho-physical condition involved in the two cases.[1]

It must be remembered that we are here speaking of purely *sensuous* feelings. As we shall see by-and-by, the pleasures of colour and tone made use of in art are not pure sense-feelings, but involve rudimentary intellectual processes, *viz.*, the comparing and "relating" of sensations. So far as feeling springs out of such intellectual processes, its conditions will obviously tend to coincide with those of cognition.

§ 6. *Feeling and Self-Consciousness.* Our discussion of the relation of feeling to intellectual activity would be incomplete without a reference to the common view that all pleasure and pain are concomitants of, and in fact arise out of, the consciousness of a modification of our subjective condition, of a heightened or lowered vitality, degree of perfection, and so forth. This theory is a development of the idea that feeling is a secondary and derivative phenomenon, being merely the outcome of the workings of intellectual elements (presentations).[2] In its extreme form, *viz.*, that pleasure and pain are *nothing but* the consciousness of furthered or hindered life, it plainly seeks to do away with feeling as one ultimately-distinct aspect of our mental experience.

That there is a certain amount of truth in the view that thus connects feeling with self-consciousness is indisputable. As already pointed out, feeling is the *subjective* side of our experience, that which belongs in a peculiar manner to ourselves. This way of regarding our feelings suggests that they are largely independent of external causes. Even the feelings which appear to arise directly from sensory stimula-

---

[1] *Cf.* above, p. 77.

[2] *Cf.* above, p. 69.

tion, *e.g.*, those of perfume, or of musical tone, are not so uniform either in the case of different experiences of the same individual, or in that of different individuals, as the presentative elements or sensations which they accompany. All feeling, as we shall see by-and-by, involves certain organic effects, or modifications of the internal vital organs, and so it comes to pass that the full expansion of a current of feeling is determined by internal organic conditions. Thus the very same sense-stimulus, as bright sunshine, which gladdens us at one moment hurts us at another. To this it may be added that our life of feeling is conditioned to a larger and larger extent as we develop by processes of internal representation (recollection, imagination). These facts, when they come to be reflected upon, lead to the habitual reference of feeling to self as subject or patient. We may thus say that feeling in general has certain marks which lead us, *when self-consciousness develops*, to include it within the sphere of our own subjective experience.

While, however, all feeling is in this way subjective in its *tendency*, *i.e.*, in its susceptibility of being wrought into the texture of self-consciousness, it is an error to say that it uniformly arises *out of this* consciousness. In the development alike of the human individual and of the zoological series pleasure and pain precede anything worthy of the name of self-consciousness. In their simplest, crudest form they appear to be as certainly the result of nervous stimulation as any psychical phenomenon can be. It may be added that, even where self-consciousness is added, the feeling is not the effect of the consciousness of furthered vitality or perfection, but the consciousness of furthered vitality is a secondary phenomenon of reflexion dependent on and prompted by the primary phenomenon of feeling.[1]

The relation of feeling to presentation has been variously conceived. In the earlier pre-Kantian stage of German psychology, before feeling was raised to the rank of an independent function, it was regarded as only a particular mode of *intellectual* consciousness (cognition of perfection and imperfection, of welfare and its opposite). On the physiological side this tendency to identify the affective and

---

[1] Horwicz well shows that feeling, so far from presupposing self-consciousness, is a factor in the development of this. (*Psychol. Analysen*, i. p. 231 ff.)

the cognitive was seen in the view of Weber that pain is a kind of sensation. The separation of feeling from presentation is due partly to the influence of Kant, partly to the physiological observations referred to above. The result of these last is seen in the hypothesis put forward by Schiff that tactual sensations and the pains that are connected with them involve separate nervous paths. This conjecture, however, has not been since established.[1]

In recent English psychology feeling has, on the whole, been sharply marked off from the intellectual element. Thus Hamilton opposes the two in his well-known "law" that feeling and cognition (and so sensation, that is, sense-*feeling* and perception) "co-exist in an inverse proportion".[2] Again, Dr. Bain marks off with great care what he regards as the "attribute" or "aspect" of feeling in sensation. Dr. Ward would separate feeling still further from sensation, viewing it not as one of its attributes or aspects but rather as a concomitant. At the same time, he appears to regard all feeling as dependent on presentation and to reject the notion that feeling as such may arise directly from certain features in the process of nerve-stimulation.[3]

CONDITIONS OF PLEASURE AND PAIN :

(A) Conditions in the Stimulus.

We have now to consider the conditions or mode of production of feeling. Here, again, we shall confine ourselves in the main to the simpler feelings, those of sense, inquiring into their nervous conditions.

Since anatomical investigation has not revealed the presence of a class of nerves whose special function is feeling, we assume that this in its simplest conceivable form depends primarily, like the sensation with which it is concomitant, on certain features of the process of sensory stimulation. We have to ask, then, first of all, with what properties of the neural process feeling seems to be connected.

§ 7. *Pleasure and Pain as Determined by Quantity of Stimulation.* The first and most obvious mode of variation of the process of sensory stimulation is quantity, and more particularly

[1] See Wundt, *op. cit.,* i. p. 114.

[2] *Lectures on Metaphysics,* ii. p. 99 ff. ; *cf.* Herbert Spencer, *Principles of Psychology,* ii. p. 250.

[3] On the relation of feeling to presentation, see Lotze, *Microcosmus,* i. p. 272 ff. ; Volkmann, *Lehrbuch der Psychologie,* ii. §§ 127, 129 ; Nahlowsky, *Das Gefühlsleben,* § 7 (p. 57 ff.) ; Wundt, *op. cit.,* i. p. 538 ff. ; Horwicz, *Psychol. Analysen,* ii. 2 ; Höffding, *Psychol.* vi. A. 2 ; Bain, *Senses and Intellect,* p. 74 ff. ; Ladd, *Elements,* p. 503 ff. ; J. Ward, article " Psychology " (*Encyclop. Britann.*), p. 40 ff.

intensity ;[1] and a very little consideration will show that this exerts an influence on the resulting feeling. In the case of the higher senses, for example, while a moderate strength of stimulus, light, or sound, is agreeable, a greater strength becomes disagreeable. The same relation holds in the case of the reflex reactions called forth by sensory stimuli. A moderate exertion of attention to sights or sounds is agreeable, a severe strain becomes fatiguing, and so disagreeable. Similarly with respect to all muscular activity. Moderate exercise of a group of muscles is enjoyable, unduly violent exercise is fatiguing, that is, disagreeable. It may be added that the prolongation of a stimulus which might for a short time be enjoyable may result in nerve-fatigue, and so a disagreeable feeling. The successive summation of stimuli seems to effect the same thing as a momentary intensification of the stimulus. Thus a moderately strong light and moderately vigorous exercise, though agreeably stimulating for a short period, grow fatiguing after a longer one.

Passing to ideational activity, a like relation appears to obtain. Apart from all differences among our representations, it may be laid down as a general proposition that the cerebral activity involved in imagination and thought is attended by some degree of pleasure, provided the effect of fatigue is excluded. Thus a rapid sequence of ideas to which attention is able to accommodate itself is exhilarating. On the other hand, the too sudden intrusion of ideas, as in certain effects of intellectual shock, or a preternatural rush of ideas which baffles and overpowers the attention, as in certain morbid cerebral states, is distinctly disagreeable. And in general, intellectual work, when unduly prolonged, grows fatiguing and disagreeable.

These facts have long since led to the formulation of a law of pleasure and pain, which may be called the law connecting pleasure and pain with quantity of functional activity, or more briefly the law of pleasurable and painful stimulation. It may be expressed as follows :—

The moderate stimulation of the central nervous substance is attended with pleasure, and the pleasure continues to in-

---

[1] Volume or extensity of area stimulated is a second dimension which also determines the quantity of the feeling. The influence of this will be spoken of later.

crease with the increase of the stimulation up to the limit of excessive or fatiguing activity, at which point it gives way to a feeling of pain.

A word or two may be added in explanation of this proposition. The reference to the central nervous elements implies that only those processes of stimulation which involve the cortex or " area of consciousness " are causes of pleasure and pain. Such central excitations include the results of the peripheral stimulation of sense-organs, and also those centrally initiated processes which underlie ideation.

Again, the expression ' moderate ' stimulation is here used in a relative, and not in an absolute, sense. It has reference to the peculiar structure and to the temporary condition of the organ stimulated. It is probable that some nerve-structures which are called on for more frequent and prolonged activity, e.g., those involved in vision, in the movements of the hands, recuperate more rapidly than others, and so allow of a longer pleasurable activity. Not only so, a vigorous condition of the organ, say, a group of muscles, or the muscular system as a whole, renders possible a greater intensity and a longer duration of pleasurable activity than a feeble condition.

As already mentioned, Wundt seeks in the case of sensation to determine a scale of intensities in the stimulus as affecting the attendant feeling. Thus, according to him, as soon as the stimulus passes the threshold and causes an appreciable sensation it begins to be pleasurable, and the pleasure goes on increasing as the stimulus is increased. At length a point or region of maximum pleasure is reached which probably answers to that medium region of the scale where the finest discrimination is possible. Beyond this the pleasure rapidly diminishes till a certain ' point of indifference ' is reached. Above this any further increase produces pain, which in its turn increases till, at the point known as the height (see above, p. 90), the maximum of pain is reached.[1]

It has been assumed above that the law of stimulation applies to every kind of psychological process, sensory and motor alike. And this certainly seems to be borne out by the facts. We can, in certain cases at least, distinguish an agreeable or disagreeable effect which is due to a sensory process, and one which is rather referrible to the motor reaction. For example, we may distinguish the agreeableness of the *sensation* in the case of a bright colour from that of the process of *attention* excited by a dull, unexciting shade of colour. As we shall see presently, the peculiar charm of low tones of colour and of certain weak sensations generally

---

[1] *Physiol. Psychol.* i. chap. x. 1. As we shall see presently, this variation of amount of feeling with intensity of stimulus is complicated by the circumstance that low degrees of stimulation produce their own peculiar mode of gratification.

is probably due, to some extent, to the special reaction which they provoke. It would, however, be a grave error to reduce the whole distinction between agreeable and disagreeable stimulation so far as it depends on quantity to a difference in the reaction.[1]

§ 8. *Impulse and its Gratification : Pains of Want.* In the above statement of the principle of stimulation no reference has been made to any impulse or disposition to activity. A correct view of our psycho-physical organisation requires us to bring in this element. Our several organs constitute not merely so many capacities for particular functional activity, but so many tendencies or dispositions towards such activity. These tendencies show themselves to some extent as original impulses, as the instinctive impulse to walk, to examine things, to play, and so forth. They are, moreover, furthered by the habitual direction of our activity. We tend and feel impelled to do what we are accustomed to do.

The effect of such organic dispositions on feeling is a double one. In the first place, the co-operation of the disposition or impulse is an important reinforcing factor of pleasurable stimulation. When strongly impelled by hunger, by the impulse to muscular activity, to reading, and so forth, the pleasure accompanying the corresponding activity is proportionately increased. Hence the common way of looking at all enjoyment as the gratification of impulse.

In the second place, the delay of such gratification gives rise to a new variety of painful feeling, *viz.*, that of want or craving. To be hungry and not be satisfied, to want to read and have no book at hand, is in itself a misery.

Such painful cravings are, to some extent, periodically recurrent, and connected with rhythmical changes of organic conditions. These periodic organically-conditioned cravings are known as Appetites. They consist of the well-known bodily appetites, hunger, thirst, sexual craving, together with other regularly recurring wants not commonly spoken of as

---

[1] This seems to be done by Münsterberg in his theory that all nerve-fatigue is a phenomenon of the motor structures. (*Beiträge zur exper. Psych.* ii. p. 95 ff.) *Cf.* the view of Wundt and Ward that the effect of quantity of stimulus on feeling is due to the production of conditions favourable or unfavourable to attention. (*Physiol. Psychol.* i. 535; and article " Psychology," *Encyclop. Britann.* p. 68.)

appetites, such as the craving for sleep, for muscular activity, for amusement. This regular recurrence of craving becomes further fixed by habits of life.

> The influence on our feelings of recurring impulse, with its correlative want or craving, is a profound one. Even in the case of the so-called "passive" sense-pleasures, as those of colour and tone, there seems something analogous to the periodic recurrence of an organic deficiency. A man who makes "the harvest of the eye" a chief source of enjoyment experiences a recurrent longing for the pleasurable stimulation of this organ. It is, however, a radical mistake to say, with Schopenhauer and his followers, that all pleasure is the satisfaction of a previously felt want. Of this more by-and-by when we take up the subject of volition.[1]

By combining the principle of impulse and want with that of stimulation, we may say that pleasure, so far as it is connected with *quantity* of stimulation, lies between two extremes, excess and deficiency, each of which is painful.

It seems possible to give this law of stimulation a teleological expression, that is, to show its bearing on the preservation and welfare of the individual. Since our organs are useful structures needed for the carrying out of certain life-functions, anything which serves to promote their efficiency is beneficial, anything which tends to destroy this efficiency, injurious. Moderate exercise conduces to efficiency. Hence the pleasures of exercise, including the gratifications of impulse which *draw* us to such beneficial action, together with the pains of craving which *drive* us thither, work to our advantage. On the other hand, the pains of excessive action and fatigue, marking the point where nervous recuperation cannot overtake expenditure, are a salutary check to injurious action.

§ 9. *Quiet Pleasures : Repose.* In the above account of the relation of feeling to quantity of stimulation we have left out of consideration an important modifying circumstance. Our greatest pleasures do not come merely by way of strenuous activity. There are the quiet pleasures, those corresponding to low degrees of stimulation, and the enjoyments of repose. A word or two on these will serve to supplement and correct the above statements.

To begin with, then, low degrees of stimulation seem to

---

[1] For a detailed account of the appetites, see Bain, *The Senses and the Intellect*, p. 240 ff. ; *cf.* Beaunis, *Les Sensations internes*, "Besoins," chap. ii. ff.

produce an effect of agreeable feeling disproportionate to their intensity. In many cases, no doubt, this is due to the circumstance that the mild stimulation is a relief from a previous excessive one, and so to the principle of change to be considered presently. But this does not explain the whole effect. Low tones and colours, slow, gentle movements, have in themselves a pleasure-value as massive forms of enjoyment. This seems due in part to the volume or extensity of sensation involved, *e.g.*, in the case of gentle movement of the body and limbs. In addition to this there is probably an indirect effect, such gentle stimulation serving in a peculiar way to further those organic processes which, as we shall see, constitute so important an ingredient in feeling.

Lastly, it is worthy of note that in the case of very weak stimuli, as *pianissimo* tones, and quiet shades of colour, there is a special activity of attention. A part of the charm of a very soft passage in music is due to the intense reaction it calls forth. The very weakness of the impression invests it with the fascination of mystery ; and this, together with the fear of losing the impression, excites the mind to a special effort of tenacious apprehension.

The delight in mere inactivity or idleness seems at first sight to convict us of a gross flattery of human nature in saying that pleasure comes from activity. There is, however, no real contradiction here. As we shall see presently, the pleasures which we call by the names idleness, repose, *dolce far niente*, are relative pleasures, presupposing, in their higher intensities at least, a contrast to a remembered previous exertion. Hours of inactivity, or idleness, fill a considerable place in those rhythmic alternations of work and rest which are required for healthy life. It must be remembered, too, that what we call doing nothing is never really a state of complete inactivity. Thus the relaxation of the severer kinds of bodily and mental work makes room for a more vigorous discharge of the vital functions. The importance of considering the effect of all stimulation on these will appear presently. The relaxation of effort is partly enjoyable because of the heightened organic consciousness. In luxurious states of idleness we may be said to consciously vegetate. In addition to this, idleness when pleasurable always involves something of play, that is, of gentle activity indulged in for its own sake. Thus it commonly includes a certain amount of agreeable sense-stimulation, as when we give ourselves up to the influence of light and sound in the country on a spring morning, or at the sea-side on a bright summer day. In the case of the educated, what is called idleness always includes further a pleasurable play of imagination or unrestrained flow of ideas.[1]

It is thought by some that in repose after fatiguing exertion the process of recuperating the exhausted nerve-structures is itself a cause of pleasurable sensation. Thus Dumont, who regards pleasure as conditioned by an equilibrium between accumulation and expenditure of nerve-energy, conceives of the pleasure of repose as arising from the beneficial restoration of the normal equilibrium between the two. Here we have a nice question of the exhaustiveness of psychological analysis. Are the pleasures of repose fully accounted for as the effect of liberated activities, bodily and mental, and of the principle of contrast to be spoken of presently ?[2]

§ 10. *Pleasure and Pain as determined by Form of Stimulus.* While intensity or strength of stimulus is thus one main factor in the determination of pleasure and pain, it is not the only condition. A difference in the form of the stimulus, answering to a difference of quality in the sensation, affects the sense-feeling also. Thus, in the region of taste, as we saw, a bitter taste is as such disagreeable in all degrees. Similarly, all degrees of roughness in sound, and all degrees of that alternate increase and decrease of stimulus constituting a beat, and supposed to form the essential ingredient in musical dissonance, are as such disagreeable. A like rule holds good in the case of a series of stimulations. In order that they be agreeable they must be arranged in a certain form. Thus a rapidly flickering light, even when not strong, may be very disagreeable. All jerky, irregular successions of stimuli are as such disagreeable.

These and other facts suggest that all modes of stimulation are not equally suited to the efficiency or welfare of our organs. There seems to be a normal mode of functional activity, and on the other side an abnormal or injurious mode.[3] In what

---

[1] *Cf.* Hamilton, *Lectures on Met.* ii. xliv. ; and H. R. Marshall, *Mind*, xvi. p. 350 ff.

[2] The law connecting pleasure with a positive functional activity and the difference between pleasure and pain with the amount of such activity is as old as Aristotle. For a historical account of this view, see below, Appendix I.

[3] So far as the same nerve-structures are engaged in the production of a variety of sensations we must of course speak of a plurality of normal functional activities.

precisely the injury consists, physiological science does not as yet enable us to say. We know that all lesions of sentient, that is, nerve-supplied, parts of the organism produce pain. Indeed, the pains of laceration, burning, and other destructive action on the tissues are one of the best recognised class. One may conjecture, then, that in the case of the unpleasant sensations now referred to, *e.g.*, of an acrid taste, or musical dissonance, the peculiar mode of nervous action tends to impair or disintegrate the nerve-substance in a way in which the normal and pleasurable activity does not. If this conclusion is established, the dependence of feeling both on intensity and on quality of sensation would be reducible to one principle, *viz.*, the distinction between a beneficial and an injurious activity of the nerve-structures involved.[1]

The attempt to refer the distinction of pleasurable and painful feeling exclusively to a difference in the quantity of the stimulus has so far failed. J. S. Mill's criticism of Hamilton's theory of pleasure and pain sufficiently shows this. The same thing is seen in Wundt's ingenious but fanciful attempt to refer all apparent influence of quality to that of intensity. He supposes that in the case of sensations, as those of bitter taste, which are disagreeable in all degrees, the point of excessive or fatiguing stimulation is reached *below the threshold of sensation*, *i.e.*, in the negative region of the scale. This idea, in addition to being open to the objection that it involves the contradiction of "unconscious" sensation, may be said to really concede the point here urged, since a kind of stimulation which becomes painfully excessive before it produces any sensation at all must be supposed to be ill-adapted *in its form* to this particular nerve-structure.[2]

(B) VARIATION OF STIMULATION.

We now come to another important condition of pleasure and pain. As already suggested, a feeling is affected not only by the nature of the stimulus at work at the time, but by the preceding psycho-physical activity. Our consciousness is not a series of detached, disconnected "states," but a continuous movement, every stage of which is modified by previous stages.

[1] For a full account of the several physiological conditions of painful feeling, see Beaunis, *Les Sensations internes*, ch. xx. p. 202 ff.

[2] On this point see Mill's *Examination of Hamilton*, chap. xxv.; Fechner, *Vorschule der Æsthetik*, ii. xliii.; Wundt, *op. cit.*, xer cap. i.; Gurney, *Power of Sound*, chap. i. § 2. Wundt's theory (already referred to), of a transition from a pleasurable to a painful degree of intensity in the case of all varieties of sensations, is criticised by Horwicz, *Psychol. Analysen*, ii. 2, p. 26.

One of the most striking manifestations of this is the fundamental importance of change or contrast as a condition of
vividness or full intensity of consciousness.[1]   And the influence
of this condition is seen yet more clearly in the region of feeling
than in that of cognition.   We have now to inquire into the
precise bearing of change, that is, variation of stimulus and
its correlative activity, on the intensity and, to some extent, on
the character of feeling as pleasurable or painful.

§ 11. *Effects of Prolonged Stimulation : Dulled Feeling.*   In
order to understand the effects of change on our affective states
we must first consider the results of prolonged *unchanging*
stimulation.   The value of change as a condition of sustained
pleasurable feeling depends on this circumstance.

To some extent the effects of prolongation are similar in
the case both of pleasurable and of painful stimulation.   The
most general effect (the exact limits of which will be considered
presently) is what we may call a weakening or *dulling* of the
feeling.   Thus, a prolonged pleasurable stimulation, as that of
the eye by a sunny landscape or stage-spectacle, of the ear by
music, and so forth, results in a gradual falling off in the
intensity of the pleasure.   The exact course of such falling
off is not as yet ascertained, but it may be said that there is
a rapid and considerable decline at the outset through the loss
of the initial freshness of impression, and then a slower and
less considerable decline till an approximation to a dull uniform
effect is reached.   A similar result shows itself in general in
the case of painful stimulation.   A large part of our physical
discomforts and mental troubles lose in intensity when prolonged.   That is to say, they sink to the level of dull or faint
psychical phenomena.

The reasons of this general subsidence of feeling when
excitation is prolonged are to be found partly in the lowered
functional activity of the nerve-structures engaged, partly in
the falling off in the reaction of attention.   That there is a loss
of intensity of feeling through lowered activity is probable from
what we know of the effects of continued optical stimulation in
exciting a less and less intense luminous sensation.   At the
same time this is not the only, nor, where the stimulus is a

---

[1] *Cf.* above, p. 175 ff.

moderate one, even the main cause.   Here the greater part
of the dulling effect may safely be attributed to the lowered
activity of attention through a decline of the stimulus of
interest.

§ 12. *Prolonged Pleasurable Stimulation : Fatigue and Monotony.*
We may now consider the more special effects of prolongation
in the case of pleasures and of pains.

As we have seen, prolonged pleasurable stimulation, if
powerful enough, issues not in a dulling of feeling, but in the
transformation of a pleasurable into a painful mode of feeling.
This is an illustration of the general principle of nerve-fatigue
when this reaches the point of inducing a conscious feeling,
viz., sense of fatigue.[1]

Even when the prolongation of a previously pleasurable
stimulation does not beget the full effect of nerve-exhaustion
(painful fatigue), it may result in a mental weariness which
arises from a *consciousness of the decline.*   This effect is seen in
all states of monotony, tedium or ennui.   Here the sense of
freshness departed and of dull, uninspiring sameness fills the
mind.   We grow weary of the drab complexion of things, and
long for a more vivid colouring.   The very fact that ennui
includes a disagreeable consciousness of time (*i.e.,* relatively
vacant time) shows that our surroundings, our doings, have
ceased to pleasurably engage our minds.   Thus ennui is a
complex feeling, involving the imagination of other and livelier
surroundings and pursuits and the *craving for change.*   Hence
it is in its more developed form a distinctively human feeling.
A sporting dog may feel a germ of it, when shut in-doors and
impulse prompts to the more exciting pursuits of the field.
But it becomes fully developed only in the case of human beings
pretty high up in the scale of civilisation.   It forms a prominent
and recurring experience in the lives of those who love excitement, and can recall it with sufficient distinctness to crave its
recurrence, and who are forced through circumstances or want
of varied interests to pass long intervals in a state of comparative vacuity.   The society-devotee in the intervals of her dis-
sipation may be said to realise ennui in its highest degree.

---

[1] Of course, the diminution of feeling, so far as it arises from lowered activity
and loss of power, may be referred to nervous fatigue also.

Tedium or ennui is closely related to the feeling of boredom. Each involves the sense of dull surroundings and the longing for change. The difference lies in the fact that when we are bored we are subjected to the pressure of some stimulus or demand for activity, and are thus positively fatigued by the circumstances of the moment. A bore is different from a merely dull person: the latter may leave us alone and allow us to overlook him, the former pesters us with his society, his conversation, and so forth, and so forces from us a measure of reluctant, because unpleasurable, activity.

The analysis of the feeling of ennui, so conspicuous a feature of modern life, has naturally been the favourite pursuit of the cynic and the pessimist. Schopenhauer has a curious theory of the matter. According to him it is, along with want or craving (the primordial state of will), one of the great categories of human misery. Want drives us to pursuit, ennui awaits us at the goal. Ennui is the miserable sense of the burden of existence. This view, as I have elsewhere tried to show, is a distortion of the facts. Ennui arises through a consciousness of monotony and a desire for a recurrence of some vivid pleasurable excitation with which past experience has made us familiar.[1]

§ 13. *Prolonged Painful Stimulation.* In the case of prolonged painful stimulation we have other special effects. It is noticeable, to begin with, that the prolongation in this case does not tend to produce the opposite of the initial effect, as prolonged pleasurable stimulation is apt to do. Unpleasant things do not become pleasant by mere extension of time. Again, any painful stimulation of a nerve-structure if unduly prolonged is apt to produce secondary effects, which tend to disguise the general effect of weakening or dulling. Thus, through the continued action of painful stimuli of considerable strength, the *injurious effect* of fatigue may increase, and may be propagated to adjacent structures, so that instead of a falling off there is an intensification of pain. This is seen in the early stages of toothache and other forms of physical pain. Something analogous to this is observable in the case of mental troubles. A recurring, worrying thought, as the recollection of something unpleasant to be done, is apt to grow worse and worse. Here it is probable that the very recurrence produces a cumulative effect, resulting in a greater persistence of the disagreeable re-

[1] See my volume *Pessimism*, pp. 94, 235. In some cases, no doubt, what is called ennui involves positive fatigue as well. The pursuer of social excitement is said to feel ennui most in the hours of reaction following excitement. It is an error, however, to identify ennui with fatigue proper. When we say we are "tired" of the monotony of our surroundings we use the word tired in a figurative sense (*i.e.*, having had enough or more than enough). The weariness of monotony comes from its stubborn opposition to our craving for fresh modes of stimulation.

minder until we can bear it no longer, and are 'driven' to fulfil the obligation. It is presumable, too, that in this case the later reappearances are attended by a *consciousness of the recurrence,* which consciousness, as an added element of discomfort, helps to strengthen the effect.[1]

§ 14. *Effects of Change.* Having thus considered the result of prolonged unchanging stimulation, we may proceed to inquire into the effects of change.

The general effect of change is to sustain the full vividness of feeling. It prevents that falling off or dulling of feeling of which we have just given an account, and secures a measure of the initial freshness and strength.

We may probably go further, and say that since all change is attended with some *consciousness of change* a transition as such is a cause of a new element of feeling which heightens the effect of the second stimulation. Thus in passing to a new pleasurable activity, as from brain-work to a game of lawn-tennis, we have in the consciousness of the transition a feeling of expansion, which as such is pleasurable. Similarly in passing from a happy to an unhappy condition the consciousness of a fall of estate supplies a secondary feeling of pain which intensifies the primary feeling.

It follows that the effect of change in intensifying feeling will vary, within certain limits at least, with its amount. Thus a very gradual change will have but little effect, and in some cases remain inoperative altogether. Gradual change only becomes an element of importance in relation to feeling when it is integrated into *a series of changes,* as in many of the effects of art (*e.g.,* crescendo and diminuendo in music). On the other hand, rapid and considerable changes, great contrasts of condition, produce a strong effect. Thus, in passing rapidly from a condition of great weakness to one of strength, of confinement to liberty, and so forth, we have the full effect of change.

---

[1] It has been said by Hartmann and others that nerve-fatigue, while tending to diminish pleasure, does not tend to deaden pain. This is to overlook the fact that nerve-fatigue produces two kinds of effect: (1) a lowering of functional activity and so a weakening of the feeling, whether pleasurable or painful; and (2) a continuance of activity *under an injurious and painful form.* The exact relation of these two effects is only very imperfectly understood. On this point see my examination of Hartmann's theory of the effect of nerve-fatigue on feeling, *Pessimism,* p. 227 ff.; *cf.* Fechner, *Vorschule der Æsthetik,* vol. ii. xxxviii.

As intensifier of feeling change is limited by more than one circumstance. Thus any change which is too sudden and violent becomes in itself a source of disagreeable feeling, so that its effect in increasing pleasurable feeling is reversed. A too violent contrast is rejected from art as harsh and disagreeable. Again, there is a certain inertia in nerve-structures, so that when once excited they tend to go on functioning. Hence a too rapid transition, so far from being pleasant, is distinctly unpleasant. When happily engaged in a bit of manual work or line of thought we resent interruption. All the higher and more complex pleasures, as those of art-contemplation, require a certain prolongation of concentrative activity for their full development, and a change which serves to deprive us of this, so far from being favourable, is distinctly unfavourable to pleasure. Lastly, it may be pointed out that a change, if too slight, will be unnoticed and so inoperative. The value of gradual change or gradation in the effects of art depends on a *cumulative* effect of many slight changes. These limits to the pleasurable effect of change are connected with the conditions of prolonged and easy attention.

§ 15. *Different Modes of Change : Rise and Fall of Activity.* Let us now consider the way in which change of activity operates on our pleasures and pains. And here we may conveniently consider first the effect of change in intensity or strength, the mode of excitation remaining unaltered ; and, secondly, of change in the nature or form of the activity.

The first and most conspicuous effect of change in the quantity of psycho-physical excitation is seen in the growing elation of rising activity. Many of our common pleasures, sensuous and intellectual, are illustrations of this effect. Thus the pleasure of passing from a dull into a bright light, of a *crescendo* passage in music, of conscious growth and advance in power, bodily or mental, illustrates the effect of heightened activity in giving us a full intense enjoyment.

It is important to note that this heightened enjoyment is not due merely to the increased intensity of the second stimulus in itself. It is not adequately represented by the difference between the two, but rather by their ratio. It involves a *consciousness of* heightened activity, that is to say, of transition from a lower to a higher level in the scale of functional energising.

It was pointed out by Fechner that there is a certain analogy between the relation of increase of (pleasurable) stimulus to that of the accompanying pleasure, and the ratio of increase of stimulus to that of sensation as formulated in Weber's Law. Thus, the more money ('fortune physique') a man has the greater must be the addition, in order that his happiness ('fortune morale') may be appreciably augmented.[1] The same may be said, perhaps, with respect to other sources of

___
[1] G. T. Fechner, *Elemente der Psycho-physik*, vol. i. ix. § 6.

happiness, as knowledge, power, reputation. The enlargement of the field of knowledge by a certain addition is more of a transition to the comparatively knowledgeless child than to the knowledgeful adult. At the same time, the geometric ratio is apt to be disguised by the complexity of the case. Thus, in the case of knowledge, a given piece of information may be said to convey more to the intelligent adult than to the ignorant child.

What holds of the rise holds also, *mutatis mutandis*, of the fall of activity. A descent from the full delight of sunshine to a comparatively dull illumination gives us, through the consciousness of contrast, a sense of loss. The smaller pleasure looks poor and contemptible after the larger. It is this circumstance which gives to exalted rank its special precariousness :—

> The lamentable change is from the best.

Lastly, what applies to pleasurable applies to painful stimulation. Increase of painful stimulation intensifies the effect through the contrast with a lesser and more tolerable misery. Also the diminution of a painful cause makes the lower degree of suffering inconsiderable by contrast with the preceding effect. Lear hardly noticed the storm through remembering the greater trouble, his daughter's cruelty :—

> When the greater malady is fix'd
> The lower is scarcely felt.

We may positively welcome the partial subsidence of a physical pain, as toothache, by regarding it as a relief.

The effects of change in amount of stimulation here briefly illustrated may be subsumed under the following principle :—

Change in the amount of stimulation increases or diminishes the accompanying feeling (beyond the point due to the bare difference between the stimuli) through the consciousness of contrast attending the transition ; this added effect varying in intensity with the *ratio* of the two stimulations.[1]

§ 16. *Change in Direction of Activity : Pleasure of Variety.* In addition to a change in the amount of a given functional activity, we have, as a great modifying condition of feeling, variation in the kind or mode of activity. This may mean, physiologically,

---

[1] It is evident from our illustrations that this principle applies to transitions not merely from one degree to another of the same activity, but from one mode of activity to another so far as they are quantitatively compared.

another mode of excitation of the same nerve-structures, or the substitution of the activity of one organ for that of another.

It follows from what has been said respecting the effects of prolonged activity of the same organ or structure, and also the recurring impulse to activity in recuperated organs, that variation of activity is one great condition of prolonged pleasure. The complexity of our organism and of the correlated activities, involving a large number of different recurrent readinesses for and dispositions to specific modes of activity, renders change of occupation a main condition of a healthy and enjoyable life. The need of such variation is further laid down in the laws of attention, the psycho-physical activity of which can only be maintained when its direction changes from time to time. The value of the principle in a properly regulated life was recognised in classical times. In modern times our way of getting relief from professional work by new modes of activity, as in mountaineering, the pursuit of art, and so forth, illustrates the recognition of the importance of the same principle.

§ 16a. *Consciousness of Change.* It is the passing from one line of activity, one sphere of ideas, to another, with the sense of freshness that attends the experience, which gives us the keenest enjoyment. Hence the ancients were right in saying that it is the variation itself which delights (*variatio delectat*).

The charm of Novelty, about which so much has been written, illustrates the same principle. It is possible that certain first experiences owe something of their delightful character to special organic conditions which never recur later on. The first greeting of bright colour by the baby-eye may bring a wave of glad feeling which is never repeated. At the same time, what is customarily called novelty, as of a first ball, a first tour abroad, a new house, and so forth, owes its charm to the transition from the accustomed to the unaccustomed. That is to say, a 'novel' experience gives us in an exceptionally full and impressive form that transition from the stale to the fresh which enters into all variation.

It may be just added that the same principle of variety and change applies not only to our pleasures but to our pains. To pass from one painful experience to another and different one has an added unpleasantness through the very element of change. This effect is most conspicuous in the case of all new

and unwonted experiences. A first trouble, whether physical as
earache, or mental as a death in the house, owes something
of its intensity, impressiveness, and retention in memory to its
very newness.[1]

As in the case of quantity, so here we may summarise the
facts under the head of a principle :—

All change in the mode of our functional activity serves to
emphasise or intensify the feeling-concomitant of the new
activity through a consciousness of such variation ; and the
amount of this added effect will vary in general with that
of the (qualitative) unlikeness in the two activities, and with
the degree of freshness of the nerve-structures subsequently
engaged.[2]

§ 17. *Transition to Opposed Feeling : Negative Pleasures and
Pains.* One other consequence of the principle of change has
to be illustrated, *viz.*, the effect of contrast in passing from a
state of pleasure to its opposite, or *vice versâ.* Such transitions,
as already hinted, are among the most impressive in our
experience. Thus the passage from the pain of craving,
as that of hunger, to the pleasure of satisfaction, from
sickness and pain to health and enjoyment, from the misery of
poverty to the delights of wealth, from the depression of a
doubting to the elation of a confident love, and so forth, is a
theme of remark in everyday life, and in fiction. Conversely,
the transition from health to sickness, dignity to shame, and
the like, constitutes a well-worn subject of pathetic emphasis.

The facts here referred to may be formulated by the simple
principle that pain and pleasure alike are heightened or intensified, or have their disagreeable and agreeable side emphasised, by a transition from and contrast to the opposite
phase of feeling.[3]

---

[1] In this case, however, the effect is apt to be disguised by the *relieving* effect
of change. A new trouble is often welcome, at the first moment, as an escape
from the weariness of a long-standing one.

[2] This last circumstance depends not merely on the dissimilarity of the two functions, but on the length of the interval of inactivity of the newly stimulated structure.

[3] If the causes of the pleasure and the pain both persist so that the two feelings
remain permanent the effect of contrast may be produced reciprocally, the pleasure
heightening the pain, and conversely, according to the direction of the thoughts.
Here, however, a new effect to be spoken of presently is apt to be interpolated, *viz.,*
that of dissonance.

A further question already touched on arises here : Can the
mere withdrawal or cessation of a pleasurable or of a painful
excitation produce the opposite phase of feeling? In other
words, are there pleasures and pains that are *wholly* negative ?

That there are pleasures and pains that seem to have
their generating condition in such a negative circumstance is
certain. We may instance the pleasure which comes from the
cessation of physical pain. The termination of acute suffering is
in itself the occasion of an outburst of joyous feeling. Similarly
the solution of an intellectual puzzle which has been worrying
us is a cause of pleasure. On the other side, the loss of a
pleasure produces pain through *a sense of loss* and the craving
which attends this.

In other cases, too, we can see that the effect is mainly due
to the transition from an opposite affective state. Thus the
pleasure of health is largely due to contrast, and is, therefore,
rarely realised, save as a transition from an actual or an
imagined experience of the opposite. The same is true of
the pleasures of liberty, one of the main examples of an
"emotion of Relativity" given by Prof. Bain. It is only as a
transition from restraint or confinement that we can be said
to enjoy freedom. Similarly certain pains, as ennui, are in the
main due to the absence of causes of pleasure.

While, however, one may thus allow that the removal of a
cause of pleasure or of pain may be a sufficient occasion for the
on-coming of the opposed phase, it is important to add that
such so-called negative feelings have in every case *one* positive
condition at least, *viz., the consciousness of the change.* An animal
that forgot its pain the very moment the cause of it ceased to
act could not enjoy the relief as we enjoy it. In the case of
all prolonged painful states this sense of escape or relief
may become in itself a delicious experience.[1]

We have then, in the circumstance of transition, *escape from*
pain, *loss of* pleasure, an intelligible cause of that secondary
mode of feeling which we call negative. It is to be added
that such negative feelings in most, if not in all, cases have

---

[1] Lamb speaks somewhere *apropos* of the lines in Coleridge's *Ancient Mariner* :—
"A spring of love gushed from my heart
And I bless'd them unaware ".
' It stung me into high pleasure through suffering.'

positive neuro-psychical conditions as well. Thus the pleasure of health and liberty has a positive stimulus in the shape of a new energetic outburst of long-repressed activity. On the other hand, it may be urged that the element of craving, which is nearly always present in the case of negative pains or losses, is a positive mode of painful stimulation.

The exact scientific explanation of negative feeling is not very clear. As hinted in the text, positive neural conditions commonly co-operate. This we have seen to be the case in the pleasures of repose (after fatiguing effort). At the same time, there seems some ground for saying that transition and contrast here produce a *quasi*-illusory effect, analogous to that known as colour-contrast. Just as the contrast of a great with a small pleasure or pain tends to nullify the latter, so the contrast of an intensely pleasurable or painful state with one wanting this element of feeling and so indifferent tends to throw this last further away, and to invest it with the appearance of an opposed feeling.

§ 18. *Change as Principle of Intellectual and of Affective Consciousness.* We have now sufficiently illustrated the workings of change or contrast in the life of feeling. In an earlier part of this work we traced its action in the intellectual domain. We saw that change or transition from one mental state to another and dissimilar state is a condition of all mental wakefulness and of the simplest mode of intellective activity, *viz.*, the consciousness of difference. It remains to point out how these two functions of the principle, the intellective and the affective, are related one to another.

Change enters into our intellective consciousness as difference, that is, definable change. We are intellectually active so far as we estimate the quality or direction, and the quantity or extent, of the differences among our successive experiences. Thus the great field for intellectual discrimination is, as we saw, the world of presentation, with the manifold and finely-graded differences in the quality, intensity, extensity, etc., of our sensations.

On the other hand, as subserving feeling, *i.e.*, its intensity or vividness, change does not involve this appreciation of definite difference. At the moment in which we are exhilarated by a considerable rise in the scale of pleasure-intensity we are not nicely measuring the magnitude of the rise. Still less in enjoying the passage from one room to another, in which a fresh colour is predominant, do we measure the amount of colour-difference or colour-interval between the two. In its bearing on feeling as such change is thus felt or realised merely as a reinforcing or vivifying condition, while its exact nature and amount remain unapprehended.

While we may thus broadly mark off the intellective and affective function of change, we must not forget that the two habitually co-operate. In those experiences which approach most nearly to a pure state of feeling the consciousness of the great qualitative contrast, pleasure and pain, is always present, as also a dim consciousness at least of a considerable amount of change. This intellective activity, which involves a measure of comparison, plays, as we shall see, a larger and larger part in the higher developments of the life of feeling. Thus in that nice discrimination of shades of feeling, which is the favourite occupation of sentimental or " subjective " natures, and which fills so large a place in modern art-criticism, we have a signal instance of the way in which feeling gathers about itself as a focus of interest a considerable amount of intellectual and even *quasi*-scientific activity (analysis, comparison, discrimination).

Dr. Bain deals with change in its bearing on intellection and feeling alike under the general head of relativity or the principle that change is a constant condition of conscious life. The influence of the principle is traced out with special reference to certain feelings called emotions of relativity, as novelty, wonder, etc. In the state of wonder he appears to find a pure effect of change in producing a feeling of "neutral excitement". (*The Emotions and the Will*, part i. chap. iv.) This view which gives to the feeling of wonder or surprise a fundamental character has a certain analogy with Mr. Spencer's theory that the simplest type or "unit" of consciousness is a "nervous" (more accurately psycho-physical) shock. (*Principles of Psychology*, part ii. chap. i.) Such an experience if conceivable might be regarded as a common root of intellectual development (along the line of discriminative consciousness) and of affective development. This point will be touched on again when we come to deal with those varieties of feeling which attend cognitive activity.

§ 19. *Decay of Feeling : Habit or Accommodation.* We have now sufficiently illustrated the effects of prolonged uniform stimulation on the one hand, and of varied stimulation on the other hand. It remains to consider the modifications of feeling which are introduced by continuous or frequent renewal of stimulation over longer periods of time. Here new influences come into view which we may describe generally as the effect of Custom, Habit, or, to employ a more technical expression, Accommodation.[1] In certain respects the effect of custom on feeling is similar to that of prolonged stimulation at a particular time : it tends to blunt its first keen edge : "Ab assuetis non fit passio ". Our permanent surroundings and manner of life tend to grow indifferent, that is, to lose all or most of their affective concomitants. This applies at once to our pleasures and to our pains. Thus we get used, that is, comparatively indifferent, to surroundings, companions, lines of activity, which, when they were new, were intensely pleasurable. In like manner custom reconciles us to what is unpleasant in our surroundings. By taking its place in our permanent environment, the distasteful, the ugly, and so forth, seems to lose much of its first repugnance.

This effect is of course only fully manifested when the pleasurable or painful circumstance is permanent in the sense of being an approximately constant element in our life. As soon as we pass from continuance to periodic recurrence we meet with the counteracting influence of freshness or variety. All forms of pleasurable activity, *if sufficiently intermitted*, retain much of their pristine freshness. The pleasures of travel, of art, and so on, when indulged in rarely, illustrate this truth.[2] The hedonic art of living includes among its foremost problems the determination of the degree of frequency at which the renewal of a pleasure more than compensates for the loss of its intensity through familiarity.

This decay or gradual abatement of feeling with permanence and custom may be supposed to involve some process of adjustment or accommodation in the nerve-structures concerned. What these changes precisely are our physiological knowledge does not as yet enable us to say. In the case of pleasurable stimuli we may, however, conjecture that the habitual excitation of a group of nerve-structures to a particular mode of

[1] On this use of the term accommodation, see Bain, *The Emotions and the Will*, p. 80.

[2] Charles Lamb writes, in one of his charming letters, that the first pleasure of landing in a foreign country "is the nearest pleasure which a grown man can substitute for that unknown one which he can never know, the pleasure of the first entrance into life from the womb ". (Ainger's Edition of the Letters, i. p. 197.)

activity, by bringing about an arrangement of its parts favourable to this activity, and so securing promptness and ease of
functional discharge, excludes vivid feeling by lowering the
conscious concomitants altogether.   In the case of customary
painful stimulation the decay in many cases at least seems to
involve a deadening of particular modes of feeling.   This effect
is observable in the blunting of sentiments of delicacy and horror
as typified in the grave-digger in *Hamlet*,[1] in the known effects of
the frequent shaming or ridiculing of children in producing
obtuseness of sensibility.   Here again we may suppose that
there is some adjustive change in the nerve-structures concerned
which renders them impervious to the particular form of process
involved.[2]

§ 20. *Counteractives of Decay.*   This general tendency of
continuance, frequent repetition, or custom, to produce a decay
of feeling is, however, counteracted and in a manner disguised
by other and more special tendencies.   To begin with, the
process of organic adjustment or accommodation just referred
to is less simple than we have supposed.   Exercise tends to
strengthen an organ, and is one main condition of organic
growth.   One important result of this is that stimuli which
by being over-exciting or fatiguing were once painful become
pleasurable.   The muscular exertion which a child really
dislikes in itself, and which he only goes through from a
feeling of pride or ambition, becomes positively enjoyable
later on when his neuro-muscular apparatus grows adjusted
to it as its normal activity.   It is possible that many acquired
likings of the palate, *e.g.*, for wine and other alcoholic drinks,
bitter condiments, and the like, involve the hardening and
strengthening of certain nerve-structures so as to render them
capable of undergoing with impunity modes of stimulation
that were at first injurious.

Other effects tending to disguise the general decay of

---

[1] Scott illustrates the same thing in his sexton, *Bride of Lammermoor*, chap.
xxiv.

[2] Dr. Bain suggests that there is a uniform course in this decay of pleasurable
excitement.   "It is rapid at first ; while after a certain time, which may be weeks
or months, but seldom years, the further diminution is imperceptible."   (*The
Emotions and the Will*, pt. i. ch. iv. § 3.)   It would be interesting, if it were practicable, to compare the course of things here with that which takes place when a
nerve-structure is stimulated continuously for a short time.

feeling are due to its increasing complication as experience advances and the mind develops. Thus through the very repetition of experience and the taking on of new adjuncts by association the tendency to decay is in a measure counteracted and disguised. The action of these processes, cumulation of traces and associative complication, in the region of feeling will be considered more fully by-and-by. Here it may be enough by way of illustrating the influence of this factor to remind the reader how a mother's love remains fresh in a measure through *the memory of* past fondlings, tendings, etc., and how it makes up for a loss of pristine sensuous intensity by an accretion of associated elements, the result of years of companionship.[1]

§ 21. *Habit and Craving.* While the direct influence of permanence or custom is thus in general to lower feeling, it has an indirect influence of another kind. What remains with us, what we habitually see, and habitually do, tends to lose its keen pleasurableness and to grow indifferent. Yet, just because it has got to be customary, and we have grown used to it, it generates an attachment or clinging of mind which betrays itself whenever it is removed. Jeannie Deans, feeling strange and lost in her London surroundings, and longing to get back to her familiar scenes, is an example of this effect. Every sudden rupture in our experience, as the loss of a familiar friend, shows the same force of custom in producing a tenacious clinging of mind. Here, then, we have an effect precisely the reverse of the blunting effect of habit. In the pain which comes to us from all severance from the habitual in our life we see custom, instead of deadening feeling, generating and intensifying it. The older and more fixed the habit, the harder is it to bear the sundering of the bond. Habit is thus a fertile source of negative pains, or the pains of craving, a source which grows more prolific as life advances. So strong, indeed, is this influence on our life of feeling that a man will become attached even to the most dreary surroundings when they have become his through long usage. It is said that men who

---

[1] It is plain that the action of memory is not wholly in the direction of sustaining feeling. We often become aware of the decline of feeling by recalling the past and setting it in contrast to the present.

have been long confined in prison have returned voluntarily
to their accustomed haunt.

It is evident that we have in the special influences just
considered elements which serve to limit the value of change
or variety as a condition of a happy life. If we were always
to abandon action at the stage at which it is unpleasantly
laborious, we should never grow to the capability of all the
higher and more difficult exercises of body and mind. The
growth of intellectual and æsthetic interests presupposes a
certain persistence in intellectual activity in spite of its temporary painfulness. This is a truth fraught with important
practical significance. The educator who fears to give a child
anything to do that is not immediately pleasurable can never
develop its higher powers.

In the principle of habit or habituation we have a still more
powerful opponent to the attractions of novelty and variety.
The new fails to delight when it involves a too great and
sudden rupture of continuity in our experience. Even a child
with all its craving for novelty is apt to break down in despair
in the midst of a social treat at suddenly waking up to the fact
that it is in a strange room and among strange faces. The
susceptibility to the charm of novelty on the one side, and to
the mastering of habit on the other, have a different ratio at
different ages and among different individuals. The art of
happy living includes a nice adjustment of these opposing
tendencies, the securing of the maximum of the pleasure of
variety without risk of suffering through a deprivation of what
has grown customary and so necessary.[1]

(C) Complication of Activities.

Thus far we have adopted the abstract supposition that
feeling presents itself as a perfectly simple phenomenon, that
it is occasioned by the stimulation of one organ or set of neural
structures only, as the organ of taste. This supposition is
probably never realised in our actual experience.

§ 22. *Diffusion of Excitation : Organic Consensus.* To begin
with, then, owing to the continuity of structure throughout

---

[1] On the whole effect of change and habit on our pleasures and pains, see Bain,
loc. cit.; Fechner, *Vorschule der Æsthetik*, i. p. 251 ff., ii. chs. xxviii. and xxxix.

the nervous system the stimulation of a particular cortical area tends to propagate or diffuse itself over other areas.[1] This effect is particularly conspicuous in the case of all markedly pleasurable or painful stimulation. Familiar instances of this are the agreeable secondary effects of pleasant rhythmical sound in exciting pleasant rhythmical movements of the limbs, of pleasurable muscular activity in promoting circulation, etc., of certain unpleasant smells, and other sensations in producing disagreeable organic sensations (nausea). The importance of this diffusion of pleasurable and painful stimulation will appear by-and-by, when we come to consider what is known as the physical embodiment and expression of feeling. Here it is enough to say that every mode of agreeable or disagreeable sensory stimulation tends to produce homogeneous secondary results, more particularly in the region of organic or systemic sensation and of muscular sensation. What is known among medical men as the reciprocal action of mind and body refers to this extension of pleasurable or painful condition from organ to organ. The influence of the bodily life in cheering or depressing the mind, and conversely of cheerfulness and gaiety, and of mental worry, grief, and so forth, in furthering or retarding the vital functions, are a conspicuous illustration of this diffusion.

This truth clearly points to a certain organic *rapport*, a kind of unconscious sympathy or " consensus " among the several structures connected. A pleasurable, that is, a beneficial, stimulation of one organ tends to a furtherance of beneficial functional activity in other organs. Similarly in the case of painful or injurious stimulation. The continuous structure and mode of working of the nervous system tend, then, to produce the result of a more or less complete participation of the whole in the varying conditions of each part. The law of our nervous organisation is that of an ideal family or state : the weal or woe of the part becomes the weal or woe of the whole.

While it is thus generally true that a wholesome or hurtful activity of one organ tends to produce a *like* condition in other and connected organs, the law is subject to a number of apparent exceptions. Certain modes of stimulation, which in themselves

---

[1] See above, vol. i. pp. 44 and 135.

and at the time yield pleasure, produce, indirectly and later on, injurious and painful results. The use of narcotics, of piquant and indigestible condiments, illustrates what is here meant. The immediate and local effect may be enjoyment, but this is apt to be dearly paid for by subsequent and diffused effects. On the other hand, many stimuli, which are in their immediate and momentary effects disagreeable, produce, indirectly, salutary and agreeable effects. The passing shock of the cold plunge may be instanced as an example of this. The disagreeable stinging skin-sensation is soon forgotten in the large increase of bodily comfort through heightened circulation, generation of heat, etc. A similar phenomenon presents itself in the rousing of healthy and pleasurable activity of muscle and brain by unpleasant skin-sensations, a phenomenon well known to the trainer of horses and of boys. We must say, then, that while there is a tendency in the organism to a condition of harmonious consensus or sympathetic understanding, the tendency is only imperfectly realised.

The secondary effects of stimulation, more especially those seen in the further-ance and hindrance of the vital processes (digestion or nutrition, circulation, and so forth), have been specially recognised and made prominent by Dr. Bain in his theory of pleasure and pain. He even goes so far as to make the organic effect the essential criterion in his definition of pleasure and pain. According to him, "states of pleasure are connected with an increase, and states of pain with an abatement, of some or all of the vital functions". (*Senses and Intellect*, p. 283.) This, it is evident, is to give a new expression to the law that connects pleasure and pain with quantity of activity. That the effects of stimuli (sensorial and ideational alike) involve these organic effects as a main ingredient is, I think, certain. Yet it seems going too far to say that this is the one essential circumstance. The plea-sure of sunshine is surely to some extent pleasure *for the eye* or optical nerve-structures, and would remain an appreciable enjoyment even if the organic effects were other than they are. Dr. Bain's theory, moreover, appears to ignore the reciprocal effect of bodily on mental conditions. An important element of the heightened pleasure of ten minutes in the playground between classes is the removal of obstacles to *intellectual* activity, the acceleration, so to speak, of the mental current. This is, no doubt, a consequence of a more active blood-circulation in the cerebral substance : yet it makes itself known to consciousness as heightened mental activity.

The harmonious agreement or consensus among different parts of the nervous system here referred to has been formulated under a biological or teleological form as a law of self-preservation. To quote Mr. Spencer, "pains are the correlatives of actions injurious to the organism, pleasures are the correlatives of actions conducive to its welfare". The facts adduced in support of this generalisation are such as the injurious effects of extreme heat and cold, both of which are painful, the general correspondence between agreeable and disagreeable gustatory sensations on the one side, and nutritious and innutritious quality on the other. Mr. Spencer goes on to reason that such a consensus between the normal pleasurable activity of the single organ and the welfare of the whole organism has been brought about by the process of evolution, since all creatures otherwise constituted, that is, so fashioned as to seek pleasure in hurtful ways, and to avoid as painful what is useful and necessary, tend to exterminate themselves. He admits, however, that the correspondence is as yet very far from exact.[1]

§ 23. *Juxtaposition of Stimulations : Conflict and Harmony.* Thus far we have considered the complication of feeling as the result of the initial stimulation and the continuity of the nervous system.[2] But complication arises in another way, *viz.*, through the simultaneous and independent production of different nerve-excitations. As pointed out above, owing to the very structure of the nervous system with its innumerable peripheral points of attack, and owing further to the large range of ideational activity in the case of the developed brain, we are exposed at any time to a multitude of stimuli, external and internal. We have now to inquire into the general effect of such multiplex stimulation on the life of feeling.

§ 24. *Pain of Distraction.* We saw above that, owing to the general form of our conscious life wherever a number of sensations or their ideational substitutes simultaneously present themselves, there is a tendency to connect or unify them as parts of a single presentative whole. If this tendency fails to realise itself, and we are subject to the stimulating action of disconnected impressions and ideas, a disagreeable feeling of confusion or distraction arises. The situation may be illustrated by the mental condition of a chairman at a public meeting when two aspiring orators rise at the same moment. Such disconnected multiplicity of stimuli baffles the effort of attention to single out, and concentrate itself upon, some one object : we feel driven in two opposite directions at the same

---

[1] See his *Principles of Psychology*, i. p. 278 ff.; *cf.* Grant Allen, *Physiological Æsthetics*, chap. ii. ; and Schneider, *Der menschliche Wille*, theil ii. kap. ix. It has been pointed out by Ward that such biological explanation of the fact of a particular arrangement of organs cannot be a substitute for a properly psychological (or one might add psycho-physical) explanation of the phenomena of feeling. (See his article " Psychology," *Encyclop. Britann.* p. 67.)

[2] Another great illustration of such complication is seen in association, the influence of which on the development of feeling will be brought out presently.

moment, and experience something of the lacerating effect of conflict.

When, on the other hand, two or more presentations are brought together as connected parts of a whole, an agreeable feeling of harmony or peaceful reconciliation ensues.   Every discovery of a connexion between seemingly disconnected things—for example, the sound and a simultaneous movement of falling water—may be said to afford a degree of this agreeable feeling of accord.   At the same time, the full agreeableness of harmony appears to be experienced only by way of contrast to a preceding state, however transitory, of dissonance.

§ 24a. *Disappointment.*   A more specialised and distinct form of conflict arises where a particular psychical element tends by the nature of its content to inhibit a second as incompatible and contradictory.   Here, through the nature of the contents, all attempt to reduce plurality to unity is futile, and the sense of antagonism and mutual arrest and destruction grows full and intense.

All disappointment of expectation is an example of this.   To have the psycho-physical mechanism preadjusted for a certain impression, and then to fail to receive the same, introduces a peculiar modification of the feeling of conflict.   The anticipatory idea and the actual presentation seem to clash and produce a sense of antagonism between the self and the not-self.   The more interesting the expectation, the less easily will it be dislodged by the subsequent impression, and consequently the keener will be the painful feeling of jar.

A weakened form of such disappointment presents itself in the effect of all that is foreign, incongruous, and 'out of place'. The immense effect of custom in fixing our ideas of what is " natural " and proper has been recognised by all thinkers.   A striking evidence of customary concomitance in making two things harmonise or fit is seen in the fact that the names of things soon lose all trace of their conventional origin, and appear to belong to the objects, and even to share in their characteristics.   The severe exactions of ceremony, the standards of morality employed by those who do not think, and the sway of custom in all forms of art, alike illustrate the truth that what goes against customary experience is dissonant.

§ 24b. *Logical Contradiction.*   We may now pass to the more distinctly intellectual or logical aspect of this principle. Here conflict assumes the form of a feeling of incompatibility and contradiction.

The state of doubt already spoken of may be taken as an example of such intellectual conflict. Here considerations in favour of a belief collide with others opposed to it, and thus a prolonged state of mental dissension is produced. Other varieties of this conflict may be found in the presentation of statements that are irreconcilable with known facts, or that contradict one another. The characteristic pain of the intellectual life which strives at the organic unity of thought is this feeling of incompatibility or contradiction.

Other forms of conflict which are analogous to this sense of contradiction may be met with in the domain of our emotional and conative experience. Thus, owing to our social instincts and feelings, we experience something closely resembling a sense of contradiction when our sentiments do not accord with those of others. The importance of this will appear when we come to consider the social feelings. Under the head of conation, again, we have many illustrations of a contradictory conflict. Thus one impulse may drive us towards a line of action, while another drives us away from the same, as when pleasure and duty oppose one another; or we may be impelled to do a certain thing, but be held back by a doubt of our ability to succeed. All such experiences, as we shall see when we come to examine them more minutely, have as their accompaniment a painful sense of jar and of rending atwain.

On the other hand, we experience a pleasurable feeling of harmony when the reality agrees with our expectation, when statement harmonises with statement, when reasons converge to a single conclusion, when others share in our beliefs and sentiments, when impulse and reason point to the same course, and so forth. In all such cases there seems to be an expansion instead of a restriction of activity. We may suppose that in all such cases of harmonious stimulation the several constituent nerve-processes involved are not only capable of being carried out smoothly without mutual obstruction, but are of such a form as to further and augment one another, so that there results a large wave of pleasurable excitation.

§ 25. *Harmony and Conflict between Feelings.* Thus far we

have been dealing with the principle of Conflict and Harmony
as an intellectual or "objective" principle, that is, as a
principle which subserves the logical unity of thought. We
have now to view it under another aspect, as a purely
"subjective" principle, that is, one which controls the combinations of feeling itself. Here we have to ask how a plurality
of affective elements behave when they come together into
consciousness.

As already remarked, there is a radical opposition between
pleasurable and painful feeling as such—an opposition which
causes the simultaneous rise of the two in consciousness to be
always attended with some degree of the effect of conflict.
The close concomitance of pleasure and pain in life is recognised
as a sort of dissonance.

It must not, however, be supposed that this sense of conflict
between the pleasurable and the painful always emerges into
distinct consciousness. In some cases it is apt to be disguised
by the fact that the two feelings are organically connected in
their origin or presentative basis. Thus certain sensations, as
those produced by particular modes of tickling, have the pleasurable and the painful so intimately conjoined that we can hardly
distinguish the two, and are disposed to regard them as
intervening or "neutral" states. Certain trains of imagination, as in recalling some pathetico-humorous incident, show the
same effect of a *quasi*-commingling and mutual counteraction of
the opposed affective constituents of a psychosis.

Again, in many instances of simultaneous pleasure and pain,
one of the feelings is so much stronger than the other as to
thrust it into the background of the sub-conscious. The present
writer knows a boy of ten who honestly believes that by
adding a small amount of syrup to very cold water this is
made less cold. What really happens here is, of course, that
the agreeable sensation given by the syrup *masks* the disagreeableness of the cold fluid.

In certain cases, moreover, when a pleasurable and a painful
feeling arise together in unequal intensity the subordinate
element, though partially realised, produces rather an effect of
contrast than one of conflict. This is illustrated in the
pleasures of sad reverie or melancholy, of tragedy, and the like.
Here there is an under-current of painful feeling which is not

fully realised as such, but rather serves to intensify the feeling of pleasure, the dominant factor in the mood. This effect of the heightening of pleasure by a sub-conscious pain, which is recognised in the theory of tragic effect of Hume and others, is well illustrated by Mr. Kinglake in his *Eothen*. The fear of the plague (he tells us), which was rife at Cairo when he was there, "operated as excitement giving rise to unusual animation". The eagerness with which he pursued his rambles among the wonders of Egypt "was sharpened and increased by the sting of the fear of death".

As pointed out above, what is called neutral excitement is frequently due to such surging of pleasurable feeling out of an under-current of painful consciousness. In the excitement of brutal spectacle, as the gladiator-combat, the bull-fight, and so forth, there seems to be a positive heightening of the total pleasure-effect by these subordinate elements of the painful. It may be added that a similar effect of contrast-heightening is seen in the intensification of pain by subordinate elements of pleasure, as in the more painful stages of grief, and in remorse.[1]

A full sense of conflict between pleasure and pain arises when the two feelings are both present in a distinct and strong form and are not so unequal in point of strength as to allow of one overpowering the other. Thus the co-existence of a harsh nasal voice with a lovely face affects us like a musical dissonance. The death of Gloucester, whose

> Flaw'd heart
> (Alack, too weak the conflict to support!)
> 'Twixt two extremes of passion, joy and grief,
> Burst smilingly,

illustrates the effect of such conflict in its more intense form.

It may be added that, since pleasure is what we desire and like to retain, any intrusion of the painful on a pleasurable state is specially resented as a dissonance. Here the opposition between pleasure and pain is reinforced by the conflict of desire with reality.

We may now pass to conflicts of a more special and definite kind among our affective elements. In many cases there is something in the whole psycho-physical condition with which a feeling is bound up rendering it specially antagonistic to, and

---

[1] On this partial blending of painful and pleasurable elements in composite feelings, see Bouillier, *Du Plaisir et de la Douleur*, chap. vi.

exclusive of, other opposed conditions. To take an example :
all quiet, soothing varieties of pleasurable stimulation, as
soft touches, slow, caressing movements over the skin, slow
movements of our own limbs, soft and slowly succeeding
sounds, put us into a certain psycho-physical mood which
renders us indisposed to respond to more exciting stimuli.
The sombre, melancholy pleasures are in this way incompatible
and dissonant with gay, exciting pleasures. On the other hand,
any new mode of stimulation of a similar kind, that is, producing a kindred affective tone, is welcomed. When gay we
look out for exciting pleasures, when melancholy we welcome
all that harmonises with this particular tone of feeling. This
influence of mood in securing continuity of a particular variety
of feeling will be illustrated by-and-by when we take up the
subject of emotion.

§ 26. *Rhythmic Sequence.* In close connexion with the
principle of conflict and harmony among simultaneous excitations as influencing feeling is that of rhythmic and arhythmic
combination among successive stimulations. A regular periodic succession of bodily movement seems to be conditioned by
the very structure of our nerve-muscular organism. Rhythmic
movement is easy in itself, and as such pleasurable, and whenever we move freely we tend to move rhythmically. Hence the
liking for it in all visible movements, as in those of a graceful
walk, of the theatrical ballet, and so forth. It is to be added
that the laws of attention require, as the condition of a pleasurable succession of sense-impressions, a certain degree of time-
regularity or periodicity. This effect of eased attention is very
conspicuous in the pleasure of rhythmic sound.

It is probable that all such rhythmic successions involve
rapid alternating preadjustments and satisfactions of expectant
attention or expectation. In this way, then, it seems possible
to connect the agreeableness of rhythmic and the disagreeableness of arhythmic series with the pleasure of harmony and
its opposite.[1]

The principle of harmony and conflict has been recognised by most psychologists as a law controlling the affective result of all compound excitation. It may
be added that in general the feeling of conflict is much the more distinct of the two.

[1] *Cf.* Ward, article " Psychology," *Encyclop. Britann.* p. 69.

Elements which are affectively disparate always produce the consciousness of conflict in some degree. On the other hand, the mere concurrence in peaceful juxtaposition of two or more excitations does not always reveal itself as a harmony. We only have a full and distinct consciousness of harmony when we distinguish the elements as (numerically) separate and apprehend the fact of their mutual compatibility and mutual support. This is best effected by a preceding state of conflict which as such separates the elements for our consciousness. Hence the feeling of harmony is most intensely realised as a transition from a preceding state of dissonance. Thus the feeling of harmony in music is greatly helped by a pre-ceding dissonance and the resolution of the same by the removal of the dissonant element. To this extent, then, the pain of conflict may be said to be the primary experience, and the pleasure of harmony be described as secondary or relative.

It is evident from what has just been said that the principle of harmony in the domain of feeling is related to that of similarity in the domain of intellect. In all fully-developed sense of harmony there is a consciousness of agreement, whether of presentative characters, *e.g.*, elements of colour, or of form, or of affective elements themselves, *i.e.*, affective quality or tone. This obviously answers to what was said above respecting the relation of contrast as a principle which governs both feeling and intellectual activity (discrimination). Hence we may say that contrast and harmony are in the region of feeling the representatives of difference and likeness in the region of intellection.

REFERENCES FOR READING.

On the distinction of pleasure and pain and the conditions which determine them, see Hamilton, *Lectures on Metaphysics*, ii. xli. ff. ; Bain, *The Senses and the Intellect*, 282 ff., and *The Emotions and the Will*, chap. i. ; H. Spencer, *Principles of Psychology*, i. § 124 ff. ; Ward, article "Psychology," *Encyclop. Britann.* 66, col. 2 and following; Wundt, *Physiol. Psychologi.*, i. cap. x. ; and Bouillier,. *Du Plaisir et de la Douleur*. Compare below, Appendix I.

CHAPTER XIV.

Having in the previous chapter discussed the general features and the conditions of Feeling, we may proceed to a consideration of its several commonly recognised varieties, such as the pleasures of colour, muscular activity, and so forth.

VARIETIES OF FEELING.

§ 1. *How Feelings are to be Distinguished.* We have seen that all our feelings are constituted by elements of pleasure and pain. Strictly speaking, therefore, there are only two varieties of feeling, *viz.*, the agreeable and the disagreeable.[1] All the various concrete feelings making up our actual experience, as those of hunger, cold, and fear, are, according to the view here adopted, *quâ* feelings, merely particular modifications of agreeable and disagreeable consciousness. What makes us speak of them as different feelings is, to some extent, the dissimilarity in respect of feeling-characters themselves (intensity, temporal course), and still more the difference in the sensational or other presentative materials with which the feeling-element is incorporated. These differences appear, until a finer scientific analysis is brought to bear on the phenomenon, to extend to and involve the feeling-concomitant itself. Thus the pleasure of quenching thirst seems to the unanalytical mind another *sort* of feeling from that of a sweet fragrance.

What is called the classification of our feelings is thus an arrangement of the psychical states or psychoses of which feel-

---

[1] Or if, with Bain, we were further to recognise "neutral excitement" as a mode of feeling, we should have to reduce all varieties to three forms.

(46)

ing proper is an ingredient as concrete wholes, and an arrangement of them primarily with reference to differences in their
presentative base, that is, sensational or ideational elements and
their groupings.[1]

That our classifications of the feelings do actually turn on
the differences of presentative character with which they are
bound up is seen in the case of the broad division adopted by
common thought and psychology between physical or ' bodily '
feelings, and ' mental ' feelings or emotions. This distinction
is obviously based on the mode of excitation of the feeling and on
the nature of its presentative substratum. A bodily feeling, as
hunger, or tickling, is directly due to a process of peripheral
nerve-stimulation, and is the concomitant of what we call a
sensation. On the other hand, an emotion, such as fear, or
anger, always involves some central (perceptional or ideational)
activity, and may, in contradistinction to a bodily feeling, be
marked off as *centrally* excited. This primary division is too
firmly fixed in common thought to be set at nought. We shall,
moreover, see presently that the division is scientifically sound,
inasmuch as the so-called mental feelings or emotions are complex feelings, into which the relatively simple " bodily " feelings
enter as constituents. Hence we shall make this bipartite
division our starting-point in the treatment of the subject. The
first class will be spoken of as Sense-Feelings, the second as
Emotions.

(A) SENSE-FEELINGS.

§ 2. *Definition of Sense-Feeling.* By a sense-feeling is
meant one that is determined by certain features in the process
of peripheral sensory stimulation, such as the intensity of the
stimulus, or its peculiar form (answering to quality of sensation). Examples of sense-feelings are those accompanying
sensations of heat and cold, taste, smell, muscular activity in
their several degrees of intensity.[2] After our numerous refer-

---

[1] This is sufficiently illustrated in the fact that the directly opposed feelings
pleasure and pain when excited by closely related ideas, *e.g.*, the beautiful and the
ugly, the (morally) right and wrong, are called by the same name—æsthetic feeling,
moral feeling.

[2] The feelings which attend muscular exercise come under our definition in so
far as they involve reflex afferent stimulation.

ences to these in illustrating the general properties of feeling, we may dismiss them with a few additional words.

§ 3. *Pleasures and Pains of Organic Life.* As already pointed out, feeling is a prominent feature in our organic sensations, those which accompany the actions of the vital organs (nutritive functions). Owing to the large extent of sensory surface involved, these feelings are, for the most part, deep and massive, rather than acute and intense. The latter are illustrated in the effects of injury to the skin, which may be definitely and narrowly circumscribed, as in the case of a painful prick.

Another feature to be noticed in the organic feelings is the greater conspicuousness of the disagreeable, as compared with the agreeable element. When we think of a feeling of digestion, we naturally represent an uncomfortable sensation. This is ex-plained by the fact that since pleasure is the accompaniment of a normal customary function, and since this is always more or less prolonged, the feeling is subject to the principle of change and accommodation. We may be deriving a certain mild voluminous pleasure from breathing under favourable circumstances, but we only note the pleasure as an appreciable quantity in passing from a lower to a higher functional activity, as in stepping out from a close room into the fresh air outside. Pain, on the other hand, being the concomitant of disturbance, is much more exceptional, and so impressive. This applies to the pains of craving, which we call appetite, for, though they recur regularly, they recur at sufficiently wide intervals of time, and with a sufficient degree of suddenness, to be eminently noticeable.

Since, as we have seen, the organic sensations are not clearly distinguished one from another, but for the most part fuse or coalesce in an undiscriminated mass of sensation, the feelings which immediately accompany them usually blend also. The result of the coalescence of the numerous elements of agreeable or disagreeable feeling at any one time constitutes what is known as the general feeling of physical well-being or health, or its opposite, *malaise.* This composite feeling forms the ground-tone of what we call the mental mood or temper of the time. As we shall see later, the organic feelings exercise in this way an important influence on the whole of the affective life. It is to be added that certain changes in the organic feelings due to modifications of respiration, circulation, secretion, temperature, etc., are now recognised as contributing important characteristic elements to our emotional states, as fear and anger.[1]

§ 4. *Feelings of Special Sense : Pleasures and Pains of Taste and Smell.* In the case of the special senses we are much better able to trace out the several elements of feeling and connect them with definite properties of the sensation. The most important of these are intensity and quality.

In the case of Taste and Smell the affective element is pronounced. As already pointed out, we commonly distinguish the sensations of these senses according to their affective aspect, as agreeable or disagreeable. It was observed further that this difference appears to be connected rather with the quality than with the intensity of the particular sensation. The pleasures of taste and smell, though intense, are limited by the absence of that range of qualitative difference, and of that scope for definitely harmonious modes of combination, which we have in the higher senses. Although there is some agreement with respect to what goes well and ill together in the matter of flavours, it would be difficult to draw up any definite rules on the subject.

The feelings connected with these senses, being well-marked and intense, occupy a prominent place in the affective life. The importance of the feelings of taste is illustrated in the fact that the names of them (bitter, sweet) have come to be representative of pleasurable and painful experience in general. The feelings of odour are known to play a considerable part in the sociality of animals, including the attractions of sex, and in the case of man readily combine with higher emotive forms of enjoyment, as in the delicate perfumes of social entertainment, and in the incense of religious worship.

§ 5. *Pleasures and Pains of Touch, etc.* Touch, while a highly intellectual sense, contributes, especially in early life, a large mass of feeling. The difference of the agreeable and the disagreeable coincides with the qualitative difference of smooth and rough. Objects which, in addition to having a smooth surface, are soft or yielding are the source of a specially luxurious enjoyment. Children revel in touching muffs,

---

[1] For an account of the feelings of organic life, see references above, p. 84, footnote [2]; *cf.* Körner, *Das körperliche Gefühl*, p. 28 ff. The feeling of *malaise* is carefully examined by Beaunis, *Les Sensations internes*, p. 185 ff.

cushions, and other soft things, and probably know a bliss in these sensations which later life cannot reproduce. Among the more exciting feelings of touch we must include that of chafing an itching surface and of tickling. These feelings are, as already suggested, probably mixed at a certain stage, becoming distinctly disagreeable when the stimulation is prolonged.[1]

Thermal sensations supply a new element of feeling to touch. Under ordinary circumstances moderate warmth is pleasant, but owing to the relativity of thermal sensations no general rule can be laid down. When uncomfortably warm we dislike contact with warm objects and welcome that with cold. The plunge of the heated body into cold water is delicious.

It would thus appear that touch, while yielding certain intense modes of feeling, does not supply a wide *variety* of pleasurable quality comparable with that of the two higher senses. Agreeable combination and its opposite have, however, a certain sway in this region. Thus the pleasure of warmth seems to comport with that of contact with soft things. Our liking, especially in early life, for furry animals, and birds, illustrates this. The combination of warmth with smoothness and softness in the human body is a well-recognised ingredient in the charm of sex.

§ 6. *Pleasures and Pains of Hearing.* In the case of the two higher senses, Hearing and Sight, we find that the element of feeling falls back to some extent owing to the greater degree of definiteness (in respect of quality, etc.) of the presentative element. At the same time, these senses contribute a large range of fairly intense feeling. It is to be noted that here the agreeable is as pronounced as the disagreeable, and in the case of sight, at least, tends to preponderate over the disagreeable. In this respect these senses show the opposite relation to that seen in the organic feelings. Owing to the wide variety of stimuli available, and the varying modes in

---

[1] The pleasurable side of touch is hardly done justice to by Ziehen when he says that the accompanying feelings of pleasure are considerably less pronounced than those of pain. (See his work, *Leitfaden der physiol. Psychologie*, p. 84.) The pains of pricking and the extremes of heat and cold, of which he here speaks, probably involve, like the pleasures of chafing an itching spot and of tickling, common sensibility. The pure tactile pleasure is best illustrated in the enjoyment of smoothness and softness. The painful aspect of skin-sensation is fully illustrated by Beaunis, *Les Sensations internes*, p. 169 ff.

which the sensations may be agreeably combined, these senses allow of a much more prolonged enjoyment than is possible in the case of the lower senses. The fact that all the arts appeal either to the ear or to the eye sufficiently attests this truth.[1]

In the case of HEARING, the difference of the agreeable and the disagreeable coincides in general with that of musical sounds, or tones, and of noises. The characteristic unpleasantness of sound is described by such terms as grating, scratching, rasping, jarring, buzzing, and so forth, terms which imply a roughness or unevenness of sound. On the other hand, all musical tone, as even smooth stimulation, is agreeable. We thus see that in the case of hearing pleasure and pain are connected respectively with regular or even, and with irregular or broken stimulation. This connexion is further seen in the pleasantness of musical harmony and in the characteristic unpleasantness of dissonance.

The several varieties of musical sensation, corresponding with tones of different pitch and of different timbre, have distinct affective concomitants. Thus tones of high pitch are more exciting or exhilarating than those of low pitch, which have a quieter and graver character.[2] Similarly, the tones of different musical instruments vary greatly in their shade of feeling. Thus we have the strong, exultant note of the trumpet, the more tender and soothing note of the flute or hautboy, and so forth.[3]

As already pointed out, all musical tone is pleasurable within certain limits of intensity. The weaker degrees of intensity are attended with a different affective result from that which accompanies stronger intensities. Soft, quiet tones are soothing; loud tones, exhilarating. With this contrast is closely connected the analogous one between slow and rapid musical successions. The art of music sufficiently illustrates

[1] On the variations of the affective element in the case of the lower and the higher senses, see Lotze, *Microcosmus*, i. p. 569 f.

[2] Stumpf tells us that a boy of five and a half years, when asked which of two tones was the higher, replied: "Do you mean which is the dull one, and which is the bright one?" (*Tonpsychologie*, ii. p. 531.)

[3] This difference in the affective concomitant of different instruments is clearly recognised by Dryden in his celebrated *Ode*.

the contrast of effect between soft, slow movements (adagio, largo) and louder and more rapid ones (allegro, presto).

§ 7. *Pleasures and Pains of Sight.* In the case of SIGHT, as in that of hearing, we find a disagreeable effect connected with excessive stimulation. The characteristic unpleasantness of this sense is that of too strong or blinding light. In visual sensation there does not seem to be a clearly-marked difference of unpleasant and pleasant answering to a difference in the quality of the sensation, as is observed in the contrast of musical sounds and noises. Besides the effect of excessive stimulation, sight only presents us with disagreeable feeling in connexion with complex modes of stimulation, as in the unpleasant impression of a flickering light, and in the combinations of colour, which may be either agreeable or disagreeable. For the rest, sensations of light and colour may be viewed as agreeable. Pleasure thus appears to transcend pain in the case of the highest and most delicate sense.

The pleasures of light are among the best recognised of human enjoyments. This is sufficiently attested by its common symbolism. In Greek popular thought light, and more particularly the sun's rays, stand for the joyousness of life as a whole.[1] The exhilarating effect of bright light on the infant shows that this feeling is largely sensuous and not dependent on associated elements. Along with the pleasure of mere light, whether supplied directly by luminous objects or reflected, there is the peculiar effect of lustre which is given by bright, smooth surfaces, as water, polished stones and wood, and which appears to depend on a complex effect of sensation, *viz.*, unequal intensities of the sensations received by way of the two eyes.[2]

The varieties of colour, like those of tone, supply a wide range of pleasant impression, and these are capable of being happily combined into rich complex effects of art. The contrast of the exciting with the quiet or grave is here illustrated in the difference of effect in the case of the vigorous colours at the warm end of the spectrum (reds and yellows) and the quiet colours at the cold end (blues and greens). It has commonly been recognised

---

[1] Antigone, in her mournful farewell to life, twice refers to the brightness of the sun.

[2] See above, i. p. 258; and, for a fuller account, Helmholtz, *Physiol. Optik*, p. 782 ff., and Wundt, *op. cit.*, ii. p. 178 ff.

that the reds have a peculiarly exhilarating and festal character, which makes them analogous in their affective aspect to the more inspiriting tones, *e.g.*, that of the trumpet. The blues, on the other hand, have been regarded by Goethe and others as producing a more tranquil, restful effect. This contrast, combined with that of degrees of brightness in the same colour, allows of many fine *nuances* of affective tone from the gloomy, half-disagreeable effect of a dull, dead black, through intermediate effects of the grave, as that of dark blue or purple, up to the gaiety of the bright yellows and reds. These differences in "feeling-tone" are fully illustrated in the symbolism of colour, the choice of particular tints for mourning, for festal revelry, for kingly pomp, etc.[1]

§ 8. *Pleasures and Pains of Muscular Action.* Finally, we may refer to the feelings which are the concomitants of *Muscular Sensations*, that is to say, the sensations excited by the various actions and changing conditions of our muscular system. To take the pleasures of muscular action first, we have a whole scale of enjoyment from the extreme of gentle movements up to that of vigorous muscular exertion. The intense delight of muscular activity is abundantly illustrated in the spontaneous or playful movements of children and young animals. There is a marked contrast of affective tone between the quiet, restful enjoyment of slow movements, as illustrated in the soothing effect of rocking the infant, and the exhilarating, exciting pleasures of rapid movements, as in the galop and other wild dances, and more generally of all vigorous muscular expenditure, as in the strenuous exertions of football and other games.

The feelings accompanying muscular action play a prominent and important part in our affective experience. As already pointed out, every process of sensory stimulation calls forth some motor reaction (varying with its intensity, suddenness, etc.), and this reaction, by altering the degree of tension in a smaller or larger number of our muscles, produces a change of

---

[1] The well-known effect of bright red in exciting and maddening certain animals is probably a case of an original instinctive reaction. It may possibly be connected with the reaction which is called forth by the sight of blood. It may be added that psychologists have been apt to attribute fanciful analogical suggestions to colours. Thus George seeks to find in the case of each of the colours an analogue in the region of taste. (See Ziehen, *Leitfaden der physiol. Psychologie*, p. 88.)

feeling.   To this it may be added that the reaction of attention, which (as we have seen) involves a muscular process, is a
further illustration of the far-reaching co-operation of the motor
element in our states of feeling.   Finally, it is to be borne in
mind that in those organic reactions which enter into our emotive
states, as joy, and anger, the muscular factor fills a very considerable place.   The feelings which grow out of varying intensities and extensities of muscular contraction are thus a main
constituent in all our affective states.

As already illustrated, muscular action, when it becomes
excessive, is a source of one of our most familiar pains.   The
sensations of fatigue induced by prolonged labour are a typical
example of massive oppressive suffering.   It is probable, as
already hinted, that this mode of painful experience involves
the nerves of common sensibility.

Lastly, it is to be noted that our muscular experience
supplies us with ample illustration of the effects of change, and
of favourable and unfavourable combination of elements.   The
transition to muscular activity after a prolonged inactivity is
one of the most exciting enjoyments of vigorous youth.   On
the other hand, repose after fatiguing exertion illustrates, as
we have seen, the effect of transition and contrast in another
way.   In the rhythmical co-ordination of a number of concurrent muscular actions, as in many forms of playful exercise,
such as rowing, riding, and skating, and in all dance-move-
ments, we see that the pleasures of muscular activity are
susceptible of a rich increase through a harmonious or
favourable co-adjustment of elements, simultaneous and successive.

§ 9. *Complexity and Alteration of Sense-Feelings.*   It follows
from what has been said respecting the uniform sequence of a
motor reaction on a sensory stimulation that our sense-feelings
are from the first never perfectly simple phenomena.   The
pleasures of the palate, of the eye, and so forth, are always
complicated by the affective concomitant of the reaction called
forth.   It must be remembered, too, that one important branch
of this reaction is the adjustive process of attention, which
itself, as moderately stimulated or overpowered, contributes an
element of the agreeable or disagreeable to the whole psychical
result.

To this it must be added that in mature life we never have a sense-feeling which is not further complicated by the presence of a representative or associated element. It is probable that the pleasures of tone and colour are considerably enriched in later life through such associative adjuncts.

It may be remarked that our sense-feelings undergo a double process of alteration as experience progresses. To begin with, there seems to be a considerable falling off in the pristine intensity of sensuous enjoyment. This is supported by the fact that many persons can recall a delicious intensity of tactile and other sensations which appears, at least in retrospect, vastly superior to later experiences. Thus Miss Martineau speaks of the exquisite delight with which, when a child of two or three, she used to touch a velvet button on a sister's bonnet. I myself can occasionally catch faint after-glimpses of a child's thrill of delight in particular colours, to which I recognise I am now a stranger.[1]

These facts seem sufficiently explained by the agencies of accommodation already dealt with, and the growing preponderance of the intellectual or presentative consciousness. As was pointed out above, we attend to sensations, under ordinary circumstances, only so far as they are signs of things which are important to us. Experience tends, in the way already shown, to invest our sensations with objective significance. We do not realise the full sensuous effect of colour owing to the instantaneous rise of these objective suggestions. Artistic training of the eye involves the stripping off of these after-growths so as to get something approaching to the pure sensuous effect which is the prerogative of childhood. While experience thus tends to dull the first keen delight in sensation, it effects another and compensatory kind of change in our sense-feelings by means of the associative process just referred to, which itself supplies added affective elements. Thus the colour blue may draw a secondary or derivate element of pleasurableness from vague suggestions of summer skies and seas, and loving eyes. How far this effect of suggestion reaches will have to be con-

---

[1] Since writing the above, I find that J. S. Mill records a quite similar recollection. (Jas. Mill's *Analysis*, ii. p. 246, footnote 47.) My own recollection, like Mill's, is that it was the saturated colours of certain flowers which specially excited this feeling.

sidered when we take up the whole subject of the influence of association on feeling.[1]

(B) Emotions.

We have now to pass to the consideration of the second great class of feelings, those commonly known as Emotions or Passions, such as joy, grief, fear, anger, love.

The nomenclature of the feelings is by no means satisfactorily fixed in psychology. The word emotion is only just beginning to receive general adoption as the name of the higher group of feelings. Hamilton marked these off as internal feelings or sentiments. The terms affection and passion have also been used by earlier writers so as to include all that we now call emotion.[2]

It seems better, however, to use the term emotion as the generic term and reserve the names affection and passion for certain modifications. Thus, "passion" is best confined to the more violent manifestations of feeling, as in the case of a passionate love or hate, which we commonly set in opposition to thought or reason. Affection, on the other hand, seems rather to point to a *recurring* feeling or fixed emotive *disposition*, as in the case of what are popularly called affections, *viz.*, human attachments. It may be added that German usage in this respect differs from ours. The term Gemüthsbewegung, "movement of feeling," answers etymologically to emotion, but it emphasises the fact of the muscular movement (expressional movement) more than our term emotion. On the other hand, the word "Affect" is used in a peculiar sense by the Herbartians and others for a sudden and violent emotional discharge.[3]

§ 10. *Emotions marked off from Sense-Feelings : Structure of Emotion.*   As pointed out above, an emotion differs from a sense-feeling in having a *mental*, or, to speak more precisely, a central psycho-physical, origin.   The pain of a prick is supposed to be the result of the afferent process in the particular nerve stimulated.   The child's fear of a dog obviously has its starting-

[1] For a fuller account of the sense-feelings the reader may consult Bain, *The Senses and the Intellect*, p. 101 ff. ; Grant Allen, *Physiological Æsthetics*, chaps. iv.-vii. ; also Wundt, *op. cit.*, i. cap. x. ; Horwicz, *Psychologische Analysen*, 2er theil, 2e hälfte ; and Ziehen, *Leitfaden der phys. Psychol.* 7te Vorlesung.

[2] "Affection" was used by Spinoza, and "passion" by Descartes as a generic term for all emotional states.

[3] On the question of nomenclature, see Bain, *The Emotions and the Will*, Appendix B.   On the use of the term "Affect" in German psychology, see Wundt, *op. cit.*, ii. p. 404.   C. Lange, in his work *Ueber Gemüthsbewegungen*, seeks to mark off certain feelings, such as love, hate, contempt, as passions (Leidenschaften), from others, as sorrow, joy, fear, anger, which he calls emotions (Gemüthsbewegungen) ; but the distinction does not appear to me to be quite clear.

point in a central psycho-physical process corresponding to what we call the percept or an idea of an object. Since, as we have seen, even a percept involves a representative element, we may say that emotions are in general marked off from sense-feelings by the presence of a representative factor.

In the second place, an emotion is characterised by a wide diffusive effect. Sense-feelings are restricted in respect of the range of nervous excitation involved. The pleasure of a sweet perfume seems mainly, if not exclusively, referrible to the olfactory nerve-structures.[1] On the other hand, an emotion of anger or terror is marked by a wide-ranging excitation, involving the voluntary muscles and the viscera (heart, respiratory organs, etc.). These diffused effects contribute reflexly a number of secondary sense-feelings which constitute an integral part of the whole emotion.

We may say, then, that an emotion is a complex psychical phenomenon made up of two factors, or, as we may call them, stages : (a) the primary stage of central excitation ; and (b) the secondary stage of somatic resonance. The first includes the sensuous effect of the initial peripheral stimulation, together with the representative elements associatively conjoined with this. Thus, in the case of a sudden fear, the primary stage includes the immediate effect of the sudden sensory stimulation, viz., mental shock, or a momentary overpowering of the attention, with vague representations of harm ; whereas the secondary stage includes all the modifications in tension of muscle, organic function, brought about by the shock, viz., loss of muscular power, disturbance of heart's action, pallor, alteration of secretion, etc. It follows that emotion is in general describable as a mass or aggregate of sensuous and representative material, having a strongly marked and predominant concomitant of feeling or affective tone.

There is reason to think that both of these factors are present in every emotion. At the same time they combine in very different proportions. Thus, for example, in the start at a sudden noise, the representative element is hardly appreciable. Such reflex emotive discharges, which, as we shall see,

---

[1] Not exclusively, because, as we have seen, every sense-feeling is probably complicated by some amount of irradiated or diffused effect.

are instinctive, occurring early in life, constitute a transition from sense-feeling to emotion. They may be roughly described as masses of sensation with the concomitant sense-feelings. On the other hand, in many of the higher feelings, as moral approval, delight in knowledge, and the like, the representative element is rich and prominent, while the reflex effect of bodily resonance becomes less distinct. Hence these may be roughly marked off as welded masses of representation, with large and prominent affective concomitants.

The question of the general composition or structure of an emotion is a some-what difficult one, owing to the great variety of phenomena denoted by the term. The older and the still common view is that an emotion is wholly mental, and that the bodily resonance is merely its physical "embodiment" or "manifestation". This conception of the matter errs by overlooking the fact that the strongest cha-racteristics of the several emotions, as anger, fear, shame, are *constituted* by the reflex sensuous feelings due to bodily diffusion, *e.g.*, of tense muscle in anger, of muscular tremor in fear, of burning cheek in shame.

§ 10*a*. *Relation of Corporeal Resonance to Emotion*. That the corporeal reson-ance does form an essential ingredient in emotion is abundantly proved by a variety of facts. Thus it is well known that in acting and mimicry we can, to some extent, take on an emotion of anger, pride, and so forth, by voluntarily assuming the mus-cular actions that form part of its expression.[1] Conversely we can repress an emo-tion in part at least by voluntarily inhibiting those actions of the "voluntary" muscles which form an important constituent in its bodily resonance. With this may be compared the fact that, when, owing to independent causes, certain organic changes are set up, these constitute a powerful predisposition to the particular emotion which has these modifications as its resonance. Thus the palpitation brought on by a fit of indigestion makes a man nervous and apprehensive of evil. The influence of what we call *mood*, that is, a sum of relatively permanent organic modifications colouring the tone of mind, is another illustration of the same thing. A cheerful mood predisposes us to joy and laughter, because the organic resonance of these feelings is already sub-excited. Similarly with gloomy, irritable, and other moods. The still more permanent changes of organic feeling produced by pathological disturbances of the vital functions are well known to be attended with marked changes in the emotional life.

While, however, we thus regard the corporeal resonance as an integral part of the emotion itself, we are not prepared with W. James to view it as the whole of the emotion. Even when representation seems to be completely absent, as in the laughter provoked by tickling, and the fear immediately excited by a sudden noise, there is the feeling due to the primary sensation itself and its central psycho-physical effect, the strange half-pleasant, half-unpleasant feeling of the titillation itself, the disagreeable mental shock. No doubt, as James seems to say, the organic resonance may, in the case of such primitive reflex discharge, be the main factor in the affective colouring of the whole emotive state ; but it is to be

[1] A like effect of induction of emotion by assumption of expressional movement seems to be illustrated in the results of suggestion on the hypnotised subject.

noticed that these are just the cases where we should hesitate to call the state an *emotion* at all. In what we are all agreed to recognise as emotion the factor of representative consciousness is present and prominent. Exactly how much each of the two factors (primary and resonant) contributes to the whole body of the feeling it is, of course, impossible to determine.[1]

§ 11. *Rise and Fall of Emotion : Emotional Persistence.* Our brief account of the composition of an emotion has led us to see that it is a process occupying an appreciable time. The pain resulting from a prick may be momentary, disappearing with the withdrawal of the stimulus. But a state of grief requires time for its full realisation. An emotion undergoes a certain rise or development from the stage of just appreciable excitement up to culmination. This course of development is determined, to some extent, by the range of resonant effect, the reflex results of which, while all occupy *some* duration, require unequal times for their realisation, and so constitute a gradual expansion of the whole emotive current. We do not become fully angry until our muscular apparatus has gone through the proper amount of characteristic action, as frowning and clenching of teeth. Similarly, fear is only fully realised when the cycle of organic effect is allowed to proceed unchecked. To this must be added that in the case of all but the primitive instinctive emotions there is the need of a certain mental occupation with the exciting cause. Fear grows through the gradual representation of the danger of what I see. Hence we often feel more afraid after the danger is over, just because the gravity of the situation is more clearly taken in.[2] Similarly, we have to mentally turn over a piece of good fortune in order to realise the whole joy of it ; to trace out, in bitter, brooding thought, all the supposed malice of an injury before we rise to the culminating point of wrath.

While there is thus a gradual rise or development of emotion, there is also a gradual fall or subsidence. A great joy, a

---

[1] On the relation of the two factors, see Bain, *The Emotions and the Will*, chap. i. ; W. James, *Principles of Psychology*, vol. ii. chap. xxv. ; Ladd, *Elements of Physiol. Psychology*, pt. ii. chap. ix. § 21 ; Höffding, *Psychology*, vi. D. i. C. Lange, in his little work, *Ueber Gemüthsbewegungen*, follows a line closely approaching that of James. (See especially p. 53 f.)

[2] This has certainly been my own experience in one or two bad accidents. Such late on-coming of fear does not, of course, apply to the instinctive forms emphasised by W. James.

fit of terror, only dies away and leaves us calm after an appreciable, and, in some cases, considerable time.   This persistence of
emotion seems most readily explained by the large range of
bodily disturbance involved.   These modifications of muscular
tension, circulation, secretion, etc., are apt to persist, and in
this way the emotive excitement is prolonged.   The influence
of the bodily resonance in prolonging a state of emotion is seen
in the fact that we often go on feeling afraid, angry, and so
forth, after the exciting cause is known to be removed.[1]

§ 12. *Influence of Emotion on the Thoughts.*   We are now in a
position to understand the well-known effect of emotion on the
intellectual processes.   This may be viewed under two aspects :
(*a*) the negative or inhibitory effect ; (*b*) the positive or promotive
effect.

The inhibitory effect of emotion springs out of the fundamental opposition of strong feeling to intellectual activity.
This phenomenon is very strikingly illustrated in the immediate results of all violent emotional agitation or shock (the
German "Affect").   The sudden arrival of a bit of exciting
intelligence, whether of a joyful character, as the inheritance
of an unexpected fortune, or of a miserable character, as the
death of a beloved friend, is apt to paralyse thought for a
while.   The mental constitution is unable at the moment to
adjust itself, to fit in the new fact with previous habits of
thought, and a sense of overpowering confusion is wont to
ensue.   Further, all intense and prolonged painful emotion
tends to retard the processes of thought by depressing or ex-
hausting the nervous system.

In the second place, emotion as cerebral excitement is in its
less agitating degrees distinctly promotive of ideation.   We
never have in our cooler moments such a swift rush of ideas
as we have in moments of emotional excitement.   This exhilarating effect is, of course, seen most plainly in the case of
pleasurable emotion, agreeably to the principle already unfolded
that pleasure furthers functional activity.   But it is not wanting in the case of painful emotion, provided it is kept to the
stimulatory pitch and is not allowed to become prostrating.

---

[1] On the duration and course of development of emotional excitement, see Bain,
op. cit., chap. i. ; and Volkmann, *Lehrbuch der Psychologie*, ii. § 129.

This being so, it is evident that the notion of the ancients, that thought is most efficient in the complete abeyance of feeling, is an obsolete error. Emotion, though in its more violent waves it may easily whelm the delicate products of ideation, is also its most potent nourisher and sustainer.

This furtherance of ideation by emotion, however, is rarely if ever impartial, and herein lies its chief drawback. It is a well-known fact that all emotion, when it is fully developed and grows persistent, tends to colour or give a particular direction to the ideas of the time. The terror-stricken man has his thoughts obstinately directed towards the terrible aspect of things. In extreme cases his mind may become permanently occupied by a fixed idea (*idée fixe*) of a terrifying character. In like manner, a feeling of personal vanity leads the mind to dwell on those features which feed or gratify the emotion. The feelings, when strong and persistent, are thus deflectors of the normal logical thought-processes They dispose us to partial, one-sided views of things, and so interfere with a courageous facing of fact and an honest search for truth.

The explanation of this selective action of the feelings on the ideational material supplied by the suggestive forces of the time is to be found in those tendencies already dealt with under the head of harmony, and bodily resonance. Every emotional state is characterised by its own affective tone; and this favours the rise of all presentations having a kindred effect, while it inhibits the rise of those which would conflict with it. A gladsome mood constitutes a particular psycho-physical attitude or disposition (of which, as we saw just now, the organic base is an important factor) specially favourable to the development of bright, happy thoughts, and unfavourable to the rise of discordant thoughts. In like manner, a fit of anger begets a special disposition at the time to single out from among all the ideas which happen to be suggested any which have a similar affective tone, thoughts of injury, annoyance, insult on the one hand, and of openings for retaliation on the other.

It is evident from this that the emotive selection of percepts and ideas has certain points of analogy with the ordinary selective action of interest. The chief difference lies in the fact that the working of interest does not presuppose a previously-excited feeling, the feeling being here excited by the object of interest itself. Thus one might say that a man was " interested " in observing the actions of another man who had once done him injury, for the reason that the sight of the person tends to revive some of the old feeling.   As a consequence of this difference, interest is always wider in its selective range than actual feeling.   Thus, in the case supposed, the interest excited would lead the man to note *all* the man's actions, and every-thing connected with him.   The selective function of an actually excited feeling is essentially directed to congruous or harmonious elements, that is, elements having a like affective tone.   It is thus closely related to an assimilative process.   We might almost describe the effect as due to assimilation of a new presentation-element through the similarity of its feeling-tone with that of the dominant feeling.[1]

Having thus briefly considered the composition and the more important effects of emotion in general, we may proceed to study the chief phases of the development of our emotional life. And here we shall naturally begin with a more detailed account of the primitive or instinctive features of our emotions, including those reflex effects which come under the head of physical embodiment, and of which one important branch is what is known as emotional expression. After examining into this side of emotion, we shall trace out the more important re-sults of experience and association in developing or otherwise modifying the instinctive manifestations.

DEVELOPMENT OF EMOTION.

§ 13. *The Instinctive Factor : Expression.*   The general cha-racter of the emotive outburst or discharge has already been sufficiently described.   It may be defined as a wide-ranging reflex motor excitation involving some, at least, of the ' volun-tary ' muscles, as well as those by which the vital actions, *e.g.*, circulation, digestion, are carried out, and, finally, the nerve-structures which are known to influence the actions of the several secreting organs, as the salivary and lachrymal glands. This wide cycle of bodily effect implicates, as already pointed out, not merely the cerebro-spinal system, but the subordinate sympathetic branch of the nervous apparatus.   This reflex diffu-sion of the nervous excitation in case of emotional stimulation is a primitive fact of our organisation.   It shows itself distinctly

---

[1] The co-operation of feeling in selectively determining the course of ideas is fully recognised by Volkmann, *Psychologie*, vol. ii. § 131, p. 334 f.

in the first weeks of life. It has much in common with those reflex movements which are brought about by congenital arrangements, and which, as we shall see later on, form one of the main rudiments of voluntary action.

What we call the expression of an emotion is merely that part of this reaction which is observable to others, and which helps us to read one another's feelings. Thus it includes, first and foremost, the actions of muscles, as those of the limbs, face, and vocal organs, which distinctly betray their effects. We read a happy emotion in the movements of the eye and mouth which constitute facial expression. Other reactions involving the organs of respiration, circulation, and even digestion may enter into the expression of an emotion. Thus the disturbance of the respiratory process in sobbing, the pallor in fear due to altered vaso-motor action, the excitation of the lachrymal gland in weeping, are among the best-recognised manifestations of emotion.

As will at once be seen, the difference between expressive and non-expressive emotive reactions turns not on any psychical difference in the concomitant (reflex) feelings, but on a difference of use or function. In the case of man and all gregarious animals, the need of expressing pleasure and pain may be said to be a paramount one, more deeply laid even than the need of expressing thoughts by a significant language. Hence the instinctive definiteness and uniformity of emotive expression among members of the same species. How far this biological or teleological view will help us to explain the ultimate origin of these expressive reactions is a point which will be considered presently, after we have examined the different directions of the emotive reflex.

§ 14. *Differences of Emotive Reaction : Pleasurable and Painful Emotion.* The bodily resonance varies conspicuously in the case of different emotions. To begin with, there are certain aspects of the reflex discharge which are determined solely by the quantity and the suddenness of on-coming of the emotive excitation. Thus it is well known that all violent emotive agitation, as a shock of joy or of grief, tends to produce a slowing of the heart's action, an effect which, in extreme cases, may actually issue in death. For the rest it may be said that all emotive excitement, irrespectively of its affective quality, produces a muscular reaction which varies, in respect both of the range of muscles involved and of the extent of their contraction, with the quantity of the feeling, and so, in general, with the height of developmental phase attained.   Mr. Spencer has shown that in the case of all varieties of emotional excitement alike there is first a contraction of the muscles of smaller calibre, as those of the fingers, toes, mouth, then a contraction of the larger muscles.[1]

While, however, there are certain features in the reflex discharge common to all kinds of emotion, it is well known that the resonance varies in character with the nature of the feeling. And here the all-important difference between pleasurable and painful emotion effects a certain differentiation in the physical concomitants.   According to the principle laid down above, we shall expect pleasurable feelings to produce in general (and within the limits of injurious shock) beneficial and pleasure-yielding concomitants.   And this is certainly what we find. Thus, as there pointed out, enjoyment furthers the vital processes.   The dyspeptic knows the beneficial influence of cheerful society and talk at table as an aid to digestion.   Pleasure enhances a woman's beauty by accelerating the circulation and adding a richer bloom to the cheek.   Among these beneficial effects of pleasure is easy rhythmical pleasure-yielding movement.   The child and the equally naïve savage express their delight by dancing, clapping the hands, or other form of purposeless movement.   The expression of joyous feeling owes part of its charm to this increased motor activity.   All these resonant concomitants of pleasure, being in themselves pleasurable, serve to reinforce the primary feeling, and to swell the whole volume of agreeable consciousness.

Painful emotion, on the other hand, when sufficiently prolonged to show its characteristic effect, is in general attended by a lowering of vital action.   This is best seen in the case of grief of all kinds, whether from the loss of a friend or of fortune, or from other cause.   There is little doubt that the organic modifications, e.g., slowing of circulation, with loss of heat, show a falling off in, or disturbance of, normal functions.

---

[1] See his *Principles of Psychology*, vol. ii. pt. viii. chap. iv.   *Cf.* Mercier, *The Nervous System and Mind*, pp. 367, 368.

The contrast is well seen in the salutary influence of laughter (when not immoderate) and the wearing and baneful effect of sobbing and crying. In addition to this contrast in the organic effects of the two, there is a clear difference in the diffused muscular effects. Painful emotion does not, like joy, call forth a large wave of rhythmic movement. Even the motor concomitant of the first stages of painful excitement seems to be rather a spasmodic, irregular, and unpleasant, than a smooth, rhymical, pleasant movement. And after this early stimulating effect is exhausted, we see a distinct falling off in muscular activity. Grief, when intense and prolonged, prostrates the body, leaving the limbs weak, and the muscles generally relaxed. Here, again, it is to be added that the various resonant factors involved profoundly tinge the affective tone of the whole emotion. To some extent, they contribute new elements of unpleasantness. Thus the feeling of smarting eye, of spasm in the respiratory muscles, of bodily chilliness, and of lowered vitality generally, make up a part, and an important part, of the misery of grief. Sorrow is sorrow just because it includes this disturbance of normal action, this prostration of our bodily energy. At the same time, some of the muscular concomitants of painful emotion, *e.g.*, the violent contractions which attend a passionate outburst of anger, probably have a *relieving* function, diminishing the initial cerebral strain in some way. Hence our common way of speaking of a fit of passion as relieving itself by such purposeless reflex discharges.

That there is a general difference between the expression of pleasurable and painful emotion seems certain. At the same time it is exceedingly difficult to define this difference otherwise than in vague terms, owing to the diversity of expression among different pleasurable and painful emotions. Thus it is evident that the pleasurable emotions, *e.g.*, love, pride, and feeling for the ludicrous, vary so greatly in their expression that it is hard to discover any general feature. Similarly in the case of the painful emotions, as fear, anger (so far as painful) and remorse. The well-marked difference between what Kant called sthenic and asthenic, or energetic and paralysing emotions, *e.g.*, anger and fear, does not, it is plain, coincide with the distinction of the pleasurable and the painful, but crosses and serves to confuse the latter.

As hinted in the text, the precise part played by the resonance in pleasurable and painful emotions is not quite clear. That, in the case of pleasurable feeling, the organic effects and accelerated movements deepen and prolong the state is certain ; on the other hand, the reflex discharge in the case of painful emotion seems partly contributory, partly alleviative.   There is little doubt that in states of anger, vexation, and so forth, the element of irregular and spasmodic muscular activity constitutes an integral factor in the whole psychosis.   The disagreeable sense of restlessness and fidget colours the feeling.   At the same time common experience tells us that the reflex discharge is a way of relieving or "letting off" our passion, and that where this reflex is inhibited by an effect of will the feeling is "hemmed in" or "pent up".   It is presumable that the motor discharge tends to relieve the preternatural brain-tension of the moment, and to reinstate the normal equilibrium.   This doubleness of effect makes it difficult to say in any given case whether unfettered expression of passion is desirable or not.   This point will have to be considered again when we come to deal with the volitional control of feeling.[1]

The fact that pleasure and pain are thus in general differentiated in their physical concomitants, and so in their expression, has an obvious teleological explanation.   It must plainly be of the greatest service to animals which associate and aid one another to be able to distinguish states of suffering and want from those of contentment and happiness in the case of their offspring, mates, and members of their herd generally.   Hence, the fact that there are well-marked cries or other sounds indicative of distress in the case of all the higher socialised animals, and that in the case of the human individual these expressions are distinct from the first days of life.   It may be added that the supreme necessity of making known states of want, injury, or danger has led to the much greater distinctness of expression in the case of painful than in that of pleasurable feeling.   This necessity entails a considerable drafting of energy to the sounding-organs, which are thus always energetically active during the early stages of misery, a fact which tends to disguise the general prostrating effect of painful emotion.[2]

---

[1] For a careful discussion of the different effects of the somatic reflex on emotion, see Volkmann, *Lehrbuch der Psychologie*, vol. ii. § 129.

[2] On the differences in the somatic resonance and the expressional movements of pleasurable and painful emotion, see Bain, *The Senses and the Intellect* ("The Instinctive Play of Feelings") and *The Emotions and the Will*, chap. i.; also, J. Ward, article "Psychology," *Encyclop. Britann.* pp. 72, 73.   A full and careful account of the characteristic manifestations of suffering and joy is given by Darwin, *The Expression of the Emotions*, chap. vi. ff.   It may be added that certain writers seem to attribute to the expression of pleasurable and painful feeling the characteristic accompaniments of *particular varieties* only.   Thus, according to Cabanis, "in pain the animal wholly recoils upon itself as if to present the smallest possible amount of surface; in pleasure all the organs seem to be in advance (*au-devant*) of impressions, they spread themselves out in order to receive them at more points".   (Quoted by Bouillier, *Du Plaisir et de la Douleur*, p. 51.)   There is, no doubt, a general tendency observable in simple states of pleasure to expansive movement, and in those of pain to arrest of movement or shrinking, and it has been pointed out that such a difference in reaction might be developed through its utility to the animal. (See Ziehen, *Leitfaden der physiol. Psychologie*, p. 94 f.)   At the same time this difference is not by any means uniformly present (*e.g.*, in the common human expression, throwing up the arms in pain).   Still less is it correct to say with Höffding that closing the eyes to the outside world is a characteristic of painful feeling.   (*Outlines of Psychology*, vi. D. 1.)   It may be, as Preyer points out (*Die Seele des Kindes*, p. 23), that all babies show pleasure by wide opening of the eyes, pain by closing and blinking the eyes, but this differencing characteristic

§ 15. *Specialised Manifestations of Emotion.* In addition to the broad contrast between the manifestation of pleasurable and of painful emotion, there are the finer differences which mark off particular varieties of emotive state, as fear, anger, love, and the rest. Each of the well-marked species of emotion has its characteristic group of reactions, of which a full and detailed account has been given by more than one physiologist and psychologist. Thus, as already hinted, fear is differentiated from other emotive states in general, as well as from other varieties of disagreeable feeling, by its peculiar organic resonance, including such familiar effects as that disturbance of the heart's action known as palpitation, tremor of muscles, pallor, certain alterations in the secretions (*e.g.*, saliva).

These different somatic resonances constitute, through the reflex sensations to which they give rise, so many distinctive emotive colourings. There is little doubt, as already remarked, that this resonant colouring is a very important characteristic of the whole emotive state. It follows that the amplitude and the range of variation of the life of feeling are primarily conditioned by the number and the perfection of the various groupings of nervous elements which underlie and render possible those several modes of reflex discharge.

In the second place, these characteristic resonances supply a differentiated language of the emotions. It is because the visible and audible part of a psycho-physical state of fear is well defined and distinctive that we are able to read one another's feelings so rapidly and so easily.

In the case of all the more primitive emotions, those which the civilised man has in common with the savage and with many of the lower animals near him in the zoological scale, such as fear, anger, love (in certain of its forms), this characteristic signature is in its main features common to all members of the species. Thus the laugh of joy, and the trembling of fear, are common to all grades of civilisation and all ramifica-

becomes, to say the least, much obscured in later life. Thus fear, a disagreeable feeling, just like admiration, an agreeable one, causes a dilation of the eye. Münsterberg seems to share in this tendency to over-simplify when he connects all pleasurable impressions with an inspiration, all painful ones with an expiration. (*Beiträge*, ii. p. 36.)

tions of the human race.   The characteristic manifestation, moreover, shows itself in early life as a strictly instinctive re- action.   The child does not learn to howl and to clench his fist, and so forth, when in a rage : these manifestations occur as primitive reflexes, the outcome of certain congenital arrange- ments in the nervous centres.

Here, again, we are naturally led to call in teleological con- siderations.   It is a matter of high utility (*i.e.*, as tending to preservation, individual and tribal) to men, and other socialised animals, to be able to express not only states of suffering gene- rally, but states of fear in particular.   There seems little doubt that in the course of human and animal evolution these instinctive markings of particular emotive states have become differentiated and fixed in the nervous organisation through the action of natural selection.

Modern research enables us to go still further, and to indi- cate with some degree of exactness the influences which have served to originate the particular characteristic expressions we possess rather than any others.   These are in the main two : (*a*) the survival as expressional sign of once useful movement or action, and (*b*) the extension of such movement by similarity of affective tone, or what has been called " the analogy of feeling," to other and kindred emotive states.

As illustrations of the first, we may take the clenching of the fist in anger, which seems plainly a survival of fighting habits.   According to Spencer, the frown of displeasure is a relic of the action of shutting out the direct rays of the sun, which was serviceable during the combats of primitive man.[1] Similarly the characteristic mean shrinking attitude of fear may be explained as the sub-excitation of the useful action of evading attack.   The second principle is illustrated in such actions as scratching the head in mental perplexity, this action having been originally serviceable in allaying

---

[1] *Principles of Psychology*, ii. p. 546 f.   Darwin also explains the frown as a survival of a useful habit, shutting out excessive light when looking at distant objects. (*Expression of the Emotions*, p. 226 f.)   This reaction may be compared with that of raising the eyebrows and expanding the pupil in states of surprise, *ibid*. (p. 278 f.). The two vision-serving movements reappear in different states of attention: the former, when we have to find an idea, or understand a knotty point ; the latter, when we seem to want to take in a new and *big* idea.

an analogous sense-feeling. Similarly the smacking of the lips by the savage to express pleasure generally may be supposed to be due to the transference of a movement originally useful in connexion with eating. Such survival of once useful actions, and their extension as expressive signs to other feelings analogous to that which originally prompted them, may be seen taking place within the limits of the individual life. Thus Preyer relates that his boy, after having suffered from eczema in the head, and got into the way of carrying his hands to this part when it was troublesome, continued to carry out the same movements when the eczema got better, *whenever something disagreeable offered itself, c.g.*, when he was invited to play and did not want to do so.[1]

It is to be added that the movements making up emotional expression are only in part instinctive. The large margin of diversity of expression observable among different races and nationalities shows that imitation and education have much to do with fixing the particular emotive dialect of each of us. This influence will be spoken of presently.

§ 15*a*. *Theories of Expression.* The discussion of the nature and origin of the different groups of movements, etc., making up the characteristic expression of the several emotions constitutes one of the most interesting chapters in modern psychology. Sir Charles Bell, in his classical work, *The Anatomy of Expression*, regarded many of our facial muscles as purely instrumental in expression, and " a special provision " for this object. Since his time the doctrine of evolution has enabled us to conjecture the mode of origin of expressional movements. Mr. Spencer has the honour of having first suggested that instinctive expressions have been evolved out of serviceable actions during the course of human and pre-human life. While this writer thus refers the distinctive elements of expression (restricted discharge) to evolution, he would explain the general diffused effect of emotion as a result of the constitution of the nervous system. Darwin, in his interesting volume, *The Expression of the Emotions*, recognises the two principles, and illustrates the transformation of once serviceable actions at much greater length. It may be added that both Spencer and Darwin implicitly recognise the action of the principle of transference by analogy, a principle which Wundt has since formulated and further illustrated.[2]

§ 16. *Special Inherited Associations.* An emotion, as already defined, is a complex presentative and affective state, consisting

[1] *Die Seele des Kindes*, cap. x. p. 140 f.

[2] For a fuller account of modern theories of emotional expression, see Spencer, *Principles of Psychology*, vol. ii. (Language of the Emotions) ; Darwin, *The Expression of the Emotions ;* Wundt, *Physiol. Psychologie*, vol. ii. chap. xxii. (where there is a good *résumé* of the theories) ; also my volume, *Sensation and Intuition*, chap. ii.

of a primary stage of excitation, and of a secondary stage of resonance. So far as we have yet considered it, an emotion is instinctive in the sense that the reflex discharge follows when the appropriate stimulus is forthcoming. Thus we say that the emotion of fear is instinctive in the sense that, when through experience the idea of danger is generated, the particular somatic reaction entering into and colouring the state follows.

But modern research helps us to go further than this. In certain cases, at least, the emotive condition is excited *independently of its customary presentative excitant*. Thus it is well established that children display fear before strangers, before dogs and other animals, and this at an age which precludes the idea of any individual experience of evil in this connexion. Here it is evident the whole emotive phenomenon is instinctive. We express the fact by saying that the child has an instinctive dread of certain animals, and so forth. In like manner the early responses of the baby to the human regard and smile suggest that a particular emotive discharge is congenitally connected with a definite kind of percept, that of the human face.

Such instinctive connexions between particular percepts and particular emotive discharges may be marked off by the description, Special Inherited Associations. A plausible explanation of such connexions is that they are the transmitted result of oft-repeated ancestral experiences. Thus the baby fears the unknown animal, because human experience through many ages has tended to connect ideas of danger with wild animals. Similarly the instinctive reaction on the presentation of the human face is explained as the echo of the results of ages of social experience with the large surplus of pleasurable suggestion which this brings with it.

This way of explaining special congenital emotive associations is, it is obvious, open to the general objection that the transmission of acquired characters has not yet been satisfactorily established.[1] At first sight, too, it seems open further to the psychological difficulty that it necessitates the transmission of *ideas* of particular forms of evil. This, however, might perhaps be obviated by supposing the presentative element to be reduced to a minimum, the feeling being merely one of vague

[1] *Cf.* above, I. p. 139. It is, of course, possible to regard such special connexions as instinctive and inherited, and at the same time to refer them to certain congenital variations in the nervous organisation of our ancestors.

uneasiness. It may be added that it is, in the nature of the case, difficult to be sure that, in referring a child's fear to inherited association, we have excluded *all* elements falling within the individual experience itself. Thus a baby's first fear of, or dislike to, a dog in a room may be due to a disturbing effect on its unstable nervous system of the rapid movements, and noise. Similarly there may be something in the very bigness and noise of the sea which excites a nascent dread. When to this disturbing effect of powerful and novel sensations is added the far-reaching effect of association through similarity, we shall find it difficult to say that a particular mode of emotive discharge is wholly instinctive.[1]

§ 17. *Effect of Experience: Modification of Instinctive Reactions.* In considering the effect of experience and education on emotion, we have first to recognise their action upon the reflex somatic discharge. The full naïve manifestations of the child and the savage become modified by the forces of education or culture.

In the first place, the early emotive discharge becomes toned down and restricted. The emotions take on a quieter form as life advances. This is more particularly true of certain unlovely or morally reprehensible feelings, as rage. This quieting effect is due to the development of conation. Passion no longer spends itself in aimless utterance, but controlled by will directs itself in the channels of useful action.[2]

In the second place, what we call education tends to differentiate the forms of emotive expression still further, substituting for the common primitive language a number of dialects, answering to different nationalities and different social strata. Thus a great deal of the pantomimic gesture of the races of Southern Europe is learnt by imitation. The characteristic movement of shoulders and palms with which a Frenchman expresses a certain mental attitude (" Don't ask me ! " " As you will ! "), and the action of pressing the side of the nose

---

[1] The tendency to refer particular directions of fear to instinct is carried to an almost absurd degree by some writers. Thus Binet seeks to prove that a baby has a special instinctive fear of falling out of its nurse's arms. (*Revue Philosophique,* March, 1890, p. 302.) Surely it is more reasonable to suppose that the feint to let the child fall produces disturbances (organic, aural, etc.) which trouble the consciousness. The existence of such special emotive instincts has been urged with considerable fulness by G. H. Schneider, *Der Menschliche Wille ;* Preyer, *Die Seele des Kindes,* p. 104, etc. ; W. James, *The Principles of Psychology,* ii. chap. xxiv. Some of the difficulties of this mode of explanation are dealt with by Bain, *The Emotions and the Will,* chap. ii. § 14 ff.

[2] *Cf.* James, *Principles of Psychology,* ii. p. 475 f.

with the forefinger by which a German expresses the sharpening
of his intellectual process to a special degree of acuteness, are
doubtless acquired actions.   The same may be said of many
of the artificial signs of those milder currents of interest and
sympathy which circulate in ' good ' society.   The characteristic
inclination of head and figure, modulations of voice, etc., by
which a polite hostess impresses on her guest a sudden
accession of personal interest is one of the *arcana* of what is
termed "good breeding".   Such acquired actions (so far as
they are genuinely indicative of feeling) tend to become assimilated to emotional movements proper, when by repetition they
grow habitual and free from the element of conscious purpose.
It may be added that, like original reflexes, they tend to some
extent to enlarge and enrich the emotion.   It may be safely
said that the stiff, undemonstrative Englishman, whose disregard of gesture is satirised by Schopenhauer, has a less
intense and rich social consciousness than the Frenchman or
Italian, an inference abundantly supported by observation of
the wide difference in their habits of life.

One other influence of experience on the somatic reflex in
emotional states must be alluded to, *viz.*, the effect of repeated
indulgence in an emotional state in fixing and strengthening
the *disposition* to that mode of discharge.   This is an illustration of the principle of habit, which, though it tends as we
have seen to dull feeling, tends also indirectly to fix and
further it by strengthening the disposition to the appropriate
motor reaction.   A child who is allowed to fall again and
again into the mental and bodily attitude of anger contracts
a stronger organic disposition to react in this way, a fact
clearly seen in the greater rapidity of the outburst, and in
the diminished strength of the stimulus requisite for calling
it forth.

§ 18. *The Presentative Factor in Emotion.*   The development
of our emotional life, while thus influenced in a measure by
modifications of instinctive reaction, is chiefly dependent on the
extension and accumulation of presentative material.   We have
already seen that an emotion contains a presentative factor.
Thus fear, love, and so forth, are excited by certain percepts,
together with the ideas which these percepts suggest.   Overlooking the possible action of special inherited association, we

see that emotion proper, as distinguished from mere sense-feeling, only displays itself when experience supplies the necessary presentative stimulus. Thus the child fears to eat his food too hot after experience has associated with the appearance of a heated substance the painful sense-feeling of a burn. The growth and expansion of emotion, its diffusion over a larger and larger range of object, its recoverability and extension in time, its differentiation into a larger and larger number of varieties or states, is due to this action of experience and association. We have now to consider this side of our emotive states more fully.

§ 19. *Feeling as Ideal.* In considering the effect of representation on emotion we must set out with the fact that when a feeling is an accompaniment of a sensation (presentative state) it reappears in a weaker degree with the corresponding representative element. Thus the pleasure attending a sensation of light, of satisfying thirst, and so forth, is revived or re-excited in a weakened form with the representation of the sensation. This reappearance of a sense-feeling through the reinstatement of the representative copy of the original sensation may be described as revived or as "ideal" feeling.

This fact that the image of a sensation is attended with a weakened re-excitation of the original feeling-accompaniment is, doubtless, connected with the nervous conditions of the representative image and of sense-feeling. It has been pointed out above that the image, say of a colour, probably has for its neural correlative a weakened excitation of the same central nervous elements that were engaged in the production of the sensation ; also that the nervous conditions of a sensation are a main factor in the production of the feeling which accompanies it. It would appear to follow that the central nervous re-excitation underlying the image will carry with it the production, in a weakened form, of the affective concomitant.

Since the revived feeling is thus organically connected with the representative image, it follows that its recoverability depends in general on the revivability of the presentative element. It is a familiar fact that the pleasures of the higher senses, tones and colours, are revived in greater proportionate intensity than those of the lower senses. As everybody knows, it is hard to recall the pleasures of the table, still harder to recall pleasurable or painful organic sensations.[1]

Now the revival of a sensation under the form of an image is already an intellectual fact involving what we call memory. Moreover, it varies, roughly at least, with the fineness of differentiation, *i.e.*, of the degree of intellectual elaboration reached, in the particular sense. It follows, therefore, that all feeling, so far as revived, is a concomitant of our intellectual life. The higher the intellectual development the greater the range for the ideal renewal or conservation of feeling.

The revival of sense-feeling is subject to a number of variations which it is difficult to explain. It is sometimes said that pleasurable sensations are more readily recalled than painful. Since painful feelings are supposed to involve injurious nervous conditions, it may be more difficult to realise these than the conditions of pleasurable feelings in the absence of the original peripheral cause. At the same time, much depends on the temporary condition ; for, as we have seen, states of dejection and sorrow dispose us to entertain painful ideas. There seem, too, to be important differences of individual temperament in this respect, some persons being more disposed to recall and dwell on pleasurable experiences, others, on painful experiences.[2]

It is a somewhat subtle point whether we should call the revived feeling itself representative or even ' ideal '. Our sharp distinction between the presentative and the affective element compels us to regard the secondary wave of feeling called forth by the image as a new feeling equally real or actual with the original, and the proper concomitant of the representative image. At the same time, it must be remembered that there is, in many cases at least, a conscious reference to the original sense-feeling. Thus in recalling a pleasurable taste, as that of a rare fruit, I may be aware that the pleasure of the actual sensation was greatly superior to the present revived feeling. This circumstance greatly limits the pleasures of memory, or imagination. Thus, as has been pointed out, recalling "happier days " may cause pain through a consciousness of the inferiority of the revival to the actual feeling. In like manner, as we shall see by-and-by, all desire implies this recognition of the inferiority of an ' ideal ' to the original ' actual ' feeling.[3]

[1] As we shall see, the organic sensations attending emotional states, *e.g.*, the trembling chilliness in fear, appear to be easily revived when we imagine the experience. Here, however, there is a reinstatement in part of the organic conditions, and so of the actual sensations.

[2] On the differences between individuals, in respect of sensibility to, and retention of, pleasure and pain, see my volume, *Pessimism*, p. 404 ff. George Eliot writes : " I don't find that the young troubles grow lighter on looking back ". (*Life*, iii. p. 144.)

[3] That, strictly speaking, there is no such thing as revived or reproduced feeling, but that what is called such is only a new and inferior production of feeling, is insisted on by more than one psychologist. (See Volkmann, *Lehrbuch der Psychologie*, § 131, p. 133.)

§ 20. *Assimilation as Conservatrix of Feeling.* In tracing out the consequences of such revived feeling, we have to ask whether the process of assimilative conservation of sensation-traces tends to the preservation of the affective element as well. Does, for example, the recurrence of a pleasurable sensation-complex, as that of a beautiful natural scene, tend to produce a deepening of the sensuous pleasure in the second case through the reproduction of the image of the first ? It has already been pointed out that repetition and custom exert a dulling effect on the sensuous feeling. The above question may accordingly be put in the form : How far is the dulling effect of repetition and accommodation counteracted by the results of assimilative cumulation ?

That a certain effect of summation is produced in these cases is indisputable. To begin with a simple illustration, the regular repetition of the rich tone of a church bell shows its cumulative effect, not merely in a closer attention to the impression, but in a sensibly deepened pleasure.

Again, in all cases of recognition of a pleasure-bringing object after an interval of separation, we can trace a like process of summation through a conscious recalling of past pleasurable sensations. Thus, in trying a favourite musical instrument after a period of disuse, we have the pleasure of tone appreciably increased by revivals of similar experiences. Enid's delight in resuming her former splendour—a splendour

> New to her
> And therefore dearer ; or if not so new,
> Yet therefore tenfold dearer by the power
> Of intermitted custom—

seems an illustration of this deepening of a present feeling through the addition of revived currents.

In certain cases, moreover, the very consciousness of recurrence contributes a new element of feeling. This applies to the deepening of horror through the repetition of a crime, as in the case of the recent Whitechapel murders, the deepening of gratitude through the repetition of a like favour, and so forth.

At the same time it must be allowed that this cumulative effect is only palpable and distinguishable in those cases where there is some distinct representative consciousness of the past experience or series of experiences. In other words, mere

automatic assimilation, in which there is no differentiation of a
consciousness of the present and of a past like experience,
appears to be inoperative in deepening feeling. Here the mere
sense of familiarity exercises, as we have seen, the contrary
effect of dulling the affective concomitant.

Similarity of presentation is thus a condition of deepened
feeling when it is consciously realised. The same thing is seen
in the curious psychological phenomenon, the effect of poetic
simile. Here we see an intensification of the feeling-element
in one presentation through an assimilative comparison of this
with another presentation having its own feeling-accompaniment. Thus the pleasurableness of the sight and sound of
foliage stirred by the wind is enhanced by the pleasurable
image of a joyous dance. It is recognised that a simile owes
its poetic value to its appropriateness, *i.e.*, the degree of kinship by means of which it brings out some aspect of the object
described ; and also to its own intrinsic charm, this latter being
in the process of assimilation transmitted to, or diffused over,
the idea of the object.[1]

That the recurrence of like pleasant and unpleasant experiences tends to pro
duce a certain cumulative result in the affective element appears to be shown not
only by the deepening of such feelings as gratitude, liking for things, places,
and persons, but by the well-known effect of indulgence in a feeling, as fear or
anger, in strengthening the disposition to that feeling. Yet it is difficult to estimate
the effect here owing to the presence of other co-operant influences. Thus we
never have assimilation without some amount of associative complication, that is
to say, addition of dissimilar presentative elements with their feeling-accompaniments ; and this process, as we shall presently see, is a main influence in the
deepening of feeling. With respect to the effect of repeated emotive indulgence it
has already been pointed out that this is in part explained by the rendering of the
reflex discharge more instantaneous.[2]

§ 21. *Feeling and Association.* As already hinted, the revival
of feeling is always, to some extent, the work of contiguous
association. That is to say, a feeling occurs in the weakened
' ideal ' form when there arises the representative copy of a sensation of which the original feeling was the concomitant. Thus

[1] This effect of transmission is analogous to that of transference to contiguously
attached concomitants to be spoken of presently.

[2] On the effect of similarity on feeling, see Bain, *The Emotions and the Will*,
p. 103 f.

it is a revived feeling when the sight of a cool stream recalls the pleasure of the bath, when the sound of a crow's cawing recalls the pleasure of a woodland scene, and so forth. Here we have to suppose that the particular feeling was in the original sense-experience bound up with certain aspects of the sensation-complex, so that the partial reinstatement of the psycho-physical state in imagination is sufficient to produce a weaker excitation answering to what we call the echo or revival of the original feeling.

This revival of feeling through associative connexion with presentative elements is a fact of far-reaching consequence, for our intellectual and for our affective life alike. The effect of this association on the former is seen in the fact that, feeling being the source of all that we call interest, the presence of a strongly-marked affective concomitant in a presentation or series of presentations tends greatly to a selective retention and reproduction of these.[1] Not only does such an affective concomitant in the sense-experience serve to fix a vivid impression, the weakened feeling which attends the uprising of the representative image serves in the reproductive stage to awaken interest, and so to secure the proper adjustive process of attention.[2] To this it may be added that the existence at the moment of reproduction of any special tendency to a particular tone of feeling will serve, as we have seen, to selectively reinstate those images suggested at the time which happen to have the proper affective accompaniment. In this way the whole course of our ideal successions or trains of ideas is profoundly influenced by the affective elements which are conjoined with the presentative.

We may now look at the effect of these associative complications of the presentative and the affective phase of consciousness on the extension and development of the latter. The action here referred to may be described in general as the gradual accretion of feeling-accompaniments about particular presentations or presentation-complexes, through and by means of successive processes of contiguous integration which connect these presentations with others that have a palpably pleasurable or painful character.

---

[1] *Cf.* above, I. p. 163.   [2] See I. p. 346 f.

The mode of action here may first be illustrated by the associative enrichment of our sense-feelings.   Let us take a particular sound, as the cawing of a rook, which, in itself, is certainly not agreeable.   This sound, in the case of those who have lived in the country in early life and enjoyed its scenes and its adventures, is well known to become a particularly agreeable one.   To some people, indeed, there is hardly any more delightful sonorous effect than that of this rough, unmusical call.   The explanation is that this particular sound, having been heard again and again among surroundings, as park and woodland, which have a marked accompaniment of pleasure, has become contiguously interwoven with these presentations, and so produces a faint re-excitation of the many currents of enjoyment which accompanied these.   It is probable that such associative integrations serve to modify our pristine sense-feelings to a much larger extent than we are aware of.   Thus the colour, cerulean blue, may take on some of its peculiar charm through association with serene, summer skies.   Alison has abundantly illustrated the workings of association in thus rendering pleasurable by association what was originally indifferent, if not disagreeable.[1]

§ 22. *Transference of Feeling.*   It is evident that this associative integration of presentations with affective elements differs from that which connects presentations one with another.   The feeling-mass, which has become conjoined with a given presentation through the medium of associated presentations, tends to appear *without the revival of the latter.*   In other words, the feeling is said to be associatively *transferred* to a new presentation.   Thus the cawing of the rook excites a pleasurable feeling directly, that is, without any distinct representative consciousness of country scenes, so that the feeling appears to belong to the sound just as much as if it were a sense-feeling proper.   Similar effects are seen in the transference of horror and other feelings to places, of dignified and undignified associations to names, and so forth.

This process of feeling-transference points to a radical peculiarity of all affective integration.   We saw above when

---

[1] Alison went so far as to say that all beauty owes its value to suggestion. This was to deny the original intrinsic pleasurableness of sensation, an error which half-an-hour passed in a nursery would have served to correct.

discussing the relation of presentation and feeling that the latter is partially independent of the former. This proposition is clearly illustrated by the phenomenon now considered. A transferred feeling is one which is excited without the mediacy of the particular presentative element of which it was originally a concomitant. The sound of a babbling brook, the perfume of a violet, give to the country-bred man an after-taste of the enjoyments of boyhood and romantic youth, without any distinct reproduction of the experiences out of which these enjoyments sprang.

The explanation of this suppression of the presentative intermediary in the case of transferred feeling seems a complicated matter. To some extent we appear to have to do here with a phenomenon of opposing representative tendencies analogous to that seen in the case of the generic image. That is to say, the multiplicity of experiences itself prevents the distinct representation of any one of the details. That this is so is shown in the fact that the present writer by dwelling long enough on the sound of the rook finds himself recalling particular scenes of early life—a case precisely similar to what, as Taine and others have shown, takes place in fixing the mind on a general name. It is possible, too, that nervous dynamics, if we understood them, would help us here. It looks as if the current of feeling (involving most probably some amount of reflex nervous discharge) supervened so rapidly as to swamp, so to speak, the representative consciousness.

It may be added that such transference takes place not only on the recurrence of (approximately) similar presentations, but on that of partially like ones. Thus M. Loti relates from his own experience how a whole Sunday morning mood, which happened, when he was a child, to synchronise with the presentation of a ray of light falling obliquely on the darkened wall of the staircase of his home as he returned with his mother from church, was uniformly revived in his later life by the sight of a similar effect of sunlight in a darkened staircase at Stamboul.[1]

It follows from this that complex mental states may form themselves, in contiguous attachment to particular objects, in which the feeling-element is predominant and the presentative remains in sub-conscious abeyance, that is to say, is only very vaguely differentiated and recognised in its constituent parts. Such a presentative-affective complex, appearing as a large undiscriminated feeling-mass, is precisely what we mean by an emotion on its presentative side. Thus the wave of feeling awakened, after an interval of separation, by the sight of a familiar object which is dear to us, e.g., our home, our favourite

[1] *Le Roman d'un Enfant*, pp. 28, 29.

book, our beloved friend, is, in its initial as distinguished from its resonant stage, the outcome of a number of confluent associated pleasures. What we mean by the development of our emotions, our permanent likes and dislikes, our recurring and characteristic fears, our feeling of pride, and so forth, is the result of such successive integrations of transferred feeling. It is this process which is most effective in counteracting that decay of feeling through accommodation of which we have treated above.

It may be well to point out that this principle of associative transference is one of the highest practical importance. To begin with, it serves to secure the persistence of feeling by extending the range of excitant. In this way we may get children to recall unpleasant experiences by associatively embodying them in certain reminders, and more particularly in special localities. Not only so, this process enables us to a large extent to create likes or dislikes for relatively indifferent objects by investing them with agreeable or disagreeable associations. Locke suggested that boys would take to books as eagerly as they take to play if study were only invested with the semblance of play. More, in his *Utopia*, shows how in his ideal community gold and silver come to be contemned by reason of degrading associations.

It may be remarked further that the same complex integration which serves to develop pleasurable and painful emotions tends to generate those mixed emotional effects which we so frequently experience. Our feeling for a locality, for a person whom we have known intimately, for a well-studied author, and so forth, is rarely an unmixed one. A tangle of agreeable and disagreeable associates results in a mixed emotion, in which now the pleasurable, now the painful factor is uppermost.

§ 23. *Differentiation of Emotion : Refinement.* As the last result of this process of presentative-affective integration, we have that growing differentiation of emotive masses which is one of the characteristic features of mental development. As an example, we may take the emergence of a feeling of anger proper out of the primitive undifferentiated baby-misery (as seen, for example, when the child is being dressed) ; and the differentiation of this early anger itself into many varieties of shade, as the feeling we cherish for a successful rival, for one who has injured us when wearing the mask of friendship, and so forth. This fine ramification of emotion is due to the ever-increasing differentiation of the integrated masses just spoken of. In the case of the more intellectual emotions—particularly, the æsthetic and moral sentiments—this differentiation reaches, in the case of cultivated persons, a specially high point. To be *differently* affected by two musical composers or two authors, to be differentially responsive to all the possible *nuances* of moral colouring in a lie, is the mark of a refined emotional nature. Here, however, as we shall see, the presentative or cognitive element assumes a much greater prominence in the whole effect, and the refinement of feeling includes a considerable increase of reflective comparison and discrimination of presentative material on its affective side.[1]

§ 24. *Influence of the Growth of Representation on Feeling.* It follows from the above that feeling becomes enlarged, spread out, as well as deepened and consolidated, by the development of representation (imagination and thought). The growth of ideation is thus a necessary condition of all the richer, more varied emotive experience.

At the same time, the effect of a growth of ideation is not wholly in the direction of sustaining and expanding feeling. As common observation tells us, the higher developments of intellectual activity constitute, along with that of will-power, already referred to, one main counteractive to the early intensity and masterfulness of feeling. The child grows calmer as he grows older, not merely and not so much by putting on the volitional screw as by learning to reflect. In this way, for example, a good deal of childish fear loses its intensity by a reflective recognition of its groundlessness. The child's outburst of anger is restricted by a recollection of previous acts of kindness from the offender; his overmastering grief is re-

---

[1] The expression "refinement of feeling," as commonly used, includes a high degree of responsiveness to weak emotive stimuli, as well as a high degree of discriminative response. Thus a refined sense of beauty is one that spies out the least obtrusive manifestations of the beautiful, as well as discriminatively appreciates its manifold forms or shades. Hence refinement of feeling may be said to answer to both "absolute" and "discriminative" sensibility in the region of sensation. (*Cf.* above, I. p. 87 ff.)

duced by the thought of the temporariness of the evil, its com-
monness to himself and to others, and like considerations.

No doubt, as we have recognised, feeling tends to suborn
representation into its service.   It may be said, indeed, that
through this selective action of feeling on ideation every emotion seeks to justify itself.   Thus the angry man selectively
fixes his attention on the circumstances which tend to awaken
resentment, *e.g.*, the deliberate nature of the injury, the want
of provocation, and the like, and which may accordingly be
put forward as logical grounds.   But just so far as representative power grows strong and independent will it exercise an
inhibitory influence on emotive impulse, complicating the situation by recalling various mollifying circumstances or aspects of
the case.   The " reasonable " man is one who is not mastered
by the emotive force of a particular idea, or of a particular aspect
of an idea, but mentally takes in and realises all the circumstances, antecedents, and reasons.[1]

§ 25. *Classification of Emotive States : Order of Development·*
It is customary in psychological works to attempt a systematic
arrangement of the emotions in which the similarities and
differences of psychological (*i.e.*, presentative-affective) character would be indicated by the proximity and the remoteness
of the several groups.   Such classifications are familiar to us
in the classificatory branches of natural science, *e.g.*, botany
and zoology.

A ground for such a classification appears to present itself
in the fixed distinctions of common thought, which not only
marks off pleasurable from painful emotion, *e.g.*, joy from grief,
but distinctively names particular modes of emotive excitation,
e.g., fear, anger, pride.

As soon, however, as we try to carry out any precise scientific
division we find the difficulties insuperable.   Material objects,
as minerals or plants, are separate things, and though the

---

[1] That intellection does tend to restrain strong emotion is illustrated in those
pathological conditions where thought is reduced to a minimum, and the feelings
have it all their own way. A similar fact meets us in the hypnotic state, where
the emotions may be excitable, even though consciousness of the external world is
suspended. Here we may suppose the reflex discharge to be more rapid and
violent, because the inhibitory action of the highest centres which underlies conscious reflexion is removed.

complexity of their affinities occasionally renders the placing of a particular form a matter of difficulty, a systematic arrangement of them in higher and lower classes is, in the main, practicable.   Far otherwise is it with those psychical phenomena which we call emotions.   As pointed out above, our states of mind as a whole are not, strictly speaking, susceptible of that kind of classification which we can carry out in the case of material objects.[1]   This is peculiarly true of those concrete states which we call emotions.   As we have seen in our analysis, they are eminently complex and variable phenomena.   What we call an emotion of fear is a changeful course of feeling which shows the greatest variations at different stages.   Owing, too, to the structure of emotion through the welding together of a mass of feeling-coloured presentative material, the particular affective complexion of the whole may vary indefinitely through variations in the mode of composition. Thus what we call a feeling of joy or grief will exhibit an infinite number of shades answering to the particular modes of presentative consciousness, and the particular currents of feeling to which these give rise.   No precise systematic arrangement can therefore be attempted.

The only thing that can be aimed at here is to mark off certain broad distinctions among emotive phenomena by help partly of differences in the presentative or excitatory factor, and partly of dissimilarities in the somatic resonance.   As already hinted, the emotions distinctively named in common life, e.g., pride, love, have a well-marked differentiating bodily manifestation.   These differences of corporeal resonance, which, as we have seen, serve to give a different colouring to the whole state, must, it is evident, be the main basis of any natural history arrangement.   This, however, will have to be supplemented to some extent by a reference to differences in the mode of excitation, or the presentative phase of the emotive state.

Having thus succeeded in marking off, roughly at least, the main varieties among our emotive states, we should proceed to deal with them in a progressive order.   That is to say, we should seek to arrange them serially, so that simple forms might pre-

---

[1] See I. p. 70 f.

cede complex forms. The ideal arrangement would be one in which we set out with the simplest emotions barely distinguishable from sense-feelings, and proceeded, step by step, to the most exalted and complex emotions. Such a serial arrangement would have to be based on a careful analysis of the feelings, with a view to comprehending their complexity, and the way in which certain simpler emotions enter as components into other and more complex ones.

This serial arrangement is possible within certain limits. We can see that there are some emotions which are complex and " representative," being indeed constituted, to some extent, by the combination, in revived or ideal form, of other and simpler emotions. Thus the sentiment of justice embodies in a representative form, and takes up into itself something of the flavour of, that ferment of anger or resentment which is the reaction instinctively following on sense of personal injury.

At the same time, a perfect serial arrangement, based on analysis, is impracticable, owing to the nature of the phenomena, as already explained. We have then to fall back on a natural history study of the *order of appearance*. Here we shall have to do immediately and mainly with the typical development of the human individual, confirming the results of our observation of this by what is known of the progressive manifestation of emotion in the development of the race and of the zoological series. This order will, it may be assumed, correspond in the main with the order of complexity or degree of representativeness.

§ 26. *Three Orders of Emotion.* Without aiming at scientific precision and exhaustiveness where these seem to be excluded, we may group the more important varieties of emotive reaction under three stages of manifestation or stadia.

(1) Under the first fall certain common unspecialised manifestations of pleasurable and painful feeling, which are best described by the current terms, Joy and Grief. These emotions, which appear at the beginning of, or very early in, life, presuppose, as excitants, merely the sense-feelings and those common forms of central reaction dealt with in the last chapter, *viz.*, change or transition, and conflict. As examples we may take the delight of the child in his bath, his fits of misery when in pain, and for a time afterwards; the first stimulating effect

of wonder at what is new;[1] the simpler forms of hope or immediate anticipation of pleasure and of aversion or rejection of what is disagreeable;[2] also the negative feelings of want or misery growing out of the cessation of a pleasure, and of relief arising from the cessation of a pain ; and, lastly, the delight of satisfied hope, and the pain of disappointment, which may be taken as rudimentary forms of the action of the principle of harmony and conflict.

The manifestation of all varieties of this joyful and of this sorrowful emotion is, in its more important features, similar. The most pronounced part of the characteristic reaction in these cases is the cycle of somatic reflexes with their effects, which, as we have seen, distinguishes pleasurable and painful emotion. It may be added that these emotive states involve the minimum of representation, and so come nearest to sense-feelings. All that they include is the representation of a pleasurable or painful experience just over by means of a primary memory-image (feelings of loss and relief), and the representation of such an experience in the immediate future through the suggestive force of present percepts (feeling of disappointed expectation).

(2) Next to these common undifferentiated forms of emotive reaction come the Specialised forms to which frequent reference has already been made. We may take anger, the specialised form of fear, and fondness or love as examples. These emotions appear in a clear recognisable form later in the history of the child than the first undifferentiated group. The child knows infantile misery before he knows anger, the misery of a suggested unpleasantness, *e.g.*, that of being dressed, of a cold bath, before he knows the peculiar state called dread, and so forth.[3] In their primitive forms, which are shared in by many

---

[1] Preyer speaks of his child showing wonder at the end of the seventh month at the disappearance and reappearance of a round fan. (*Die Seele des Kindes*, pp. 22, 23.)

[2] This feeling must be distinguished from the specialised reaction of fear already referred to. It can only be described as a feeling of discomfort or vexation, accompanied by signs of rejection, which arises when anything unpleasant is suggested.

[3] This is true of the human individual at least, and seems true of certain animals, as the dog, which give a fairly clear expression to states of joyful feeling.

of the lower animals, these specialised emotive reactions involve very little of the representative element. Thus an infant's display of tenderness when he touches his mother's cheek,
or puts his arms about her neck, is, like the "fondness" of a kitten
for its mother, and the reciprocal feeling, hardly distinguishable
from a (complex) sense-feeling (that of soft touch, warmth,
etc.). As pointed out above, the specialised reactions here referred to are instinctive. As shared in by all the higher
animals, these specialised instinctive emotions may be marked
off as Animal emotions.

(3) As a third order of emotive states we have a group of
feelings characterised by a high degree of prominence of the
representative or ideational element. They may, roughly at
least, be marked off as Human feelings in contradistinction to
the instinctive emotions common to man and animals just
referred to. They appear later than these in the evolution of
the race and of the individual. In their fully-developed form
they presuppose a considerable degree of representative power
and also a certain range of emotive experience as supplied not
only by the sense-feelings but by the earlier and instinctive
forms of emotion themselves.

These Representative emotions, as they may be called, fall
into two clearly-marked sub-divisions. (a) Of these the first is
Concrete representative emotion, or that mode of feeling which
arises through an imaginative reinstatement of the original
causes of an emotion. All "ideal emotion," as it has been
called, e.g., the secondary emotions of anger, love, fear, excited
by mere *ideas* of their objects, would fall under this sub-group.
There is, however, one mode of such representative feeling so
well marked in its characters and so important that it may be
singled out as the typical example of this variety of emotion.
This feeling is Sympathy, that is, the imaginative entering
into others' feelings through recallings of our own similar
experiences.

(b) The other sub-division of the representative emotions
may be marked off as Abstract, because in their clear form they
involve as a constituent factor an abstract idea. These are the
feelings which grow up about and colour the idea of truth, of
beauty, and of duty. These are more complex and later in
their development than sympathy, and, as we shall see, they

presuppose a certain development of the sympathetic feelings. They may be briefly described as Abstract Sentiments.[1]

It is natural to inquire whether the order of development of the emotions just indicated accords with the general requirements of evolution. The most important of these needing to be considered here is that those modes of feeling manifest themselves first in the development of the race and consequently in that of the individual which are most closely connected with, and most directly subserve, life-preserving functions. To some extent we can see this to be so. Thus, as already hinted, it is of material consequence to self-preservation in the case of all the higher animals, whose young pass through a period of infantile helplessness and require parental tendance, that all varieties of painful experience at least should manifest themselves by certain common signs. Again, we can see to some extent why, owing to their bearing on the preservation of the individual or of the species, those specialised instinctive reactions which we have called animal should be developed so early. Thus it is clearly of more direct consequence for purposes of life-preservation that the child should have the instinctive feeling of anger or resentment, the essential factor in the fighting instinct, than that he should have a moral sense. There is, accordingly, every reason to suppose that the latter feeling was acquired much later in the history of the race, as it certainly shows itself later in the history of the individual.[2]

REFERENCES FOR READING.

On the nature of emotion, its development and varieties, see Bain, *The Emotions and the Will;* Herbert Spencer, *The Principles of Psychology*, vol. i. pt. iv. chap. viii., and vol. ii. pt. viii. chaps. ii., vi., and vii.; Ladd, *Elements of Physiol. Psychology*, pt. ii. chap. ix.; and W. James, *The Principles of Psychology*, ii. chap. xxv. The reader of German will do well to consult further Volkmann, *Lehrbuch der Psychol.* ii. §§ 127-132 ; J. W. Nahlowsky, *Das Gefühlsleben ;* and C. Lange, *Ueber Gemüthsbewegungen.*

[1] On the order of emotive development in the animal world, consult Romanes, *Mental Evolution in Animals*, chap. xx. On the relative strength of the several kinds of emotion in the lowest known races of mankind, see H. Spencer, *Sociology*, pt. i. chap. vi.

[2] On the various modes of classifying the emotions, see below, Appendix J.

＊

CHAPTER XV.

A detailed natural history description of the several types of emotion must not be looked for here. We shall content ourselves with taking a rapid glance at some of the main distinctions among the earlier and primitive forms in order to pass to those higher distinctively human types of feeling which educationally and otherwise claim a fuller notice.

(A) GENERAL FEELING OF HAPPINESS AND MISERY.

§ 1. *Characteristics of General Feeling.* The general or un-specialised manifestation of pleasurable and of painful emotion need not detain us long. The causes have been dealt with in Chapter XIII., and the general character of the reactions in Chapter XIV. A word or two must suffice by way of completing the account of this group.

As pointed out above, the most prominent characteristic of these emotions is their common feature of agreeableness or disagreeableness, and the absence of differentiating psycho-physical concomitants such as we find in the case of the emotion of fear or of anger. This shows itself in the resonant discharge and the expression. Joy at seeing a friend, at the unexpected sight of a beautiful thing, and so forth, is pleasure more than anything else, and it manifests itself as such. Similarly with respect to sorrow or grief, such as that excited by mere loss or disappointment. This general form of mani-festation is more distinct in early life when certain expressive

＊

(88)

manifestations, more particularly laughter and tears, which in later years come to be restricted as signs of specialised emotive states (*e.g.*, feeling of the ludicrous, wounded self-love), are common modes of utterance of pleasurable and painful emotion.[1]

It is to be added that while the main characteristic of the group of emotions here referred to is this common agreeable or disagreeable complexion, they have as a subordinate factor certain more special and distinctive features, which grow more clearly manifest as development advances. Thus, the emotive state attending a small recurring annoyance with its teasing, worrying effect, its spasmodic and abortive motor discharges, has a different tone or emotive colour from that induced by a heavy loss (*e.g.*, of fortune, reputation). The difference in the form of presentation involved, with the connected difference in the attitude of attention, would alone differentiate the pleasure of a glad surprise from that of relief from pain, and the gentle melancholy enjoyment of retrospection from the exciting pleasure of anticipation or hope.

Lastly, it must be borne in mind that even in these most general manifestations of happy and of miserable consciousness we have a certain commingling of elements. Thus, the state of suffering commonly indicated by 'grief' (due to some weighty trouble as to loss of friend or wounded self-love) is marked off from those forms of misery which involve a feeling of bitterness and the hardening effect of defiance by an accompaniment of massive pleasurable sensation. The softened or melting attitude, with its effusion of tears, and so forth, not only provides a vent or relief, but supplies agreeable concomitants. Hence the paradox, that we enjoy and nurse our grief, and that we speak of sorrow as a luxury.[2]

---

[1] On the part played in early life by laughing and weeping as common manifestations of pleasure and pain, see Darwin, *The Expression of the Emotions*, chaps. vi. and viii. *Cf.* the full account of the early manifestations of pleasurable and painful feeling in the first years of life by Preyer, *Die Seele des Kindes*, cap. vi. The characteristic physiological accompaniments of joy and of sorrow are described by C. Lange, *Ueber Gemüthsbewegungen*, p. 19.

[2] The relieving, soothing effect of tears in states of misery is set down by Mr. Spencer to its action in lessening the congestion of the brain. For a careful investigation of the causes and the accompaniments of weeping, see Darwin, *The Expression of the Emotions*, chap. vi. and chap. viii. (*Cf.* Bain, *op. cit.*, chap. vii.)

(B) SPECIALISED INSTINCTIVE EMOTIONS.

Passing, however, from these common forms of emotive reaction, we may seek to indicate the chief characteristic differences among the specialised forms of instinctive or animal emotion. Here it is obvious we may be guided by certain broad and well-marked differences in the somatic embodiment, including the manifestation or expression. While keeping as closely as possible to the psychological point of view and considering the psychical differences in the emotive states, we shall endeavour to include the biological aspect as well.

§ 2. *Distinction of Egoistic and Social Feelings.* We may sub-divide these primitive instinctive emotions according as they have to do exclusively with the individual or embrace others as well. In this way we get the current distinction, egoistic (or selfish) and social feelings; or, since the social feelings are other-regarding as distinguished from self-regarding, egoistic and altruistic feelings.[1] Fear (of personal evil) would be an example of the former, love or tenderness towards offspring an example of the latter.

It is important to inquire how far this distinction is a psychological one. The so-called egoistic (animal) feelings, as fear, do, no doubt, in their crude form, develop before the social, and to this extent are psychologically distinguishable. Again, in their higher forms in the human consciousness, they involve the idea of self, and are thus marked off by the presence of a distinctive presentative element. On the other hand, the main ingredient in social feeling proper, *viz.*, sympathy, is, as we shall see, a close reproduction and simulation of egoistic or individual feeling. Thus, to sympathise with another's suffering, fear or anger, is to be in a psychical condition closely resembling that of the corresponding personal feeling.

From all this it follows that the distinction of the egoistic and the social feelings is in the main biological and ethical. It is because of the profound difference in the significance and purpose of the feelings, as subserving the preservation of the individual, and that of the community or race, that the distinction has become fixed in psychology. In order to illustrate

---

[1] So, Herbert Spencer, *Psychology*, pt. viii. chap. vi. ff.

the two groups of feeling here distinguished, we may take a brief glance at one or two of their main varieties.

§ 3. *Fear : Its General Characteristics.* The best type of an egoistic emotion having no necessary reference to others is fear, which in its more intense form is known as terror. A word or two in addition to what has been said on this emotion may suffice to fix its general psychological character.

Fear may be defined as the particular emotive reaction which takes place through a sufficiently vivid and persistent representation of a possible pain or evil. It is the peculiar psycho-physical attitude taken on by the organism when confronted with danger or threatening evil.[1]

The well-known somatic symptoms, *viz.*, cessation of motion, crouching bodily attitude, heart-beating, pallor, trembling of the surface muscles, etc., point to the disturbing and lowering effect on the organism. The only stimulation of muscular activity seems to be in the direction of the higher sense-organs, and especially the eye. These organs, together with the connected apparatus of attention, are, as is well known, in a state of rigid fixation in confronting an object of terror.[2]

Fear is an emotion which illustrates in a peculiarly striking way the need of including the resonant factor. The attempts of the earlier associationists to account for fear as a mere product of association—as in James Mill's doctrine that it is the idea of a painful sensation associated with the idea of its being future—were wholly inadequate through ignoring what J. S. Mill called the animal (*i.e.*, the bodily) element in our feelings. As he pertinently remarks, the fear of flogging has physical demonstrations quite dissimilar to those of the flogging itself, so that the effect of fear cannot be wholly set down to the idea of the pain.[3]

The organic substratum of fear constitutes a well-known disposition to the feeling. Thus we are all prone to fear when ' low,' that is, wanting in nervous tone. A fit of heart-palpitation from indigestion may excite a range of vague apprehensions. Children's special liability to fear is connected with the weakness of the organic functions in their case, and their little power of resisting disturbance.

[1] Since all vivid imagination is apt to beget a vague form of expectation (see above, I. p. 487), it follows that fear may be excited by any vivid suggestion of evil as possible to ourselves. Thus, according to Aristotle, fear is an essential ingredient in the effect of tragic spectacle.

[2] There is reason to suppose that in extreme cases of fascination, as when a man is attacked by a tiger, this factor in terror, by becoming the predominant one, not only reduces the misery of the state, but may even lend it something of the complexion of pleasurable excitement.

[3] See his editorial note in his father's *Analysis*, vol. ii. p. 234 ff.

The connexion of fear with organic disturbance is strikingly illustrated in the miseries of hypochondria.[1]

§ 4. *Development of Fear.* Fear appears early in the life of the child, as it seems to appear low down in the zoological scale.[2] At first it is hardly differentiated from the vague condition of mental shock or " start " which attends all sudden and powerful stimulation, and more particularly loud sounds. As we have seen, fear probably appears in the vague form (*i.e.*, without any distinct representation of a particular kind of evil) in connexion with presentations, *e.g.*, of strange animals, which have contracted no associations from individual experience, and which derive their emotive force from special inherited associations.[3]

Experience is, however, the chief determining factor in the evocation of fear. The child is as conspicuously fearless in many directions as he is fearful in others. The infant Scott found during a thunderstorm lying on his back on a knoll, clapping his hands at the lightning, and crying out " Bonny, bonny ! " at every flash, is a charming example of this. At the same time, the timidity of childhood is seen in the readiness with which experience invests objects and places with a fear-exciting aspect, in its tendency to look at all that is unknown as terrifying, and in the difficulty of the educator in controlling these tendencies.

It follows that since fear is excited through the suggestion of pain or evil by some sense-presentation, or the imaginative representations of this, its development is furthered, in a manner, by intellectual progress. The greater the representative power the more numerous and varied the suggestions of possible evil. Just as a good deal of endurance means want of sensibility, so a good deal of courage means want of imagination. It is in this way that we come to fear so many more things, *e.g.*, loss

---

[1] For a careful account of the physical manifestations of fear in man and in animals, see Darwin, *The Expression of the Emotions*, chap. xii. The author supplies excellent photographic illustrations of the more exaggerated forms of terror in the insane. (*Cf.* Bain, *The Emotions and the Will*, chap. viii ; also Prof. Mosso's monograph, *La Paura*.)

[2] See Romanes, *Mental Evolution in Animals*, p. 342 f.

[3] For the facts going to prove inherited fear, see Preyer, *Die Seele des Kindes*, theil i. kap. vi.

of wealth and of reputation, as life advances. Nevertheless, as pointed out above, intellectual culture, while thus tending to the spreading out and to the differentiation into more varied forms of primitive timidity, tends, at the same time, to greatly reduce the early intensity of fear. This it does by substituting knowledge for ignorance, and so undermining that vague terror before the unknown to which the child and the superstitious savage are a prey, an effect aided by the growth of will-power and the attitude of self-confidence which this brings with it.

§ 4a. *Biological View of Fear.* The facts here briefly described are capable, to some extent, of a teleological explanation. Fear is commonly said to be useful by holding us back from what is destructive. It is, however, to be noted that as a fully-developed state of emotive excitement which paralyses, or at least depresses, motor activities, fear, so far from being helpful, is distinctly injurious. This paralysing effect is of use only as checking a line of action which is injurious—as, for example, the child's impulse to play with a candle flame; it fetters, instead of setting free, the activity required for all self-protective action (shunning of danger, escape). The child and the animal may be too frightened even to run away from that which terrifies them.[1] Owing, moreover, to the peculiar effect of fear in keeping the attention rigidly directed to one object or idea, it exerts a disastrous influence on the intellectual processes, incapacitating the subject alike for precisely estimating the nature and amount of the threatening evil, and for devising means of escape. Here, then, we see a mode of reaction which can only be set down to the primitive constitution of the neuro-muscular organism, and cannot be explained on teleological grounds. The influence of development, on the other hand, by reducing the intensity of early fear, and transforming it into a calmer feeling, and a motive to action, does undoubtedly tend to secure for it this functional utility ; and it may be said that the object of education with reference to fear is to direct it to right objects, *i.e.*, real, as distinguished from imaginary or fictitious evil, and to transform it into an efficient deterring force.

§ 5. *Feelings of Anger, Combat, etc. : General Characteristics of Group.* As a second main variety of egoistic emotion, we may take the feeling variously called anger, resentment, and antagonism. By anger is here meant the peculiar emotive reaction called forth by the conscious suffering of injury or an infliction of pain. It is eminently an "animal" passion, shared by sentient creatures generally, as an incident of the struggle for life

---

[1] A remarkable case of fascination directly opposing the self-conservative instinct is to be seen in the so-called "curiosity" of deer, monkeys and other animals leading them to approach, instead of flying from, their destroyers. (See Darwin, *Descent of Man*, pt. i. chap. iii.)

and of combat. Unlike fear, anger has a reference to others, but, so far from subserving their interests, it is directly injurious or destructive. Hence it is frequently described as anti-social. Looking at the psychological character of the feeling, we find that it resembles fear in having a painful experience as its base or starting-point. But the differences are greater than the resemblances. To begin with, fear has to do with an impending evil, anger with one that has actually been experienced. Again, fear is throughout a painful feeling, a state of misery or wretchedness. Anger, on the other hand, is a mixed feeling, with a very distinct and intense element of pleasure. We smart under the sense of injury, but we find a keen enjoyment in the irate outburst itself.

This last point brings us to the differentiating somatic accompaniments. Whereas fear is lowering and prostrating to the muscular energies, anger is exciting. It is essentially an "active" emotion, that is, one venting itself in energetic muscular action. Thus, after describing the well-known effects of rage on the circulation and respiration, Darwin writes :—

"The excited brain gives strength to the muscles, and, at the same time, energy to the will. The body is, commonly held erect, ready for instant action. . . . The mouth is generally closed with firmness, showing fixed determination. . . . Such gestures as the raising of the arms, with the fists clenched, as if to strike the offender, are common." [1]

This brief description of the outburst may help us to understand how it comes to have so intense a pleasurable element. To begin with, the heightening of the activities of the respiratory and circulatory organs, causing a special reflex augmentation of cerebral activity, is itself a cause of agreeable excitement : we seem preternaturally alive when ' roused ' by anger. It is, however, in the energetic reaction of the voluntary muscles, and the sensations to which they give rise, that we must look for the chief pleasurable ingredient. The full gratification of angry passion is the retaliative injury (e.g., the blow, the insulting word). What we call the sweets of retaliation are the concomitant of this conscious infliction of injury. Hence anger is in general pleasurable so far as this injury is foretasted, as

---

[1] *The Expression of the Emotions*, pp. 240, 241.

in the partial clenching of the fist. It is only when we are conscious of being restrained in this retaliative action, as in states of "impotent wrath," that the pleasurableness of the re-action is lost in a new feeling of disappointment or conflict.

§ 5a. *Pleasure of Retaliation.* The exact nature or source of the pleasure of retaliation, and, generally, of the delight in inflicting pain or cruelty, is a subject which has given rise to much discussion. In the case of the human being and the animal this pleasure is closely bound up with the fighting instinct. The baby's crushing of a fly can hardly be said to be conscious cruelty. It is in resenting injury, in the situation of resistance to attack, that the characteristic flush of angry enjoyment or malevolent passion shows itself. Now this situation gives rise to other feelings, and more particularly to the vivid consciousness of transition from a state of danger to one of strength or superiority. To strike our assailant is to disable him, and so not only to remove a cause of fear but to substitute the sweets of power. Hence we find Dugald Stewart deriving the pleasure of cruelty from the feeling of power. This pleasure of power must, it is evident, include that of the strenuous exciting muscular action itself. It is in and through successful combat that we realise our bodily strength, the ultimate base of all power. It may, indeed, be said that there is an instinctive impulse to resist all assault in this strenuous way, an impulse traceable far down in the animal scale, and forming a main ingredient in what is called the instinct of self-conservation. Whether, however, this exhausts all the pleasure of the angry outburst is doubtful. The appearance of a disposition to torture where there is no attack and no combat, as in the cat's toying with its victim, and man's delight in tormenting, and seeing tormented, helpless creatures, is difficult to account for on this view. There is some reason to say that man has derived from his animal origin a certain bloodthirstiness or love of cruelty, and that though this is rigidly circumscribed by law and custom it is allowed a certain moderate gratification in all cases of resentment of injury, whether to ourselves, to our friends, or to the community to which we belong.[1]

§ 6. *Development of Anger.* Anger shows itself in the case of the child in close connexion with physical pain. It becomes differentiated from mere physical suffering by the vague con-sciousness of another's action, *e.g.*, when forced by the nurse to take its bath, or to be dressed. The extension of the feel-ing by the child and the savage to injuries not inflicted by living things may be explained by the primitive impulse to endow all material objects with something analogous to will. Even grown-up persons find themselves resenting the painful arrest of the foot in locomotion by a stone or other obstacle as a per-sonal attack very much as the child resents it.

---

[1] This is Prof. Bain's view. (See *The Emotions and the Will*, chap. ix.; *cf. Mind*, i. p. 285 and p. 429.) Prof. W. James calls in a special hunting instinct to explain the pleasure of brutality. (*Principles of Psychology*, ii. p. 412.)

The feeling becomes expanded and differentiated into a variety of shades through the progress of experience. Thus the situation of competition for the limited pleasures and good things of life is a powerful excitant of the passion. In this way the unlovely feelings known as envy and jealousy which play so large a part in early life are developed. Again, in the fighting propensity so strong in boyhood we see a signal display of this anti-social feeling. This propensity is congenital and derived from the instinct of combat common to all animals (especially when the food-supply is deficient), and more particularly to the stronger and more combative male.

In later life the passion, owing to its unloveliness and social injuriousness, is in its more turbulent forms subject to peculiar restrictions. It remains, however, throughout life one of the strongest and most deeply-rooted of our feelings. A good deal of the excitement of ambition springs from the gratification of the combative propensity, the delight of triumphing over adversaries. The pleasure of surpassing others exceeds that of surpassing ourselves mainly through the addition, in the former case, of the element of veiled malice. A curious scene of the pleasures of successful attack and humiliation is the fashionable drawing-room, which, as modern satirists, both literary and pictorial, remind us, hides under its smiling complacencies a considerable fund of malignity. It takes on a worthier form in all refined antipathy, which is the resentment of injury to our higher æsthetic and moral feelings. Under this form it is differentiated into a number of *nuances*, the gratification of which constitutes one of the complexities of the modern novel as compared with that of the last century with its one strong typical villain. The feeling of anger attains its highest development in the state of moral indignation, in which the sense of personal injury is lost in the feeling of our solidarity with the community.

§ 6*a*. *Teleology of Anger*. From this account of the feeling we can see that it has an obvious teleological explanation. As already pointed out, anger in its primitive form is resentment of attack, and so one of the most widely diffused manifestations of the instinct of self-conservation. A child without the impulse of retaliation would be unfitted to defend himself in the battle of life. Owing, however, to the special social conditions under which the individual lives in a civilised community, the instinct of resentment loses some of its original utility. As an anti-social feeling, anger tends to our own injury in a state which connects the well-being of the individual with that of the community. Thus a violent disposition to angry outburst brings us into dangerous collision with law. On the other hand, so far as anger is sympathetically extended, so as to become a disposition to resent injury to others, the family or the community, it acquires a new teleological significance as race-preserving.

§ 7. *Self-Feeling : Pride, etc., General Characteristics.* We may now pass to an emotion which is much more circumscribed in the animal world, and more distinctively human, than either fear or anger. This is the egoistic feeling *par excellence*, the feeling, or rather the group of feelings, which becomes organically connected with the presentation or idea of self, and which may be described generally as self-complacency, and its opposite, self-displeasure. Although, if we were following the order of development strictly, we should have to postpone the consideration of this feeling until we had dealt with the instinctive form of social feeling, it will be more convenient to give a brief account of it here, and then discuss social feeling as a whole.

The stimulus of the feeling of self is the presentation of some aspect, feature, or concomitant of the ego or self, recognised as such, that is to say, as a part of ' me ' or as ' mine,' which has a distinctly agreeable or disagreeable character. More particularly the feeling is excited by the presentation or idea of something in the ' me ' or its belongings which is regarded as useful, worthy, which raises our value, or the opposite. Thus it is a self-feeling when a child is pleased at the sight of his own fresh-washed face, or is proud of his exploits in games, or when, on the other hand, he has the disagreeable feeling of self-depression, confusion, or the blush-feeling under the gaze of a stranger.

It is evident from this definition that we have here to do with a feeling of a much wider range, and much more difficult to demarcate, than so well-marked and comparatively invariable an emotion as fear. Self-feeling is as varied in respect of its presentative or excitatory factor as the presentation or consciousness of self is varied.

With respect to the somatic reaction, we have here, in addition to the general distinction of pleasurable elation and depressing misery, certain more specialised manifestations. Thus there is the characteristic bodily attitude of pride (of which the strutting peacock and turkey are often taken as the emblem), the erect carriage of trunk and head, as if the proud man were trying to enlarge himself.[1]   On the other hand, there is the well-marked reaction of blushing, which is confined to man, and seems to appear first in the third year of human life. This reaction, of which Darwin has given us a most careful account, is distinctly a disagreeable experience, involving a painful and crushing consciousness of self (under another's gaze), and attended with a further unpleasant effect in what is called confusion of mind.   Another and closely-related reaction is that of shyness (covering the face, etc.).   This, however, it is obvious, is closely akin to, if not a modification of, childish timidity.   Much the same may be said of the characteristic attitude of humility, the crouching pose of body, which, as Darwin has observed, is the direct antithesis of the elevation which goes with pride.

§ 8. *Development of Self-Feeling.*   We may now pass to the development of the chief phases of this many-sided emotion. As self-consciousness begins with the presentation of the bodily self, we may expect the self-feeling to display itself first with respect to the body.   A trace of bodily pride or self-complacency may, perhaps, be seen in the care with which many animals clean and spruce their hair, feathers, etc.   An infant, on beginning to envisage the body as self, may be supposed to contract a certain feeling of self-complacency in connexion with the touch and sight of it.   The intense delight which young girls are capable of realising from this source has recently been illustrated by the frank confessions of Marie Bashkirtseff.[2]

The feeling attaching to the bodily self, as ' nice ' or ' pretty,' is analogous to the tender feeling the child has for its mother. Markedly different from this is the shade of self-feeling attaching to exhibitions of active power.   The boy rejoices in his strength, and has in the realisation of his bodily powers a delightful consciousness of self, as expanding.   It is this mode of the self-feeling, the delight in exhibiting or realising our active powers, which is more specially referred to by the term pride.

---

[1] This, as Darwin observes, is clearly indicated by the expressions, ' haughty ' (*i.e.*, high), " puffed up " with pride, and so forth.   (See *Expression of the Emotions*, p. 263.)

[2] See her *Journal*, translated by Mathilde Blind.

8a. *Pride and Power.* This aspect or self-feeling connects itself very closely with what some writers treat as the distinct emotion of power. It is obvious that, if pleasurable self-feeling comes from the conscious realisation of some attribute of the self, it will be chiefly stimulated by our active efforts. It is the man who does much, who plans and achieves on a large scale, that has most of the enjoyment of pride. Conversely, inactivity, impotence, bring with them a painful consciousness of a compressed self. The feeling of power in its more intense forms is differentiated from a pure self-feeling through a special element, the delight in superiority. So strong is the impulse of combat or rivalry in human nature that a man's achievements are apt to yield him an imperfect pleasure when they do not sensibly add to his feeling of lordly triumph over others. The boy who is proud of his strength is pretty sure to begin to hector it over his comrades. All boasting or " bragging " is indeed a form of insolence, that is to say, intended affront, disrespect, or humiliation. Now this feeling of superiority, as we saw above, is a modification of the anti-social feeling of malignity. All arrogance, contempt of inferiors, and the like, are forms of anger, and involve the rudiment of an impulse to destroy or crush. This explains the close association of the feeling with the fighting propensity, the fact that the luxury of conscious superiority is enjoyed by preference both among men and among women in relation to those who were once their superiors or equals (pleasure of triumph) or are near enough to them to be thought of as possible rivals.

It may be added that while the emotion of power is thus accounted for in the main as a compound of self-exaltation and malignity, it takes on a special colouring in such forms as disdain, contempt, and so forth. The characteristic bodily manifestation here, movements of the nose and mouth, some of which, as Darwin has shown, stand in organic connexion with particular sensations (*e.g.*, foul smell, nausea), contribute special sensational elements, partly disagreeable though in the main agreeable, to the whole emotion.[1]

§ 9. *Self-Feeling and Self-Consciousness.* The development of the feeling of self proceeds *pari passu* with the growth of self-consciousness. We may trace the effect of this development in different directions. Thus the expansion of the bodily self by the incorporation of clothes, serves, as is well known, to effect an irradiation of the self-complacency on to these.[2] As the area of our belongings enlarges under the influence of experience and association, so does our self-feeling dilate or spread itself out. In this way our surroundings, our home, our books or other implements of work, our friends, and so

---

[1] On the psychological nature of the emotion of power, see Bain, *op. cit.*, chap. x. On the expression of pride, arrogance, and contempt, see Darwin, *op. cit.*, chap. xi. The shading off of common anger into contempt is curiously illustrated in the expressional movement of the sneer (uncovering the canine tooth), which is at once, as Darwin remarks, a sign of angry defiance and of contempt. (See pp. 249 f. and 255.)

[2] See Lotze, *Microcosmus* (Eng. transl.), i. p. 592 ff.

forth, come to take on the glow of self-complacency. We take them up into the private region of our personal belongings, and feel proud of them. Again, the growth of the higher and more representative self-consciousness tends to expand and at the same time to equalise the feeling. A steady and relatively permanent consciousness of intellectual or moral worth based on representations of past achievement becomes more and more marked, giving to the man his habitual bearing of pride or the opposite. The feeling becomes further refined by the discrimination of a lower and a higher self, and the enjoyment of self-complacency through the realisation of the latter, *in spite of* the depressing effect of the former.

It may be added that the whole effect of experience and social discipline is to tone down the feeling, or at least to check its early manifestations. A wider contact with others is commonly supposed to teach us the limits of our importance, and in any case the code of good manners requires us to hide all self-inflation and to assume the external aspect of a reasonable or modest self-esteem.

§ 10. *Self-Feeling and Others' Opinion : Love of Approbation.* Thus far we have considered the growth of self-feeling with little reference to the individual's social relations. But the slightest attention to the circumstances in which the feeling is excited, and to its characteristic manifestations, will show us that it involves the presence and co-operation of others, at least in the capacity of spectators of our doings. A word or two must suffice for elucidating this side of the subject.

As was pointed out above, the child's self-consciousness becomes distinct by help of others' actions and words, which tend to direct its attention inwards upon itself. We may expect, then, to find that what others say will have a good deal to do with the early developments of the feeling of self. The first clear indications of self-consciousness, *viz.*, blushing and shyness, are directly excited by the look or word of another.[1] This feeling of shame might almost be defined as a disagreeable consciousness of another's and a superior's scrutiny. It is a mode of timidity in relation to what others will think of us. This reference to external opinion becomes more distinct in all those modes of self-feeling which fall

---

[1] See Darwin, *loc. cit.*, p. 326 f.

under the head of love of display or vanity, love of praise or approbation, reputation, and so forth. As is well known, this feeling
is very strong, reaching to the intensity of a passion not only
in the child, but in his prototype, the race-child or savage.[1]
Here it is evident that the gratification (or dissatisfaction) arises
through others' regard for, or estimate of, oneself. The feeling
is still, properly speaking, the feeling of self. The child is
realising its own skill, or other good quality, *by means of* what
others say. And the reason why it needs this objective stimulus
in order to enjoy its self-feeling is that it is unable to carry out
independently of this any clear process of self-reflexion and self-
estimation. At the same time, the enjoyment of others' good
opinion and praise involves an additional element, *viz.*, the
pleasure of favourable regard, an element that arises out of the
instinctive sociality of the child, his disposition to heed what
others think and say of him, and, more particularly, that impulse
to " look up " to his elders, which is the main condition of all
education. We may then say that the love of approbation is
a semi-social, or, to use Mr. Spencer's language, an ego-
altruistic feeling.

The pleasure taken in others' approbation must be distinguished from that
connected with the giving of pleasure to another, though the two feelings shade
off one into the other, and are nearly always combined in various proportions. A
child is not in the same mental state when it is enjoying its mother's praise *as
praise*, and when it is agreeably conscious of having given her pleasure. The first,
the love of approbation proper, springs from the survey of the ego or self, and is
an egoistic feeling, although it has a subordinate social character also : the second,
as we shall see presently, is a purely social feeling.

The later developments of the self-feeling free themselves,
to some extent, from this dependence on others, and so there
arises the distinction of an independent self-esteem, which may
maintain itself even in the face of what others think of us.
This detachment shows itself in the weakening of the desire to
impress others or to make a display, which desire is a main element in vanity. It is illustrated in the character of Guy Man-
nering, of whom Scott says that he " was too proud a man to
be a vain one ". This liberation of the feeling from its early
social props is, however, at best only a partial one. Few men

---

[1] On the huge vanity of savage races, see Spencer, *Sociology*, i. p. 71 f.

would set much store by their own good opinion of themselves if this received no external support. We are counselled by philosophers, ancient and modern, to despise others' opinion, reputation, and the like, and to live content with an inner self-satisfaction.[1] Such a counsel, however, if taken literally, would lead to all the ridiculous extravagance of obstinate self-conceit, a phenomenon most strikingly displayed in the delusions of the insane. Regard for others' opinion is a mark of the normal social man ; and true independence here consists not in flouting all extraneous opinion as valueless, but in discriminating whose opinion is worth consulting, and whose is to be disregarded because unenlightened, or biased.

§ 10a. *Teleology of Self-Feeling.* Our brief account of the self-feeling may suffice to show how it has its roots deep down in the instinct of self-conservation. If strength and beauty alike are the outcome and the signs of health, the setting, store by these must, it is plain, tend to the preservation of the organism. Thus the girl's instinctive feeling for rosy cheeks, and the boy's pride in his biceps, may be seen to have a teleological significance. To this it may be added that all gratification of self-feeling, so far as it springs out of successful life-effort, has a like utility. To derive pleasure from the recurring consciousness of growing power, knowledge, wealth, and so forth, is to find our gratification in connexion with what is useful, and tends to the confirmation of life. Lastly, it is evident that the self-feeling, so far as it depends on and reflects others' feeling towards us, is an aid to individual self-preservation *in a social state.* The child that cares nothing for what others say is doomed to a dangerous collision with the will of the community. It may be added that a mutual regard for others' opinion is one of the chief guarantees for the security of the social structure itself. Here we see the social character or tendency of the feeling, its connexion with the maintenance of the community, as well as with that of the individual.[2]

§ 11. *Instinctive Social Feeling : General Characteristics.* Having now taken a survey of the more important develop-

---

[1] No writer has urged this with greater force than Schopenhauer. (See especially *The Wisdom of Life*, translated by T. B. Saunders, chap. iv.)

[2] The nature of the self-feeling is variously conceived by different writers. Prof. Bain appears to regard the emotion of self as in the main a reflexion of feelings primarily directed towards others. (*The Emotions and the Will*, chap. xi.) This view is criticised by W. James, who points out, after Horwicz, that many things which offend us in others do not offend us in ourselves. (See his whole account of the feeling, *Principles of Psychology*, i. pp. 305-329 ; *cf.* Horwicz, *Psychol. Analysen*, 2 theil, 3te hälfte, § 11.) The teleological aspect of the feeling is dealt with by H. Spencer, *Principles of Psychology*, vol. ii. pt. viii. chap. vii. ("Ego-Altruistic Sentiments "). The term self-love has been used in a peculiar sense by Butler and other moralists for the regulative principle presiding over and duly co-ordinating the particular affections of self, *i.e.*, egoistic feelings.

ment of Egoistic feeling, we may pass to the opposed group of Social feelings. As already pointed out, the distinction between the two groups is not purely or mainly psychological, but teleological or ethical, a fact which appears at once in the attempt to define social feeling. This cannot be done by saying that it is a feeling which has the presentation of another sentient being as its proper excitant and 'object'; for then anger, contempt, and the like, would be social feelings.[1] We can only frame a definition, so as to include the well-recognised varieties of the feeling, by adding a teleological reference. Social feelings are those feelings which are excited by and gather about the presentation of other sentient beings, *in so far as they are supposed to be favourable to others, and to maintain the social structure or community.*[2]

As thus defined, social feeling has a wide range, appearing far down in animal life in the family attachments developed in connexion with the rearing of offspring, parental, and more especially maternal, attachment, and in a secondary measure attachment to mate ; further, in the gregarious or herding propensity, that love of society which keeps many species of animals in permanent association. It is, however, as manifested in man that we shall be more particularly interested in it here.

§ 12. *Two Ingredients of Social Feeling.* A very little examination of the social feelings, love of society, of co-operation, personal attachment, and so forth, shows us that it has two ingredients, or that it springs from two roots : (*a*) In the first place, there is the liking for others, growing out of the pleasure or satisfaction which the presence or companionship of others brings, or the bare feeling of Attachment, an emotion that has in its more concentrated forms the characteristic reaction of fondness or caressing.[3] (*b*) In the second place,

---

[1] Anger or malevolence is indeed often spoken of as a disinterested, that is, non-self-regarding feeling along with sympathy or benevolence ; and, psychologically regarded, they have a close affinity.

[2] The social feelings as thus defined include those which subserve the biological end, race-preservation, and more particularly parental love.

[3] I use this term in preference to "tender emotion," the phrase employed by Prof. Bain, as this carries with it a much more distinct reference to the common altruistic accompaniment and provocative, *viz.*, a compassionate or kindly feeling.

there is the feeling of Sympathy, or the sharing or entering into the feelings of others. The former ingredient has an egoistic basis. It springs partly out of gratifications instinctively derived from others, *e.g.*, maternal satisfaction, the gratification of sex, and partly out of transferred products of association, *e.g.*, the liking for those who have given us pleasure in various ways. The second ingredient is the pure altruistic element in the social feeling.

Although the ingredients are never, perhaps, found in perfect isolation, and commonly interact and blend in a very intricate manner, they may be studied to some extent apart. In the lower animal forms of social feeling sympathy, though present, appears in a vague and circumscribed form. In order that this feeling attain any considerable degree of development representation must have acquired a certain vividness, clearness and stability. Hence it is only found in well-developed form in man, and among civilised men. We may then consider social feeling first of all in its aspect of bare attachment as a feeling common to man and animals, and then give a fuller account of the distinctively human element of sympathy.

§ 13. *Feeling of Attachment or Fondness.* The simplest manifestation of attachment is that of a pleasure or satisfaction when with others, and a correlative pain or dissatisfaction when bereft of their society. This feeling is common to all animals that form the temporary but close attachment of the family, or the more permanent attachment of the herd.[1] In the case of certain animals, more particularly our domesticated ones, it takes on the form of a liking for the society of others outside the species. This form of attachment is very strikingly illustrated in the dog.[2]

In the case of the human being, the first display of the feeling is the early and partially-instinctive attachment to the mother or other nurse. The relation involved in the sustenance of the child, a relation only a degree less close than that

---

[1] How strong such animal attachment may be is shown by the fact that birds and other animals have been known to die from grief at the loss of their mates. (See Romanes, *Animal Intelligence*, p. 270 ff.)

[2] The attachment of domesticated animals to man is not the only case of the extension of fondness beyond the limits of the species. Many curious stories are told of animals making a chum of a member of a foreign species.

of the fœtus to the maternal organism, constitutes in itself the chief source of the feeling. Along with the supply of nutriment there goes that of warmth, support, or propping, which again is a continuation of the fœtal dependence. This first instinctive or sensuous attachment of the child grows into what we call fondness by the complication of this instinctive feeling with numerous "ideal" or transferred feelings, the product of the many pleasurable sensations, including those of the eye and of the ear, of which the mother is the source.

The mass of feeling thus formed constitutes a true emotion, and one which has its own distinctive or specialised manifestation. Its proper excitant is the presentation of the mother, more particularly the visual presentation of the face; or, later on, what serves as its substitute, *e.g.*, the sound of the voice. This presentation, provided the necessary conditions, intermission, or change, etc., are fulfilled, excites a peculiar emotive reaction, which, in addition to the general manifestation of pleasure (the smile, general excitation of movement), has a special element which, in its most general form, may be best described as cuddling or nestling. This reaction in its common animal form, which is seen not only in man but (with obvious differences) in many birds and in mammals, is at first the physical symbol of the attachment. Later on it develops into a more distinctively human form, *viz.*, caressing touches, soft strokings, and embraces. The completed reaction supplies a wide range of quiet sensuous enjoyment, so that the feeling is a massive pleasure. This early reaction differentiates itself later on into the well-known manifestations of the lover, and the out-goings of pity or compassion.[1]

§ 14. *Expansion of Feeling of Attachment.* The emotion here briefly described is irradiated in varying degrees of intensity on the child's companions. It is sometimes said that there is an instinctive tendency in the infant to react pleasurably, more

---

[1] For a full, detailed account of the organic accompaniments of the human feeling of fondness, see Bain, *op. cit.*, chap. vii. ("Tender Emotion"). The writer points out that the characteristic utterance of "tenderness" with its relaxation of muscle contrasts with that of anger, and he suggests that among our uncivilised ancestors the mood of tenderness may have been induced to some extent as a reaction from the fierce, tense attitude of combat. The organic connexion with touch is seen in the expression of friendly feelings by savages. (See Tylor, *Early Hist. of Mankind*, p. 51.)

especially with the responsive smile, on the sight of the human face; yet this is far from being certainly established.[1] However this be, other instinctive tendencies aided by experience serve to develop attachment, liking, or fondness for others. Thus the child's sense of weakness and dependence, his desire for others' approbation, the experience of the larger and more varied field of pleasurable activity opened up by the initiation of others, and his own imitative propensities, all conspire to make him seek society and to develop a fondness for others.

Later on, as his mind expands and he is able to discern other qualities in those with whom he lives, and more particularly mental qualities as knowledge, kindness, sincerity, the early fondness becomes enriched by the infusion of higher emotive elements, and especially æsthetic and moral admiration. This recognition of indwelling personal quality becomes a more and more distinct element in all permanent attachments, including filial fondness, sexual attachment, and the more widespread feeling of friendship.

From this brief sketch we may see that human beings are objects fitted to attract us in many and various ways, and to develop within us a strong pleasurable emotion (liking, fondness). It is obvious, however, that the full intensity of this feeling depends on others' friendly or kindly manifestations towards us. The child's largely non-sympathetic and egoistic 'love' is the response to the mother's sympathetic or altruistic love. Social relations, and the measure of sympathy and good-will which are necessary to their maintenance, are thus presupposed. We have now to turn to this central and all-important clement in the social feeling, viz., fellow-feeling or sympathy.

(c) REPRESENTATIVE EMOTIONS.

(1) CONCRETE REPRESENTATIVE EMOTION : SYMPATHY.

§ 15. *General Nature of Sympathy.* By Sympathy, in the most extended signification of the word, is meant feeling

---

[1] Such a reaction would be an illustration of a special inherited association. The observations of Preyer on the first development of smiling tend to show that the responsive smile only appears when the imitative age is reached, and that even then it is not always forthcoming. (*Die Seele des Kindes*, p. 185.) It is to be noted that the general infantile amiability soon becomes checked by the development of a new instinctive feeling, viz., shyness or timidity before strangers.

excited by the manifestation of a like state in another, or, as we may call it, concomitant feeling. Thus, when the expression of grief depresses, it may be said to be an effect of sympathy. In its higher and more complete form sympathy is "fellow-feeling," that is, feeling with and for another, which involves an imaginative "intuition" or realisation of his affective state.[1]

In its more extended sense sympathy is not confined to man. We see, indeed, among many species of animals not merely concomitant feeling but the germ of true fellow-feeling. The out-goings of parental compassion among all the higher animals are a considerable set-off to their general callousness.[2] At the same time, sympathy, in all its higher and more distinct or articulate forms, is pre-eminently a human emotion. It appears, indeed, in its richer and more varied outflow, even in man, only among the civilised and "humanised" branches of our species, and constitutes an essential and fundamental element in the higher social culture.

The reason of this limitation of sympathy to the higher stadia of human development is to be found in the nature and conditions of the feeling. These will be expounded more fully presently. Here it is enough to say that, since sympathy with others is only possible when we imaginatively represent and realise their affective states by help of our own similar experiences, it can only reach a considerable development after a certain accumulation of emotive experience through the gratifications and disappointments of the more instinctive emotions, and when the general representative power of the mind attains a certain strength. It is to be added that the wider and more varied manifestations of sympathy presuppose that type of social life, with its closer and more systematic co-operation, and its humanising institutions and education, which is to be found only in civilised communities.

---

[1] The idea of concomitant feeling passing into conscious feeling with others is suggested by the etymology of the term (σὺν and πάθος). (Cf. the German Mitgefühl.)

[2] The manifestations of sympathy among animals occur in close connexion with those of fondness, that is to say, towards members of the family group, and those of the herd or flock. For examples among birds, see Romanes, *Animal Intelligence*, chap. x. p. 270 ff.

The process of sympathy in its most general form may be described as the revival in a lower degree of strength of an affective state through the medium of a presentation answering to a part of its reflex discharge or manifestation. Thus depression on seeing a person in tears is due to the presence of a presentation corresponding to a part of the manifestation of pain or grief as experienced in our own case. Since this presentation presumably acts by exciting a faint degree of the corresponding reaction in our own organism, all sympathy is closely akin to imitation. In its lower and more instinctive form, more particularly, it might be described as reinstatement of feeling through imitative motor discharges. In the higher forms, where the responsive feeling depends on more elaborate processes of representative imagination, this imitative factor, though still present, plays a more subordinate part.

The feeling of sympathy appears under a special or restricted, and under a more general or comprehensive form. Chronologically, both among animals and in man, it is first developed in close connexion with special feelings of attachment, and more particularly those binding together parent and offspring, during the more or less prolonged duration of the family, and the members of a gregarious band or herd. It is here that we see the manifestations of animal sympathy ; and it is within like limits that sympathy displays itself in the lower stages of human culture, and in the earlier period of individual development. The wider and impartial manifestations of sympathy belong to the higher stages of civilisation and human culture.

With this preliminary account of sympathy to guide us, we may now proceed to trace its principal developments.

§ 16. *Contagion of Feeling : Sympathy and Imitation.* The simplest manifestation of sympathy is to be found in the phenomenon known as the contagion of feeling. This shows itself among all animals which are brought into habitual contact, and subjected to similar external conditions, mothers and their offspring, and members of a herd of gregarious animals ; witness the imitative communication of distress from chick to hen, or of fear from one member of a flock of sheep to others. It shows itself, too, very early in the life of the child, and is, indeed, a characteristic of low development in the case both of the race and of the individual.

Here, it is evident, we have to do, primarily and mainly, with an imitative reproduction of expressive movements, in-. cluding those by which sounds are produced. The animal that bleats, bellows, and so forth, in response to a similar cry of his herds-mate, takes on something of the corresponding feeling by taking on one important factor of the whole organic reaction. Similarly when children or savages are affected by the contagion of laughter, or of fear.

Now we have seen that the reaction characterising a particular emotion is, to some extent, instinctive and unacquired. It follows that animals and children would show a tendency to this contagious or imitative sympathy as soon as they were able to connect the sound (or other sign) as presented by another with its analogue in their own case, that is to say, as soon as they acquired the power of imitation. This power, as we shall see, is attained by the child in about the fourth month, and there seems no reason to suppose that the signs of others' feelings are reproduced before this date.

The connexion between sympathy and imitation has been noticed by more than one writer. Thus Adam Smith illustrates sympathy as follows : " The mob, when they are gazing at a dancer on the slack rope, naturally writhe and twist and balance their own bodies, as they see him do, and as they feel that they themselves must do, if in his situation ".[1]

Mr. Spencer advances the ingenious hypothesis that sympathetic response to manifested feeling arose among gregarious animals through a simultaneous subjection of the members of the herd to common experiences, e.g., alarm. This common experience would beget an association between the feeling in the case of any given member and the corresponding sound as uttered by other members. In this way sympathetic responses would be independent of imitation, and might become instinctive through hereditary transmission.[2] Possibly this is true of the lower animals, but in the case of the child sympathetic reaction is not instinctive, but waits on the development of the imitative impulse.[3]

§ 17. *Sympathy Proper : Fellow-Feeling.* We may now pass to sympathy in its more complete form, *viz.,* the imaginative realisation of another's feelings *as his,* or the participation in a

[1] *The Theory of the Moral Sentiments,* chap. i. (" Of Sympathy ").

[2] *Principles of Psychology,* ii. pt. viii. chap. v. § 505.

[3] See Preyer, *op. cit.,* chaps. xii. and xiii.

pleasurable or painful experience consciously referred to another. Here, it is evident, we need more than the instinctive imitative response just dealt with : we require a representative consciousness sufficiently developed to allow of an apprehension of another sentient creature as such.

This power is only developed among animals, and even among the lower races of man, within certain narrow limits which have already been indicated. There seems no reason to doubt that among all the higher animals, under the concentrating and intensifying influence of the maternal instinct, the power of representing others' states of satisfaction, want, or injury, attains a sufficient steadiness and distinctness. The mother among mammals, and even among birds, shows all the signs of understanding and' entering into the changing states of her progeny.[1] Among uncivilised races of mankind the strong tribal instinct may develop a similar power of genuine fellow-feeling. At the same time, as already indicated, sympathy appears only sporadically among animals and savages, and for its fuller and more varied manifestations we must look to civilised communities.

§ 18. *The Process of Sympathy.* The general process in this higher sympathy or fellow-feeling may be described as follows : (*a*) Certain external presentations, as cries, looks, gestures, are noted. These produce directly by imitation a faint resonance in the spectator. (*b*) But this instinctive factor is inhibited by a more complicated central activity, answering to the conscious reference of the signs observed to another as their subject or agent, and to the imaginative interpretation of the manifestations. This last may be quite simple and instantaneous, as when a mother hears her child cry and sees it prostrate on the ground. In other cases it may be complicated and prolonged. This applies to all sympathetic participation in new, untried experiences, *e.g.*, the exaltation of a friend, a loss of fortune by another, and so forth. Here there is a process of constructive imagination, a building up of an imaginative representation of the new by the help of old and analogous experiences of our own. (*c*) Finally, we have as the outcome of such more or less prolonged central representative activity the sympathetic re-

---

[1] On the power among animals of recognising the mental states of others, see some good remarks of Romanes, *Mental Evolution in Man*, pp. 197, 198.

action, that is to say, the excitation in a weakened form of the whole round of organic processes involved in the feeling. This organic reaction will, of course, include those manifestations which are specially marked off as expression.

(d) One other factor remains to be spoken of. Sympathy, though mainly an emotion, is also a stimulus to action ; and so close is the connexion that we cannot attempt any description of it without referring to this. In sympathising with another we desire to augment his joy, or to assuage his sorrow. This tendency to benefit the subject is particularly manifest in the soothing, alleviating actions that characterise pity, which, as we shall see, is the earlier and stronger form of sympathy. The expression of true sympathy is thus more than a mere emotive reflex ; it has something of the character of a voluntary action. Hence the expression 'feeling of benevolence,' which is used by moralists as interchangeable with sympathy.[1]

The prominence of the representative factor in sympathy accounts for the common attribution to it of intellectuality. To sympathise with another is to understand him, to think out his situation, and its effect on his known character, tastes, and desires. Hence we never understand those who lie wholly outside the circle of our sympathies.

While sympathy and intellectual apprehension are thus closely related, they are not identical. In each case there is the representation of another's mind or feeling, but the mode of representation differs. In sympathising with a person we are occupied with his feelings as such, and are ourselves in a state of resonant feeling ; in understanding him we are intellectually active, fixing our attention on the *relations* (causal, etc.) of his mental state. Hence we can often understand passions and impulses, *e.g.*, a homicidal hate, without a full sympathetic realisation of them.[2]

§ 19. *Relation of Sympathetic to Personal Emotion.* It is well known that sympathy in many cases comes near in point of

---

[1] *Cf.* the emergence of a conative element in anger or malevolence—the half-volitional impulse to strike. It is to be observed that active helpfulness is the most prominent element in the rudimentary sympathy of animals. This is strikingly illustrated in the thoroughly practical sympathy of ants. (See Romanes, *Animal Intelligence*, pp. 46, 47.) Our expansive human emotion looks like an efflorescence on a primitive stem of practical helpfulness.

[2] On the relation of sympathy and understanding, see Leslie Stephen, *The Science of Ethics*, chap. vi. § 2.

intensity to the actual experience. This specially applies to all participation in *emotive* states, as fear, joy, indignation. To see another in danger is to experience a fear hardly distinguishable from the actual personal feeling. On the other hand, our fellow-feeling with another's physical pleasure or pain is, as a rule, much less vivid and further removed from the actual experience.

The reasons of this have already been suggested. Bodily sensations, and so their affective concomitants, are only revivable in a feeble degree. Just as I cannot recall with any satisfying distinctness the pleasures of a good dinner, so I cannot enter into these pleasures when I see another enjoying them. An emotion, on the other hand, being already representative in part, is revivable in very much of its original intensity. I can imagine myself injured till I feel very much the same as if I were certain that the injury had been inflicted. Hence the strength of my sympathetic indignation. To this must be added the important circumstance that the revived emotion, by engaging the same plexus of organic structures as the original, and so producing sensuous feelings similar to those of the actual experience, is to a considerable extent a re-experience of this last.[1]

§ 20. *Effects of Sympathy.* We may now consider the concomitants and effects of sympathy. Here we shall consider first the state of the subject of the sympathy, and then that of the object or recipient.

Sympathy is representative emotion, and as such an extension of our life of feeling. This is of course most obvious when the feeling we are entering into is agreeable. To witness or to imagine others' delights is a means of enjoyment. Even selfish persons like to surround themselves with happy faces. Those who are badly off in respect of the luxuries of life find a certain solace in dwelling on the delight of those who possess them. Similarly women feel their lives expand by entering sympathetically into the larger interests of men.

When the exciting cause of sympathy is a painful experience the case is somewhat different. The sympathetic representation of another's sorrow is as such painful. Hence the selfish

---

[1] On this characteristic of Ideal Emotion, see Bain, *op. cit.*, chap. v.

man dislikes the sight of suffering and shuts it out as much as possible. It is only the humane man or woman, who as such is interested in others' sufferings, that will give so close an attention to their manifestations as to enter into their painfulness. Yet, while sympathy with pain is primarily painful, it has in its later stages distinctly pleasurable ingredients. That characteristic reaction which we call the outflow of pity or compassion, with its gentle look, its low-toned voice, its soft touch, is a massive complex of sensuous enjoyments. It is a matter of common observation that sympathetic natures find a rich satisfaction in this outpouring or "gush" of pity through the manifold channels of looks, words, and so forth. Mr. Spencer goes so far as to speak of the "luxury of pity".

It is to be added that all complete manifestation of sympathy is pleasurable so far as it involves a representation of its beneficial effect on the recipient. This idea of benefit is already present in a vague sub-conscious form in the pitiful reaction.[1] And it becomes more distinct in the practical benevolent man, who, not content with the 'luxury of pity,' that is, the effusion of mere sentiment, sets about some beneficent action.

The nature of the special emotive reaction known as pity or compassion has given rise to much discussion. One chief part of the problem of tragic effect as Aristotle defined it was to account for the pleasure attending the gratification of our pitiful impulse. One of the most striking and suggestive attempts to account for the rich massive enjoyment in compassion is that made by Mr. Herbert Spencer. He deals with compassion or pity as a specialised instinct evolved in the course of human and animal life by the action of natural selection, and primarily developed as a parental impulse towards the small, feeble offspring.[2]

Let us now turn to the recipient of manifested sympathy. Whether we be the subject of joy or of grief, the consciousness that another is in sympathy with us causes us pleasure. As Bacon puts it : ". It redoubleth joy and cutteth grief in halves ". This effect is the outcome of the principle of harmony already

---

[1] Present, but not easily susceptible of analytical detachment. Let the reader ask himself how much is pure self-enjoyment, how much conscious promotion of another's pleasure in ordinary baby-fondling, or in lovers' caressings.

[2] On Mr. Spencer's attempt to make the parental feeling towards infantile helplessness the germ of all compassion and tenderness, see his *Principles of Psychology*, ii. pt. viii. chap. viii. § 532; *cf.* Bain, *The Emotions and the Will*, chap. vii. p. 124 ff.

dealt with working on our social instincts and propensities. To find others in agreement with us is to have a restful sense of concord or unity, whereas to find them in disagreement is to have a disagreeable sense of conflict or division. So valuable is this consciousness of others' sympathy that even a great suffering becomes in a measure transformed by its presence into a mixed state of half-painful, half-pleasurable grief.

It may be added that in general we look for sympathy more in the case of painful experiences than of pleasurable ones. Pleasure is as such sufficiency, whereas pain is defect, and need. It was pointed out above that the distinct expression of pain seems to be more widely diffused in the animal kingdom than that of pleasure, and to come before the latter in the development of the child. This paramount claim on sympathy in states of pain or sorrow has led to the correlative development of the sympathetic impulse in this direction. The mother's sympathy is displayed in the alertness with which she responds to every utterance of pain in her child.

Adam Smith points out that "we are still more anxious to communicate to our friends our disagreeable than our agreeable passions, that we derive still more satisfaction from their sympathy with the former than from that with the latter, and that we are still more shocked by the want of it".[1]

Mr. Herbert Spencer argues that, since common causes of pleasure act upon us far more frequently than common causes of pain, the tendency to respond sympathetically (i.e., imitatively) to manifestations of pleasurable feeling, e.g., laughter, is stronger than the impulse to respond to the manifestations of pain.[2] This seems to me to be very questionable. There may be some special anatomical reason for the infectiousness of laughter. The reaction of laughter is, as everybody knows, one of the most rapid and irresistible. But it is surely an error to say that the tendency to respond sympathetically to a laugh is stronger than the tendency to respond sympathetically to an expression of pain *of equal intensity*, say a shriek. As pointed out above, the expression of pain is in general antecedent (in the case alike of the zoological series and of the individual) to that of pleasure, a fact shrewdly turned to account by the pessimist. At the same time, culture and the rules of good breeding tend to the habitual repression of the signs of pain, whereas they directly encourage the moderate manifestation of pleasurable feeling. The same causes tend to produce a certain artificial strengthening of the sympathetic impulse in the case of all agreeable manifestations. None the less the response to pain in what is called pity seems, as Mr. Spencer appears in another place, already alluded to, to recognise, to have been the primitive manifestation of true sympathy.

[1] *Moral Sentiments*, chap. ii.
[2] *Principles of Psychology*, ii. pt. viii. chap. v. § 511.

§ 21. *Conditions of Sympathy : The Sympathetic Bent.* From this brief account of sympathy, together with its concomitants and effects, we may deduce its more important conditions. As already observed, the development of true altruistic feeling presupposes the social state, civilisation, and what we mean by the educative action of a civilised community on its individual members. Assuming these primordial conditions to be realised, we may proceed to point out more special conditions.

To begin with, then, sympathy, in all its higher, more expansive, and more refined forms, presupposes a sympathetic bent or disposition. There are well-known differences between the two sexes, and among different individuals in this respect. Some people, especially women, live more than half their life in the sympathetic realisation of others' feelings.

The more important factors in this special disposition seem to be the following : (1) First of all, the sympathetic person is necessarily the sensitive or emotional person, with quick and varied susceptibility to pleasure and pain. In order to feel *with*, we must be able to feel, and to some extent have felt, *without* others. A cold, unemotional temperament is incapable of sympathy.

(2) The sympathetic disposition includes a special interest in the external manifestations of feeling, leading to quickness and fineness of observation. This condition is in part contained in the first ; for when another's situation and aspect awaken the echo of a personal emotive experience, this affective element itself gives an interest to the presentation. Yet the two conditions are by no means the same. We may see a person in a situation of great happiness or great misery which reminds us of a past personal experience, and yet not be excited to a true sympathetic response. Indeed a very strong tendency to dwell on our own affective experiences is distinctly antagonistic to sympathy. This last involves the co-existence with a warm emotional temperament of an " objective " attitude of mind, a bent of the attention away from self-regard towards a certain group of appearances in the external world. And this special direction of the attention is precisely what we mean by the sympathetic, or, to take its higher and larger form, the humane interest. There is a special interest in the manifestations of feeling, in the dramatic play of facial feature and limb, just as

*

there is a special interest in flowers, in the song of birds, and so forth. This peculiarity is oddly illustrated in the proverbial blindness of trained scientific observers to the manifestations of feeling by their wives or other companions.

This special interest is developed first of all, and to some extent, through the workings of the egoistic instincts, and more particularly, perhaps, the fear of human wrath.[1] In many persons it never rises much above this childish interest in noting how others feel *towards ourselves*. But in a few it expands, becomes detached from its parent-germ, and grows into a disinterested love of watching the display of human passion *on its own account*. This, again, may take on the form of an artistic or dramatic interest in expression, as powerful, touching, beautiful, and so forth ; or, if coupled with strong active instincts of helpfulness, becomes the sympathetic interest proper.

(3) As a last factor in the sympathetic attitude, we have a certain liveliness of imagination. The connexion between sympathy and constructive imagination has already been touched upon. Dulness of imagination, even where there are the emotional temperament, and special observation of affective manifestations, is a fatal obstacle to all the more extended and more finely ramified sympathy.

§ 21. *Special Directions of Sympathy : Feelings as Unequally Realisable by Sympathy.* Having considered the fundamental conditions of sympathy as laid down in the nature or temperament of the subject, we may glance at the main circumstances which determine the particular lines or directions of discharge of the emotion. Whatever our capacity for sympathy, we are not equally disposed to participate in all the feelings that may happen to manifest themselves. Some manifestations are in themselves, and apart from the nature of their particular subject and his relation to ourselves, much more readily entered into than others. One reason why sympathy with the pleasures of others is so difficult and so rare is that it is apt to be choked at its birth by envy. I never yet heard of the child who could look on sympathetically at another child's

---

[1] Hence the quickness of the child in discriminating real from sham anger. (See B. Perez, *The First Three Years of Childhood* (Eng. transl.), p. 219.)

enjoyment of a feast.[1]   Again, many feelings are difficult to enter into, because their expression is ugly and repellent.   This applies not only to all the unpleasant manifestations in the lower regions of bodily appetite, but to certain emotional displays, as those of sulkiness, peevishness, and the like. Æsthetic and moral culture tend to narrow the range of our sympathies by branding many manifestations of feeling as unseemly and degrading.   Thus resentment, though a torture to its subjects, is, as Adam Smith remarks, not a fit subject for sympathy.[2] Conversely, education and moral culture tend to strengthen sympathy in directions which it does not naturally follow, as in kindly consideration for old age, infirmity, poverty, and so forth, the painfulness of which is not easily imaginable by the young and happy.

§ 22. *The Sympathetic Rapport.*   Let us now turn to the determining conditions as residing in the particular nature of the object of our sympathy and his relations to ourselves.   Here we may confine ourselves to the common case of reciprocal or mutual sympathy, and ask ourselves what are the conditions most favourable to the development of this sympathetic *rapport*.[3]

(1) In the first place, then, a certain community of nature and of feeling, as also of the emotive experience which this conditions, is necessary to a lively interchange of sympathy. Thus we find sympathy developing by preference between persons of the same age ("Crabbed age and youth cannot live together"), of a similar grade of culture, of similar interests and tastes, and so forth.   The need of a certain similarity of experience is seen in the proverbial callousness towards the poor of people who have never themselves known the miseries of poverty.

While, however, a certain degree of similarity of temperament and experience is necessary to close sympathy, it does

---

[1] Unless, indeed, his humane feeling has previously been excited by the spectacle of the child's hunger.

[2] See his admirable and subtle discussion of this point, and generally of the aspects of passion which make it a fit object for sympathy. (*Moral Sentiments*, sect. 2, chap. iii.)

[3] Not that sympathy is always a reciprocal affair.   Many a mother has lavished sympathy on a selfish boy without calling forth any response save a greedy demand for more.

not follow that perfect similarity is favourable to the feeling. It has already been pointed out that sympathy is valuable to its subject as a revival and as an extension of his affective life. Now such extension is realised best of all where, with a certain general agreement in the mode of feeling, there go considerable differences in actual experience. The often-remarked tendency to choose a friend and a lover of a ' complementary ' disposition is explained in part by the love of contrast and novelty, and in part by the desire for new directions of emotive expansion. A main charm of all growing friendship is the successive discovery of new points of sympathy, the gradual and mutual approximation and assimilation of emotional nature, which again is due to the reciprocal stimulation of new modes of sympathetic or reflected feeling.

(2) Next to this subjective condition, affinity of emotive nature, there is an objective condition, *viz.*, habitual proximity, and a common subjection to like external influences. The effect of this condition in promoting sympathy is seen in the instinctive fellow-feeling developed among gregarious animals, and also in the peculiar strength of the feeling which binds together the mother and her offspring.[1] Through such proximity, together with the common exposure to life's beneficial and injurious influences, and the conjoint co-operative actions which it brings with it, the habit of feeling together would be developed, and this would tend, in a measure, to promote true fellow-feeling. Such a relation of daily companionship and co-operation would, moreover, supply a common fund of experience to serve as a basis for sympathy. Lastly, this community of situation and experience would secure, in some measure, that observation of the outer manifestations of joy and grief from which all sympathy and all understanding of others set out.

These conclusions are borne out by observation. People who are least attuned to fellowship by nature somehow find

---

[1] This would, of course, be favoured by actual bodily contact. In the uterine life the fœtus may be said to be a part of the maternal organism, and to share in the intra-organic sympathy or consensus. (See above, p. 37; *cf.* Höffding, *Outlines of Psychology*, p. 247.) Even after birth the habitual carrying of the child by the mother would be favourable to a direct transmission of the movements caused by feeling, *e.g.*, the start of fear, to the offspring, and conversely the emotive movements of the child would communicate themselves to the mother.

themselves, when living in the same house, entering into one another's pleasures and pains. So powerful, indeed, is the influence of propinquity that those who began by disliking one another acquire, after a certain period of daily association, the habit of looking out for, and entering into, one another's displays of feeling. A like result is seen in the overcoming of antipathy and growth of sympathy towards a foreign people, which commonly follows from a residence in their midst.

(3) A third, and still more special, condition of the partial concentrated sympathy we are now dealing with is that restrictive influence already referred to under the head of fondness or attachment. Liking, attachment, opens the sluices of sympathy ; dislike, antipathy, hatred, effectually closes them. As we have seen, the first manifestations of sympathy, alike in the animal world, in the life of the human race, and in that of the individual, are concomitants of special attachment (parental, gregarious, filial, etc.). A child is sorry for his mother long before he is sorry for any outsider.[1] Family attachment is the matrix out of which primitive sympathy emerges. Even in later life, when sociality grows wider, the effect of personal attachment, and more generally of liking, on sympathy is everywhere apparent. Witness the lively outpourings of sympathy between girl friends, and between lovers, and our special readiness to give sympathy to those who have in any way pleased us.

The tendency to sympathise with those dear to us may be explained, to some extent, as the result of the powerful influence of fondness or love on the attention. Where we are fond our attention is held captive. The lover 'hangs' on every look and word of his mistress. The very use of the word ' regard ' for love is suggestive of this observance. Yet this is not the only circumstance which favours the development of sympathy in the case ; for, were it so, we should be specially disposed to sympathise with a person whom we hate or whom we fear, both of whom, it is plain, excite a close and fixed attention. For a final explanation of the connexion of liking and sympathy we must look to the fact that love, that is, fondness, and humane

---

[1] Early childish sympathy usually finds a main vent in pity for its animal pets. Here, too, the instinctive connexion of sympathy with fondness is seen. For illustrations of this childish affection for animals, see Perez, *op. cit.*, p. 75 ff.

concern for the object of this feeling, are, in the primitive and more instinctive form of the passion, one organic whole, and that in all the higher transformations of the emotion this organic unity still asserts itself in the impulse to follow up our liking with a kind of gratitude, the desire to sympathise with and to benefit.

§ 23. *Impartial Sympathy: Feeling of Humanity.* The highest development of sympathy is seen in the detachment of this kindly impulse from the restrictive bonds of personal liking, and in its expansion into a widespread, universalised human sentiment. What we call the growth of sympathy in a people or in an individual means this progressive expansion. Thus the child grows more sympathetic when he extends his fellow-feeling beyond the narrow limits of the family circle to those outside, who do not so powerfully appeal to this feeling by reason of the conditions of proximity, common experience, and personal attachment just spoken of. In like manner, as a human community rises in the scale of culture, and grows more humane, its members shed their sympathy over a larger and larger area, gradually transcending the limitations of family and tribe, so as to include those who were once objects of an unmitigated malevolence.

This higher manifestation of sympathy takes on a more humane or beneficent, and a more intellectual, aspect. The expansion of the humane sentiment with culture is seen in such phenomena as the extension of our benevolent concern to oppressed and suffering members of the human family in remote parts of the earth, and at widely different levels of culture, in the recent rapid growth of sympathy with those classes which have in the past excited feelings distinctly antagonistic to it, *viz.*, the "masses" and the criminal class, and lastly, in the widespread, and in some cases exaggerated, feeling of beneficence towards the lower animals. In all this we see a larger and larger irradiation, as knowledge and representative power develop, of the primitive form of human sympathy, that is to say, fellow-feeling in the common elemental experiences of pain and pleasure.

Concurrently with this development of humane feeling there goes the growing *interest* in everything human, in all varieties of character and life, that intellectualised sympathy which prompts the traveller to study and understand a new tribe, which incites the historian to imaginatively re-construct the feelings and doings of the extinct generations, which makes us all interested in a new and original individual, and in new national types, whether we meet with them in real life or on the page of fiction. This intellectual interest in others, though, as already pointed out, not identical with sympathy, or benevolence, is closely allied to it, and, as we see in the history of communities and of individuals alike, the two feelings tend to develop concurrently, and to react one upon the other.

It is evident that in the case of sympathy we reach a feeling much less obviously related to the ends of individual and race preservation. In its primitive instinctive form (maternal, family, gregarious sympathy) it has, no doubt, been an important factor in the conservation of species, and in the maintenance of the tribe or community. But in its modern forms it has passed far beyond these teleological requirements. In truth, it has been recognised by the modern evolutionist that our whole system of charitable institutions, hospitals, poorhouses, and the rest, so far as they tend to preserve the diseased, the old, and generally the inefficient, is opposed to the tribal end as biologically conceived.

REFERENCES FOR READING.

On the varieties of instinctive emotion the reader may consult Bain, *The Emotions and the Will*, part i.; W. James, *Principles of Psychology*, chaps. xxiv. and xxv. The reader of German may also refer to Nahlowsky, *Das Gefühlsleben;* Horwicz, *Psychol. Analysen*, 2er theil, 2e hälfte. On the subject of sympathy further reference may be made to Adam Smith, *The Theory of the Moral Sentiments;* Herbert Spencer, *The Principles of Psychology*, vol. ii. pt. viii. chap. v.; Volkmann, *Lehrbuch der Psychol.* vol. ii. § 136 ; and Höffding, *Outlines of Psychology*, vi. c.

CHAPTER XVI.

Having in the preceding chapter given a brief account of the more important varieties of instinctive emotion, and examined into the characteristics of concrete representative emotion, or sympathy, we have in the present chapter to consider somewhat more fully the highest group of emotions, *viz.*, those which we have marked off as abstract sentiments.

(c) REPRESENTATIVE EMOTIONS: (2) ABSTRACT SENTIMENTS.

§ 1. *Characteristics of Abstract Sentiments.* Sympathy, though a representative feeling, is eminently concrete. In order to sympathise there must be a real living object, or at least its fictitious simulation. The excitants and objects of sympathy are concrete psychical states in particular individuals. At the same time we recognised a tendency in its later and more subtle form, the feeling for humanity in general, to take on something of a general or abstract character, that is to say, to attach itself to certain common aspects or attributes of human nature.

This tendency in feeling to transfer itself from a concrete to an abstract form of representation shows itself still more clearly in the case of the emotions now to be examined. Philosophers have long since familiarised us with the notion that there are three main types of objective worth or of ideal end, which are valid for all minds alike, and answer to the three main directions of mental activity. These are: Truth, the objective correlate of intellection; Beauty, the objective correlate of feeling

(in its purest form) ; and what the ancients called the Good, and
we moderns usually envisage under the narrower aspect of the
Right, the ideal end of human action or endeavour.   These are,
it is obvious, ideas of an unusual degree of abstractness.   Yet
with each of these abstract ideal conceptions there is connected a peculiar feeling.   Thus we commonly speak of the
love of truth or knowledge, the feeling of the beautiful, and
the sentiment of duty.   These feelings may be marked off as
the logical, the æsthetical, and the ethical sentiment.

The general characteristic of these sentiments is that they
are highly representative, the outcome of many processes of
combination and transference, and as such far removed from
the instinctive emotions, and that they are wanting in the
energetic manifestation, and in the well-marked organic resonance of these last.[1]   A man may occasionally glow with an
enthusiastic love of beauty, or even of that colder deity, truth,
but for the most part the affective element in these psychical
states is subdued through the preponderance of the representative factor (imagination, thought).   The term " sentiment " as
distinguished from " emotion " appears to indicate this quiet,
contemplative character of the whole mental attitude.

Since these feelings in their fully-developed form presuppose
considerable representative power, and some progress in
abstract thought, it is obvious that they come late in the
evolution of the race and of the individual.   It is only among
the few highly-cultured peoples that they become distinct at
all.   To this it may be added that, as feelings attaching to
objects of common worth, they are pre-eminently an outgrowth
of social life and common interests.   In the case of the community and of the individual alike, the love of truth, the
sentiment of duty, and the feeling for the beautiful, are only
developed when the sphere of egoistic feeling has been trans-
cended, the sympathies awakened and deepened, and so a
large conjoint emotive satisfaction rendered possible.   This
dependence of the abstract sentiments on the higher sociality
and a refined sympathy, though sufficiently clear in the case
of each of them, is specially striking in that of the ethical
sentiment.

[1] No doubt, as W. James points out (*op. cit.*, ii. p. 470), there is a somatic
factor here too ; only it is less pronounced.

(A) The Logical or Intellectual Feelings.

§ 2. *Characteristic of Logical Feelings.*   By the logical, or, as some call it, the intellectual sentiment, is meant in its most general signification that group of feelings which accompanies the intellective processes *as such*, and which culminates in the love of knowledge or of truth.

We must not, it is obvious, look in this feeling, at least in its ordinary manifestations, for any of that emotive excitement or agitation which we find in the case of the instinctive emotions, or in that of sympathy in its more passionate form.   The logical sentiment, just because it is feeling bound up with intellectual processes in their fully-developed form, is a quiet affective state.   The general opposition between strong passionate feeling and intellection again meets us here.   As pointed out above, all intellectual activity under favourable conditions is warmed by a certain glow of agreeable feeling.   Yet it is comparatively rarely that the feeling rises to the volume and intensity of an emotion.   It will, however, be our aim to define so far as practicable the conditions which determine the more marked manifestations of agreeable and disagreeable feeling in this domain.

It may be said in general that intellectual activity only yields a considerable feeling of enjoyment when the common conditions dealt with in Chapter XIII., *viz.*, quantity of activity, contrast and relief, and harmonious adjustment of opposing elements, are present in a marked degree.   The carrying out of the intellectual processes, the intenter acts of observation, the search for ideas, and so forth, are commonly accompanied by disagreeable concomitants—a feeling of strain, of difficulty, of obstacle to movement.   Hence, though the quest of truth has its own peculiar delight, this comes mostly as a reward for a previous endurance of disagreeables.

With respect to the special manifestations of intellectual feeling there is little to be said.   The agreeable and disagreeable modes of consciousness here referred to are examples of that common manifestation of joy and sorrow, elation and depression, already dealt with.   It is only in those cases where a special attitude of attention, and as the concomitant of this a particular group of muscular effects, are involved that we get a characteristic physical embodiment. Thus in the look of wonder, the peculiar attitude or pose of curiosity and of wrapt attention, the well-known manifestation of mental perplexity, and the correlative expression of mental relief, *viz.*, the sigh or intensified expiration after the holding of the breath, we have reactions which connect themselves in a particularly close way with the intellectual life.

§ 3. *Feelings of Surprise and Wonder.* Perhaps the best starting-point in dealing with the sentiment for truth is the attitude or feeling of surprise. This phenomenon in its simplest form reaches, indeed, far down in human and animal life. Thus it has been pointed out by Preyer that surprise is one of the first emotions which are distinctly manifested by the child.[1] It is noteworthy, moreover, that Descartes and more recent writers deal with the feeling of wonder as one of the simplest and most fundamental forms of emotive manifestation.[2]

The feeling of surprise in its simplest form arises from the sudden presentation of something unexpected, that is to say, for which the attention is not prepared. The immediate effect of such a presentation is mental shock or disturbance, involving a derangement of the mechanism of attention and a sense of mental vacuity or stupor. In this its immediate effect it is distinctly disagreeable, and this disagreeableness is clearly manifested in the bodily reaction following the intenser forms of shock, and also in the readiness with which surprise passes into a vague form of fear.

The effect of the disturbance is to call forth a self-preservative reaction, *viz.*, the intensified look by help of which the strange, unexpected object becomes clearly apprehended. This intensified attention, necessitated by the absence of all pre-adjustment or "pre-perception," gives the characteristic ex-

[1] Preyer recognised it distinctly in the twenty-second week. (*Die Seele des Kindes*, p. 108 *seq.*) See further Romanes, *Mental Evolution in Animals*, p. 344.

[2] Descartes deals with it as the first of the emotions. (*Les Passions de l'Âme*, art. liii.) As already pointed out, Bain takes wonder as a typical case of "relativity". (*Op. cit.*, chap. iv.) Wundt, too, appears to regard surprise as a primitive form of emotive reaction or "affect". (*Physiol. Psychol.* ii. cap. xviii. p. 405.) The bodily expression of surprise (astonishment) is carefully described by Darwin. (*The Expression of the Emotions*, chap. xii.)

pression.[1] Through this special reaction mental activity is increased, the strange object is carefully inspected, imagination is stimulated by associative suggestion, and in this way the momentary shock leads on to the agreeable feeling of relief, of self-readjustment, and of intellectual mastery.

If, instead of being merely unexpected at the moment, the object is strange and unfamiliar, the feeling of surprise passes into the more prolonged state of wonder or astonishment. Here we have new affective elements, growing out of nascent processes of reproduction and comparison. Thus when we are astonished at a strange or rare phenomenon, as an eclipse, we are affected by the fact of novelty or of rarity.[2] This realisation of something new and extraordinary is in itself exhilarating, and, under favourable circumstances, we find wonder manifesting itself as a distinctly pleasurable elation involving an energetic and prolonged reaction of attention. That this is so seems proved by the fact that people eagerly seek the stimulus of the marvellous in nature and in art, and find a certain gratification in wondering even where, as in the case of preternatural wickedness, the object is qualitatively repulsive ; and, further, by the existence of such expressions as " love of the marvellous ". On the other hand, the sight of what is strange, especially if it is, at the same time, big and suggestive of power, is apt to excite apprehension or fear. A child shows the recoil of timidity rather than the joyous greeting of wonder at the first approach to the vast, many-voiced sea.[3]

The bare feeling of wonder is not an intellectual emotion, though closely related to it. Since the pleasurable excitement depends on the strangeness of the phenomenon, it may readily oppose the process of understanding, that is, assimilative comprehension. This is what happens whenever the love of marvel and mystery so intoxicates the vulgar mind as to lead it to resent a scientific explanation of those natural phenomena and

---

[1] For a careful description of this in man and animals, see Darwin, *The Expression of the Emotions*, chap. xii. ; *cf*. Ribot, *Psychologie de l'Attention*, p. 39 ff.

[2] Descartes says : " Wonder (l'admiration) is . . . caused, first of all, by the impression one has in the brain *which represents the object as rare, and consequently worthy of being closely (fort) regarded* ". (Quoted by Ribot, *loc. cit.*, p. 39.)

[3] A vivid description of a child's first alarming impression of the sea is given by Pierre Loti in his reminiscences of childhood. (*Le Roman d'un Enfant*, chap. iv.)

human actions (*e.g.*, of the mystery-man or conjuror) which excite the passion of wonder, an effect abundantly illustrated in the history of superstition.

At the same time, the feeling of wonder, through the preternatural reaction of attention or mental fixation called forth, is, when not intense enough to intoxicate the mind, favourable to inquiry. The impulse to " take in," assimilate, or comprehend becomes specially excited in presence of what is strange and foreign to our minds. The transition from ' This is strange ' to ' I do not understand it,' and from this last to ' What is it ? ' seem natural and inevitable. And, as a matter of fact, we find, in the evolution of the race and of the individual, curiosity developing into a strong and effective impulse as an action on some new wonder-exciting presentation. This close relation of wonder to inquiry is illustrated in the overlapping of the meanings of the words wonder and curiosity, and in the saying that all philosophy, that is, pure speculative inquiry, had its origin in wonder.[1]

§ 4. *Feeling of Intellectual Pursuit.* The intellectual feeling proper, that is, the feeling attending the process of intellection, only grows distinct when curiosity or the desire for knowledge is sufficiently developed to prompt to a prolonged search or pursuit.[2] To begin with, all energetic intellectual activity is under favourable conditions, that is to say, provided it is not unduly impeded, attended with the agreeable consciousness of exertion. To cast about for an idea, to search out among previous knowledge for some analogy which may throw light on a new experience—these and like operations are, to the vigorous mind, a source of enjoyment. This fact is amply proven by the liking of intellectual men for the pastime of witty talk, and clever youths for conundrums and other forms of guessing pastime. A certain amount of difficulty, moreover, tends to sharpen the edge of desire, and to add to the zest of the pursuit.

At the same time, the whole enjoyment in this case is

---

[1] The difference in the effects of wonder when this is intense and overpowering, and when it is moderate, is a particular illustration of the general principle that feeling in its higher intensities is unfavourable, in its lower intensities favourable, to intellectual activity.

[2] The active element in this pursuit, the impulse of curiosity, will be dealt with later on.

dependent on an assurance of coming success. As already
pointed out, intellectual discovery is nearly always accompanied
by disagreeables, such as the sense of perplexity at the outset, and
the temporary baffling of inquiry. The delight of the intellectual
man comes through the surmounting of these obstacles and
the joyful transition out of the bewilderment and impotence of
ignorance into the serenity and assurance of clear knowledge.
This feeling of relief is anticipated throughout the quest. From
the very outset we have a vague foreboding of the direction in
which our goal lies ; and the agreeable foretaste of attainment
grows (under favourable circumstances) more vivid as we move
onwards.[1]

In this excitement of intellectual pursuit we have a general
mode of feeling, that common to all prolonged and directed
activity, bodily as well as mental. In the satisfaction, too,
which comes through attainment we have a common type of
gratification (discharge of impulse, attainment of object).[2]
We have now to investigate a more distinctive intellectual
element in this experience, *viz.*, the peculiar feeling which
accompanies the solution or explanation. This feeling may be
called the pleasure of assimilation.

§ 5. *The Pleasure of Assimilation.* The exercise of each of
the intellectual functions, discrimination and assimilation, has
its own affective accompaniment. There is a gentle satisfaction
in discerning the precise difference between two objects. At
the same time the workings of assimilation, as the unifying,
simplifying process, the essential factor in all comprehension
or understanding of things, constitute a much more intense
and noteworthy enjoyment.

To wake up to a resemblance between two things hitherto
kept apart in the mind is always an agreeable experience.
When the interval between them is great, when our everyday
associations and habits of thought tend to keep them in remote
categories, the assimilating of them may bring a peculiar thrill
of delight through what has been called the 'flash of identity,'
an effect which has the exhilarating tone of a glad surprise.

---

[1] Hence Lessing's saying, that, if the Deity imposed the alternative, he would
prefer the pursuit of truth to its actual attainment, remains a splendid paradox.

[2] Prof. Bain deals with this feeling in its general aspect under the head " Emotions of Action—Pursuit," *op. cit.*, chap. xiii.

The charm of wit and of poetic simile is in part explicable by this principle. In like manner the assimilative comprehension of an object or fact which has at first looked strange and baffling brings an appreciable delight. In the higher processes of scientific thought, classification, induction, this pleasure of similarity in diversity is accompanied by a further gratification, *viz.*, the feeling of relief or easement, which is due to the fact that the discovery of similarity supplies a connecting bond for a multitude of diverse and previously-scattered fragments of knowledge.[1]

> The pleasure of assimilation is only appreciable when the difference makes itself felt. It is essentially the accompaniment of the discovery of similarity *in* or *under dissimilarity*. Hence the everyday automatic assimilation and classification of objects give us no appreciable element of the feeling. If we knew the exact nervous conditions of the assimilative process we might be able to see how the sudden rush of the nervous current in the case of the first assimilative comprehension of unlike things supplies a special condition of pleasure.

§ 6. *Feeling of Logical Consistency.* All assimilation of what was strange and isolated gives us a feeling analogous to that of reconciliation or harmony. This feeling of resolved conflict or dissonance in the intellectual domain becomes more specialised and pronounced as soon as facts, statements, ideas, are compared one with another, and the consciousness of consistency, mutual implication, logical dependence, and their opposites grows distinct. As reference has already been made to these feelings under the head of harmony and conflict, a word or two by way of additional illustration must suffice.

A case already referred to is that of doubt and its resolution. Owing to the complexity and variation of experience the indications presented in any particular case may point in a dubious or uncertain way. Thus the juxtaposition of heavy clouds with silvery interspaces leaves me in the conflict of doubt as to whether it is going to rain or be fine. In like manner, as we saw, the slipping away of ideas from our firm intellectual clutch excites doubt by suggesting non-reality. The removal of such conflicting indications, the restoration of an idea to its full vividness and stability, by bringing relief from doubt, and

---

[1] On the peculiar pleasure accompanying the assimilative process, see Bain, *op. cit.*, chap. xii. 2 and 3 ; W. James, "The Sentiment of Rationality," *Mind*, iv. p. 317 ff.

supplying the experience of belief in its fresh intensity, is the source of a deep delight.

A somewhat different case of logical antagonism and its happy resolution presents itself where a statement appears to clash with some previous knowledge obtained through our own experience or from others, so that we are disposed to reject it as improbable. Here the peculiar feeling of contradiction is excited. Where this happens, and where the opposition is afterwards removed, we experience a sense of relief, of harmonious adjustment.

In the organisation of knowledge into a consistent whole the reconciliation of contradictory elements plays a large part. Thus, owing to the complexity and subtlety of nature's processes, facts often appear to our incomplete observation to be antagonistic one to another, and to refuse to come under one connecting principle. The joy of a great discoverer, of a Newton or a Darwin, includes the consciousness of mastering these seeming contradictions. So, again, since organised knowledge is knowledge valid for all, the work of scientific research has frequently to reconcile discrepant ideas and views of things. The value of the large all-embracing theory of a subject, the product of the big mind, is that it takes up into itself the partial views of smaller minds.

It may be added that the feeling for consistency and inconsistency is the test of the truth-loving mind. It is curiously wanting, not only among the lower races of mankind, but among the majority of so-called educated persons. It is evident that a fine feeling for congruity will only arise as the concomitant of a severe logical self-discipline. To trace out all the little veiled discrepancies in our ideas about things involves a painful effort, and is only undertaken where inconsistency is felt to be a real and substantial misery.

§ 7. *Feeling and Intellectual Conviction.* We have so far supposed that intellectual feeling is the concomitant of clear discernment. It is to be remarked, however, that the intensity of feeling in the intellectual domain frequently manifests itself where the representative or intellective process is incomplete, vague, or sub-conscious. This fact meets us in the common forms of the " feeling " of conviction or belief. In many cases this shows itself as an anticipation of clear intellectual insight.

We often recognise a proposition to be true, an action to be right, before we can distinctly set forth the grounds of our conviction. Here the feeling of satisfaction is the concomitant of a half-developed intellectual process. In other cases, however, the feeling of conviction appears as detached from the intellective process. Thus what is known as faith, in its antithesis to illumined reasoned conviction or knowledge, is specially marked off as a feeling. The mother *feels* sure that her son will have a brilliant career, the religious man *feels* the reality of the objects of his faith, and so forth. Here it is evident we again encounter the opposition of feeling and intellection, an opposition strikingly exemplified in the faith that defies reason : " Credo quia impossibile ". In such cases, however, it is to be noted that conviction loses its properly intellectual character altogether and takes on the independence of an emotion. The lover's faith is his passion dominating his intellectual processes. All the more ardent sorts of conviction thus fall outside the logical domain ; and intellectual, that is, clearly thought out, belief, though containing an element of feeling, must be carefully marked off from this emotional faith.[1]

The antithesis of feeling and clear intellectual insight in the domain of conviction is emphasised by more than one German writer. Nahlowsky gives on the whole an excellent account of the matter in his little volume, *Das Gefühlsleben* (§ 16). At the same time it is an error to make all intellectual feeling the accompaniment of obscure intellection. The carefully reasoned conviction of the scientific mind has its own mode of joyous satisfaction. In truth the intellectual or logical feeling only deserves its name when it appears as the concomitant and dependent of clear rational thought.

§ 8. *The Love of Knowledge as Active Impulse : Curiosity.* We have in the above account of the intellectual feeling confined ourselves as far as was possible to the affective aspect of the phenomenon. At the same time the very reference to the common expression, scuriosity, love of knowledge, and the like, implies that the feelings considered have a close bearing on action. As already pointed out, the keen pleasures of intellect arise through active exertion, the pursuit of knowledge, the search for facts or truths, the effort to unify and reconcile, and so forth. This active phase can of course only be understood

[1] *Cf.* above, I. p. 491 f.

fully when we come to deal with the process of voluntary action or conation. A word or two on the subject must at this stage suffice.

Curiosity or the desire for knowledge arises in the case of the animal world, the human race, and the individual out of the pressure of practical needs. What the animal, the savage, and the child are mainly occupied in finding out and understanding is that which bears on their preservation and welfare, *e.g.*, the whereabouts of nutriment, the indications of foes, the properties of things that can be turned to uses of protection, warmth, and so forth. At this stage there is, properly speaking, no estimation of knowledge *for its own sake.* Knowledge is valued for its practical utility only.

The germ of a disinterested, a purely speculative appreciation of knowledge is to be found in the curiosity at the new, and more particularly the marvellous, or that which excites wonder, dealt with above. It is commonly said that the impulse to inquire, to examine things, especially new and strange objects, is instinctive in man. There may not improbably be an instinctive element in the case, but, if so, it is at first exceedingly weak. It has often been remarked that the savage is strangely incurious. Travellers will show him perfectly new products of civilisation, *e.g.*, mirrors, without exciting even surprise. It was pointed out above that wonder does not necessarily pass into curiosity, and we find that astonishment in the savage and in the child often stops short of inquiry. The true impulse of curiosity only begins to show itself when knowledge has developed sufficiently to generate the expectation of further knowledge, and more particularly to suggest the existence of a ' why ' or a cause.[1]

When this point is reached and the impulse to inquire is strong enough to prompt and sustain prolonged and energetic action it becomes enriched and strengthened by the varied feelings of pleasure attending intellectual activity. In this way there emerges a more distinct form of the love of knowledge, as illustrated in the desire to inquire about, and to understand things.

---

[1] *Cf.* H. Spencer, *Data of Sociology*, p. 97 ff., who illustrates and explains the absence of *rational* surprise in savages.

Even after the speculative impulse is thus developed it remains in organic connexion with, and closely limited by, special interests, as practical aims and æsthetic tastes. We see in the case of the child how the free inquisitiveness which appears fitfully and sporadically in such questions as 'What is this?' 'Who made that?' soon condenses into a special interest in a particular domain of fact, *e.g.*, the horse and his ways, railway- or ship-lore, and so forth. Here we see the development of a love of knowledge, but only of particular kinds, such as are related to his previous knowledge and to his interests, natural and acquired. Even the scientific man who shows the speculative feeling in so intense a form is often surprisingly narrow in the range of his intellectual interest. The general or *abstract* sentiment, a pleasurable interest in *all* or *any* new ideas, and a veneration for truth in all its forms, is in fact a kind of fiction. A few great " universal " minds, as an Aristotle, a Diderot, a Goethe, may have a feeling approximating to this, but in most cases the speculative interest in all its stronger degrees is limited to a narrow range of ideation corresponding to our particular interests and preferences, practical and æsthetic, which are partly instinctive, partly the product of our life-circumstances.

(B) THE ÆSTHETIC SENTIMENT.

§ 9. *Definition of Æsthetic Feeling.* By the æsthetic sentiment is meant that pleasurable feeling which accompanies the perception of the beautiful in all its modes or forms of manifestation. It may also be described as the pleasures of taste. Since, moreover, the feeling is excited in its greatest fulness and in its most perfect purity from other feelings by the fine arts, it may be spoken of as the fine art sentiment. This pleasurable mode of feeling has for its opposite the disagreeable emotion excited by what is ugly, in bad taste, and so forth, an effect which it is the concern of the fine arts to avoid.

It is at once evident that in the æsthetic feeling we have to do with a sentiment of a highly representative and refined character, one which only appears in its distinct and fully developed form among civilised communities, and in men of culture. We will now consider its distinctive characters, those peculiarities which differentiate it from the other sentiments standing at the same psychological level.

§ 10. *Characteristics of Æsthetic Pleasure : The Mode of Excitation.* The enjoyment of what is beautiful is plainly a different feeling in its mode of origin and its psychical features and accompaniments from the feeling for knowledge or truth. It is a fuller, or deeper pleasure, freer from disagreeables, and more of a luxury. In seeking knowledge we are aiming at something more or less directly useful, and we find the pursuit in certain of its stages arduous and even painful. In giving ourselves up to the beauty of a natural scene, to the charms of music, and so forth, we have done with all thought of utility, and are seeking enjoyment as children seek it, for its own sake.

These distinguishing characters may be connected with the way in which the feeling is excited or gratified. All beauty appeals to us through one of the two higher senses, sight and hearing. Thus in looking at a landscape, or a picture, in listening to music or poetry, we are, it is evident, having our minds stimulated through one of these higher sense-channels. And the pleasure produced is due altogether to the particular grouping of sense-impressions or presentative elements supplied, together with the objective suggestions or implications of these, such as the health suggested by a rosy cheek, the force expressed by a cataract, an impending cliff, and so forth. In other words, the pleasure arising from an impression of beauty is wholly contemplative and disinterested, that is, free from reference to self and its concerns. Thus a mother's delight in looking at her child, so far as it depends on the consciousness of its being hers, is, it is obvious, excluded from the category of properly æsthetic pleasure.

§ 10*a*. *The Beautiful and the Useful.* It has been the custom, from the time of Plato, to sharply mark off the beautiful from the useful. Knowledge and duty are matters of utility, necessary to the welfare of the individual and of the community. Beauty is, so far as its utility is concerned, a thing we could dispense with : it is an adornment of life, a pure luxury. According to the pregnant suggestion of Schiller, enlarged by Herbert Spencer, the æsthetic function is analogous to play in being the overflow of a redundant activity or energy not needed for the necessary or life-ministering functions. An individual and a community, so far as they cultivate beauty and art, are dispensing energies of sense-organ and brain not used up in the preserving and securing of life. Hence the familiar fact that art is the resource of leisure hours, that it only flourishes at periods where a community is strong, secure, and materially prosperous, and so forth. We may say, then, that the love or pursuit of art has no teleological significance. Indirectly, like play, it may contribute to health, vigour, and efficiency through its refreshing or *recreative* effect : but this must be regarded as accidental. One is tempted to say that the gods must, in a generous moment, have snatched beauty from the greedy hand of necessity, and made it over to much-oppressed mortals as their one pure source of delight.

§ 11. *Psychical Peculiarities of the Æsthetic Feeling :* (a) *The Contemplative Attitude.* With this general description of the nature of the æsthetic faculty and its place in human life, we may proceed to examine more closely into the psychical peculiarities of the feeling involved.

(a) Perhaps the first and most striking peculiarity of æsthetic enjoyment is to be found in its *contemplative* character. By this is meant the fact that when under the spell of a beautiful object we are in a peculiar attitude of outward-directed attention. In surveying a picturesque scene, in following the development of a musical theme, our mind is under the intrinsic charm of the object, is acted upon by its immediate affective influence. The whole pose of mind contrasts with those commoner modes of perception in which personal and practical interests are predominant. It is purely objective, that is, addressed to the object in itself, whereas these run off to the relation of the object to the self ; or, to express the difference otherwise, we may say that the contemplation of beauty is passive, whereas the common modes of viewing things are essentially active, including the stirrings of desire.[1] Hence the difficulty the practical man finds in taking on the æsthetic attitude. Again, as purely contemplative, the æsthetic mood is free from all that turbulence of passion which belongs to those forms of emotive perception by which egoistic or animal passion is excited, *e.g.*, the sight of danger, of an enemy. Hence the difficulty the emotional mind brooding on its subjective personal concerns finds in rising to the æsthetic point of view.

This contemplative attitude is, it is obvious, closely related to the intellectual attitude of observation or scrutiny. Both

---

[1] It is this thought which is expressed by Schopenhauer when he says that in art-contemplation we escape the misery of willing. (See my *Pessimism*, p. 97.)

are objective, *i.e.*, involve the direction of attention outwards to presentations or objects as such.   Yet the two modes of perception differ in important respects.   In all the finer intellectual inspection there is, as we saw, a special strain of the attention in discriminating details, and so forth.   Further, intellectual observation is essentially an active state, involving effort, desire, and the pursuit of an end.   Æsthetic contemplation, on the other hand, is free from this peculiar strain of attention, and it involves no desire for an unrealised object or end.   It is essentially a falling in with what is actually before us, a relatively passive submission to the affective forces residing in the object as it presents itself.[1]

(*b*) *Purity and Richness of Æsthetic Enjoyment.*   Having touched on the characteristic difference of the æsthetic attitude, we may pass to one or two prerogatives of the pleasures involved.

To begin with, then, the enjoyment of the beautiful is, among all our pleasures, the purest and the richest in respect of the variety of its elements.   The purity of the delight is connected to some extent with the particular channels of sense employed.   The higher senses, as compared with the lower, are, as we have seen, free from disagreeables.   Not only are they wanting in such unpleasant antecedents and consequents as the desire and the satiety which mar the enjoyments of appetite, they are relatively weak in painful elements.   There is nothing in the region of visual sense-feeling corresponding to the disagreeableness of certain tastes and smells.   In the case of tones we have, no doubt, in the effect of musical dissonance something distinctly and intensely unpleasant.   Since, however, this effect only becomes conspicuous in that tone-world which art has created for us, this as well as other disagreeables is effectively excluded, or made use of merely as a subordinate element of transition and contrast by the selective function of art which seeks to give us the fullest and most perfect enjoyment.

With this purity of delight there goes special richness or

---

[1] As we shall see presently, intellectual activity enters into æsthetic effect, but only so far as it contributes to a properly æsthetic pleasure.   Even where the two are most closely interwoven, as in recognising nice adjustment of means to end, the intellectual recognition of the relation and the æsthetic appreciation of its pleasing aspect, as pure spectacle, are two distinguishable ' moments '.

variety and complexity of pleasure. This is due in part to certain peculiarities of the two æsthetic senses. Both sight and hearing have, as we saw, an extensive graded scale of sensuous quality (colours, tones), and this series of qualitative differences supplies an important element in æsthetic effect. In addition to this, both visual and audile sensations admit of an indefinite variety of pleasing groupings, harmonious distributions of colour- and form-elements, rhythmical arrangements of tones. Lastly, the impressions of the eye and of the ear will be found to have the fullest and richest significance or suggestive value. This is clear enough in the case of sight, the impressions of which never fail at once to convey, to the adult mind at least, a definite objective meaning, more or less deep according to the spectator's previous knowledge. In the case of tones, this depth and richness of meaning are supplied by their *expressive* function, as signs of feelings, aspirations, and the like, in other minds.

(c) *Expansiveness of Æsthetic Pleasure.* Another prerogative of the æsthetic feelings is their *expansiveness* or susceptibility of prolongation. Being wholly disconnected with appetite, and practical needs generally, they are possible at any moment if only the appropriate object present itself. That is to say, they are in a peculiar sense frequently renewable pleasures. Moreover, being enjoyments arising through the two senses which are most susceptible of prolonged stimulation without the disagreeable result of fatigue, or considerable decline in the intensity of the pleasurable concomitant, being, further, accompaniments of that mood of contemplation which is free from the strain of effort, the æsthetic pleasures lend themselves in a pre-eminent way to prolongation through judicious variation of the stimulation. The contemplation of beautiful scenery can be extended indefinitely if there be only the requisite feeling and the needed leisure. Listening to music, though to many it has something of strain in it, may nevertheless be the untiring occupation of a whole evening.[1]

---

[1] That looking at pictures and listening to music are fatiguing must be a commonplace to a modern Londoner. All such strain is, however, accidental so far as the proper art-effect is concerned, as may be seen in the fact that a perfectly-cultivated sense and mind would feel nothing of the strain. Amateurs are fatigued

(d) *Shareability of Æsthetic Pleasure.* As a last distinguishing attitude of æsthetic pleasure, we have its high degree of
shareability. As objective feelings bound up with the contemplation of objects, and having no reference to personal interest
or need, the pleasures of beauty and art are, of all pleasures,
the most sociable, the most susceptible of a wide and impartial
distribution. The pleasures of self-gratulation, and even of
friendship, are restricted to one or a few. The pleasures of
art are free from such restrictions.

This prerogative of æsthetic pleasure again points to certain
peculiarities in the sense-channels employed. Both sight and
hearing convey to us impressions from objects at a distance.
Hence they can be stimulated in the case of a number of persons by the same objective stimulus at the same moment, as
when an assembly watches the same dramatic spectacle, or
listens to the same musical performance. This peculiarity—
which strongly marks off the two highest senses from touch
with its limiting condition of contact with the organism—
serves to give these senses something of their distinctive
' objectivity '.

A direct result of the peculiarity here spoken of is that the
pleasures of art are, in a pre-eminent manner, susceptible of
sympathetic co-enjoyment. We may experience them together
and in connexion with the same objective presentation, and the
impulse to enhance our pleasure by sympathetic interchange
is here subject to no check through the prompting of a lower
feeling of envy. Thus the pleasures of art constitute the
purest and most valuable of all our social or common enjoyments.

§ 12. *Constituents of Æsthetic Enjoyment :* (a) *Sensuous Element.* As already suggested, æsthetic pleasure varies in the
mode of its origin. Now it is referrible more particularly to
some sensuous element, *e.g.*, bright colour, a rich musical
timbre, now to a perception of relations among sense-elements,
and so forth. These elements or constituents in the æsthetic
effect are rarely found in isolation. Our enjoyment of a beautiful
natural scene, or of a picture, is a complex affective pheno-

by art-contemplation just because the works are strange to them, that is to say,
because they have not the culture which would allow them to enter into the restful
attitude of pure æsthetic contemplation.

menon in which we can analytically distinguish various contributory feelings. These constituents may be conveniently
grouped under three heads, *viz.*, the sensuous or material element, the relational or formal element, and the representative
or ideal element.

(*a*) By the sensuous or material element in a beautiful
object is meant the pleasurable aspect of the sensations involved.
It will thus include the agreeable sense-feelings of the two
senses concerned. The enjoyment of bright light, of lustre, of
the various gradations of colour, of the linear elements of form,
also of musical tone, and of the allied effect in articulate sound,
form a fundamental portion of the delight in beautiful objects.
The fine arts, painting, music, and the rest, seek, so far as their
limitations allow, to give us this sensuous charm in its highest
intensity and purity.

(*b*) *The Formal Element in the Beautiful.* By the relational
or formal element is meant the agreeable effect due to certain
modes of grouping the sensuous elements. The æsthetic value
of such grouping depends on a union of the two principles
already pointed out, (*a*) variety, or (in its intenser degrees)
contrast, and (*b*) harmony, or peaceful co-ordination of diverse
elements. Such a satisfying harmonious arrangement of elements may have to do with relations of sensuous quality. Thus
in visual art (decorative and imitative painting) considerable
attention is paid to what is called the harmonious distribution
of colours. Similarly in music, and, in a less marked degree,
in poetry, much of the effect of good combination turns on a
due attention to the affinities of sound, whereby, along with a
sufficient variety of elements, a certain harmony or unity of
effect is secured.

In addition to this arrangement of sensuous material according to its own qualitative affinities, there is a pleasing distribution of it in the forms of space and time. This gives us what is
commonly known as the formal element, space-form, time-form.
Thus a considerable ingredient in the pleasure of beautiful
visible things, including natural objects, as crystals and organic
structures, as well as the products of visual art, is referrible to
an arrangement of form-elements, more particularly lines, which
supplies the eye with a certain variety and freedom of movement, and, at the same time, secures unity of design under its

various aspects, linear continuity (arabesque, contour), proportionate division of a whole into parts, and symmetrical arrangement about a central element (point, axis).[1]

The value of such arrangement is in part due to the satisfaction of the conditions of easy attention and of clear perception.
Thus, as already suggested, the delight of rhythm in music and
verse arises, to some extent, through the ready readjustments
of attention to a regular succession of sounds, and the accompanying satisfactions of expectation. So the value of pleasing
space-form, *e.g.*, symmetry, depends on the ease with which
the details are comprehended as parts of a simple scheme. At
the same time, according to what was said above, we must
not confound the æsthetic enjoyment of form with a properly
intellectual gratification. The æsthetic aspects of form,
balance, regularity running through manifold detail, are *intrinsically* pleasing without reference to their objective significance. We are in one attitude when distinguishing and recognising a crystal, a shell, a plant as an object of a particular
group of characters, in another attitude when appreciating its
formal aspects as æsthetically valuable.

(c) *Associated or Ideal Element.* In the two constituents
already considered we have been occupied with the presentative
features and relations, or what has been called the "direct"
factor in æsthetic impression. We may now pass to the *re*-presentative, or the "indirect" factor. This includes that part of
the æsthetic effect which arises from association, suggestion, or
the play of imagination.

It has been pointed out that, as experience and knowledge
grow, sense-presentations take on more and more of significance
or suggestiveness. The pleasures of art are largely based on
this circumstance. It produces its voluminous feeling of
delight, because it can count on the reawakening through the
stirring force of sense-impressions of whole psycho-physical
tracts which we mark off subjectively as ideation and "ideal"
feeling. In this way, as we have seen, colours and tones themselves become enriched with ideal concomitants (pleasures of

---

[1] For a fuller account of this formal element in beautiful objects, see my article,
"The Pleasures of Visual Form," *Mind*, vol. v. p. 191, and *Sensation and Intuition*,
chap. viii. (Aspects of Beauty in Musical Form) ; *cf.* E. Gurney, *Power of Sound*,
chaps. iv. and v. ; and G. T. Fechner, *Vorschule der Æsthetik*, vi.

suggestion). The objects of nature, *e.g.*, the delicate and fragile maiden-hair fern, the Alpine crag, the curved vale, and so forth, awaken a fund of suggested idea in the instructed mind. So far as such imaginative activity is pleasurable, art makes full use of it. Lastly, it is through their suggestiveness that objects of beauty stir our emotive promptings, pulsations of wonder, of joy, of luxurious pity, of humour. The well-known emotive effects of music depend on the nascent re-excitation of ideas, impulses, and feelings, through the expressive features of tones and their combinations.

This indirect or·associative factor is not equally prominent in all æsthetic effect. Thus, to most persons, the beauty of a flowering plant is largely a matter of pleasing colour and form. On the other hand, the pathos of a crumbling cottage, the sombre sublimity of the fjord, is nearly altogether a matter of imagination. The value of the indirect factor becomes particularly prominent in certain modes of æsthetic impression. Thus in what is known as relative or dependent beauty, that is, the pleasing effect on the eye of fitness or adjustment of means to end in tools, in the architectural disposition of material, and the like, we have an effect which depends primarily on an imaginative realisation of the use of the object.

The indirect or associative factor is seen in a peculiarly distinct manner in all varieties of æsthetic effect which depend on the use of a symbol or a representative sign. Here, it is evident, the feeling excited is due to the mind's discernment of the reality represented through the representative medium. This constitutes a large factor in art-effect. In the imitative arts, as painting, there must, it is evident, be a considerable ideal reinstatement of the object represented, the ratio of the indirect to the direct factor increasing as the fulness of sensuous presentation declines, as, for example, in passing from imitative painting to delineation in black and white, or from the pictorial and 'plastic' arts generally to poetry.

It may be added that, in the peculiar æsthetic effect of all art as distinguished from that of nature, we have a further associative element, *viz.*, the suggestion of the productive intention, power, and skill of the artist. To look on any product of human hands is to have some idea of the craftsman's purpose, and of the means he has selected in order to realise this.   Relative beauty always involves this idea of purpose and conscious adaptation of means to an end.   Here we see that a new and *quasi*-sympathetic ingredient enters into the æsthetic effect.   To enjoy a work of art is to enter sympathetically into the artist's aims and their fruition.   In music this intuition takes on the form of a sympathy with the emotion expressed by the tones.   This view of art as expressive of idea and emotion will, of course, be more prominent and definite in the case of the spectator-artist, whose technical knowledge enables him to intuit the brother-artist's precise intention.   Yet it is a universal and a primitive feeling.   As we shall see presently, art-beauty is appreciated before nature-beauty, just because the interest in human doings is stronger and more deeply laid in our nature than the interest in nature's spectacle.   In truth, the modern æsthetic delight in nature is itself to a considerable extent a reflexion of this feeling.   We view nature's scenes and movements as products, and admire the creative and expressive spirit behind.[1]

While we may thus trace out the chief constituents in the impression of beauty, we cannot be said to have reached a final psychological analysis of its several component elements.   The æsthetic feeling, through the very richness of its sensuous and ideational elements, is one which in a peculiar manner defies analysis.   Not only so, the large part played by *harmonious* arrangement in objects of beauty obviously tends to circumscribe the range of analysis in this case.   Hence the prejudice, especially among sentimental and "appreciative" artists, amateurs and critics, against any attempt at analysis of art-effect as a thing profane.   Such an objection, though absurd when put in its extreme form, is justified to some extent by the shallow kind of philosophical or psychological explanation of beauty often put forth, as in the common doctrine that all beauty is essentially formal in its nature, residing in certain relations of contrast, proportion, etc., which are best brought under the principle or law, unity in variety, or uniformity in diversity.   That this principle formulates some of the more important limiting conditions of æsthetic effect is indisputable.[2]   At the same time it offers no *adequate* explanation

[1] On the range and importance of the associative element in æsthetic effect, see G. T. Fechner, *Vorschule der Æsthetik*, ix. and following, who shows in an interesting way how it varies from art to art.   According to Mr. H. Spencer, heredity plays a considerable part in this associated element, as in the feelings excited by beautiful scenery and music. (*Principles of Psychology*, vol. i. pt. iv. chap. viii. § 214 ; and vol. ii. pt. viii. chap. ix.)   Compare Darwin's interesting speculation as to the way in which music acquired its emotive effect, *Descent of Man*, pt. ii. chap. xix. ; and E. Gurney's chapter on "Association," *Power of Sound*, chap. vi.

[2] This has been shown with great wealth of illustration and in a most interesting manner by Fechner, *op. cit.*, vi.

of the phenomenon. As has been well argued by E. Gurney, a beautiful melody obeys this law just as much as, and no more than, the most commonplace correct one.[1] The attempt of the 'formalists' to account for all æsthetic effect by certain laws of form is now recognised as insufficient. As pointed out above, even the forms that please us most, *e.g.*, the favourite proportion of lengths in a cross, or a rectangle, probably owe their worth to association ; that is to say, to the fact that they answer to, and dimly suggest, recurring customary forms in actual life. However this be, it is plain that association, that is, dim recallings of feeling-coloured experiences, individual and possibly also racial, constitute an important part in the rich emotive effect of beautiful things. This fact in itself must always serve to prevent the appreciation of beauty from becoming a distinct intellectual 'intuition,' and to invest it with something of a dreamy and *quasi*-mystical character.[2]

§ 13. *Development of the Feeling for Beauty.* The sentiment here briefly described and analysed emerges gradually into recognisable distinctness alike in the case of the individual and of the race. To begin with, the disinterested contemplative feeling for the beautiful has to detach itself from more elemental feelings which primitively sustain it. Thus the germ of a feeling for beauty is detected in that care for bodily appearance which, as we saw above, is betrayed by certain animals as well as by man, and which is genetically related to the instinct of self-preservation. The savage love of beauty is largely a joy in personal adornment (colouring and marking of the skin, feathers, etc.).[3] Next to this egoistic matrix the æsthetic feeling finds a powerful stimulus and support in the race-preserving, and more particularly the sexual, emotions. It has been argued by Darwin that the selection of the male in the animal world has been profoundly influenced by a rudimentary sense of beauty in the elective female.[4] This hypothesis, taken with the undoubted fact that certain birds show an appreciation of

[1] See the able chapter on "The Relations of Reason and Order to Beauty," in his *Power of Sound*, chap. ix.

[2] An analysis of the constituents of beauty resembling the one given above will be found in Mr. Spencer's *Principles of Psychology*, pt. viii. chap. ix. With this may be compared the psychological treatment of the subject by Dr. Bain, *The Emotions and the Will*, chap. xiv. ; by Grant Allen, *Physiological Æsthetics ;* and by Fechner in the work just referred to. For a brief account of the different theories of beauty in their psychological aspect, see below, Appendix K.

[3] For an account of the savage fondness for personal ornament, see Darwin, *The Descent of Man*, pt. iii. chap. xix. ; Tylor, *Anthropology*, chap. x.

[4] *Ibid.* pt. ii.

colour and form in the adornment of their love-bowers,[1] and with the prominence of animal music at the wooing season,[2] suggests that the feeling for beauty stands in a peculiar organic correlation with the amatory passion. And this conclusion is supported by the observation that in man the age of puberty is commonly attended with a considerable expansion of æsthetic or *quasi*-æsthetic sentiment.

The gradual development of the æsthetic feeling into an independent sentiment is further seen in the fact that the artistic impulse is first called forth in close connexion with and dependence on the production of utilities. The æsthetic feeling is not strong enough to prompt the uncivilised man to shape a beautiful form for its own sake : he produces it as an ornamental finish to some useful object, for example, a weapon. This shows that the æsthetic interest is still germinal, and sustained by stronger interests, delight in the sight of what is serviceable.[3]

Lastly, this transition from the status of a dependent to that of an independent feeling is seen in the priority of the feeling for art as human workmanship to the feeling for nature. It has often been pointed out that the æsthetic feeling for nature in its purity is a modern growth. This sentiment as exhibited by the ancients was an enjoyment of cultivated and embellished nature, that is, as transformed by human hands, and as resonant of the sounds of human life. They had no appreciation of the picturesqueness of wild and lonesome scenery. Indeed, the feeling for nature's wild solitudes is hardly older than Rousseau. This shows that the sense of beauty is at first restricted, being interwoven with, and needing

---

[1] Darwin, *op. cit.*, pt. ii. chap. xiv.

[2] Darwin, *ibid.* pt. ii. chap. xiii., and pt. iii. chap. xix. Mr. H. Spencer has, however, recently brought forward facts which go to show that the connexion between music and the sexual impulse is less close than Darwin supposes. (See *Mind*, xv. p. 450 ff.)

[3] It follows that all appreciation of " relative " beauty is a lower form of æsthetic admiration in so far at it needs the support of suggested utility. It may be added that the relics of the earliest known attempts at pure art-production, as drawings of animals on bone, bark, etc., sculptures of figures, suggest that the artistic impulse was at first reinforced by religious feeling. (See Tylor, *op. cit.*, p. 301.)

the support of, the feelings and interests which centre in man himself.

While we thus see a gradual self-detachment of the æsthetic sentiment from more elemental or primitive feelings, we note also a gradual development of it in respect of richness or complexity, of variety of manifestation, and of refinement. The growth of the feeling of beauty follows, to some extent, the order of our analysis. Thus, in the evolution of the race and of the individual, the feeling for bright colour precedes the delight in symmetrical form, and the ideal element, so far as it involves experience and reflexion upon this, appears latest of all. Again, the æsthetic feeling grows in complexity as the mind develops on its intellective and emotive side. Thus a finer, more discriminative eye brings with it a larger and more various enjoyment of colour-effects, a growing power of comparison and measurement secures a fuller and a subtler appreciation of proportion, and widening experience invests objects with a richer suggestiveness. Lastly, this growth of æsthetic responsiveness, aided by a deeper reflectiveness, has for its result that refining of æsthetic enjoyment or differentiation of it into a number of distinct shades of tone which, as we saw above, is one characteristic feature of emotional development.

§ 14. *The Sentiment of Beauty and its Congeners : Feeling of the Sublime.* One consequence of our analysis of æsthetic effect is to show that there is no perfectly uniform type of emotive experience in art-enjoyment. The feeling with which we regard a perfect bit of classical sculpture or architecture, with its simple severe proportions, its dignified and graceful flow of contour, is markedly different from that with which we view a tragic spectacle. The same term beauty seems, indeed, scarcely applicable to effects so different. In the one, the dominant feeling is admiration for what we all recognise as beauty of form ; in the other, though present, this perceptual element is submerged in the deeper emotive current of hero-worship, compassion for stricken greatness, and the like. Common language shows a decided tendency to mark off such differences of æsthetic effect by special terms, as pretty, graceful, picturesque, grand, pathetic and touching.

Among these varied effects there are two which stand out clearly enough in the mode of their excitation and their characteristic reaction to have been commonly treated apart. These are the feelings of the Sublime and the Ludicrous. A word or two on these must complete our account of the æsthetic sentiment.

The feeling of sublimity is that peculiar emotion which is excited by the presentation or ideal suggestion of vastness, whether in space or time (Kant's "mathematical" sublime), or physical or moral power (Kant's "dynamical" sublime). It is thus differentiated in its mode of excitation from the feeling of beauty in the narrower denotation, *viz.*, the appreciation of beautiful form. With this there goes a certain difference in the reaction involved. The enjoyment of beauty expresses itself in a serene contemplation, a quiet gladness of look, in which there is no strain beyond that of delightful surprise itself. On the other hand, the feeling of sublimity has, in its physical embodiment, much more of active tension, of that elevation and expansion of the corporeal frame which goes with the enjoyment of power.

A little examination tells us that the feeling of the sublime, though less complex than that of beauty, is by no means quite simple. Being excited by the presentation or the idea of the vast in space, time or power, it involves as an initial effect a vague sense of bafflement of faculty. Thus, when confronted with the myriad-starred canopy of night, or the cycles of geological evolution, the mind seeking to take in and comprehend so vast a range and multiplicity of object is checked and defeated. Hence the kinship of the feeling with religious awe, the emotion excited by the infinite, and the incomprehensible. Not only so, there goes with the feeling in all its earlier and cruder forms a touch of fear. It is probable that the sense of the sublime, like that of beauty in its narrow sense, is primarily controlled by human interest, and so directed exclusively towards great and striking manifestations of man's power, and is only later on directed to nature on the ground of the analogy of its phenomena with human operations. Now extraordinary displays of man's power would, it is obvious, tend to excite in the breast of the spectator a mixed feeling, in which a vague form of dread has a prominent place ; and it is probable that this ingredient is always present in a disguised form in the emotion of the sublime.

While, however, the feeling of the sublime has thus a painful "moment," this is by no means its dominant element. The true feeling of sublimity, that which expresses itself in the characteristic bodily attitude described above, is a transition out of momentary bewilderment, sense of feebleness, and vague dread, into wondering delight at realisation of the vast proportions presented. Here the very contrast to the everyday scale of things excites an emotion of glad surprise. At the same time, a new and peculiar effect emerges in the realisation, by a sufficient effort of perception and imagination, of the elating and joyous effect of the grandeur itself. Such imaginative enjoyment of a sublime spectacle always includes a sense of personal elevation, or imaginatively realised consciousness of power, whence the close similarity in its bodily manifestation to that of the feeling of power or superiority. We represent more or less distinctly an amplification of our own powers to the magnificent dimensions of the object contemplated. This is most clear when the object regarded is a display of human prowess, as that of the classical Hercules and Jupiter, the Miltonic Satan, Bonaparte, and the other examples in literature and in history. In following their titanic achievements we enjoy a reflected, borrowed or *quasi*-sympathetic consciousness of power. But the same thing is true also of nature's sublimity. Thus, in enjoying the sight and sound of a great cataract, or of a storm-driven sea, we take on something of the glorious sense of its unconquerable might, the feeling being intensified by the rebound from a weak, craven tendency to dread. Similarly, the large space we view from a mountain-peak, and the long procession of the ages, fill us with delight by yielding us an imaginative expansion of our own powers.

Writers on the sublime have usually laid emphasis on one or the other of the two 'moments' here distinguished. Among those who accentuated the effect of awe is our own Burke, who, in his well-known work, *The Sublime and Beautiful*, regarded the essence of the sublime as the terrible operating either openly or latently. On the other hand, a number of writers have referred the effect to a sympathetic sense of power. Thus Longinus and others found the secret of the spell in an inward glorying or sense of greatness. Among modern writers the pleasurable effect of elation or expansion has been made prominent by Dugald Stewart, who, basing his argument on the etymology of the word (*sublimis*, lofty), contends that the sense of exaltation attending an elevated position is the fundamental form of the feeling.

Dr. Bain, taking the sublime of force, Kant's "dynamical" sublime, as the fundamental form, regards the feeling as a sympathetic consciousness of power. (*The
Emotions and the Will*, chap. xiv. § 27 ff.) Midway between these we have writers
who recognise at once a painful and a pleasurable element in the feeling. Thus
Lucretius describes the effect of a sublime object in the words :—

" Me quædam divina voluptas,
Percipit atque horror ".[1]

Hamilton, following Kant, also recognises a pleasurable and a painful "moment"
in the feeling. In the early evolution of the sentiment, which probably had its origin
in fear of human power, the painful element was more conspicuous. The truly
æsthetic sentiment of delightful reverence may thus be viewed as the transformation
of a disagreeable into an agreeable feeling through the elimination of the gross
element of personal fear.[2]

§ 15. *The Sentiment of the Ludicrous : (a) The Excitants of
Laughter.* Still further removed, in respect of its characteristic
tone, from the effect of the beautiful than the sentiment of the
sublime is that of the ludicrous. Here, it is evident, we have
to do with a feeling of a lower level. Quite unlike the dignity
of beauteous form, and the majesty of sublime spectacle, is the
littleness, the valuelessness for all purposes save laughter itself,
of what we call the amusing, the grotesque, the comical.
The feeling excited has its familiar and sharply-defined physical
embodiment, the unmistakable gamut of laughter, from its
nascent, half-developed form in the smile up to the hearty,
boisterous explosion. This neuro-muscular element is always
present in the feeling of the ludicrous. We are never affected
by the comicality of a thing without experiencing 'a tickle' in
the throat, that is, having a partial excitation of the laughing
movements ; and this muscular element gives much of the characteristic tone to the effect. The feeling or rather the group
of feelings which thus express themselves are elemental and
naïve, and found far below the level of art-culture. The child
laughs before he is stirred with the sublimity of things. Laughter is loud among savages, and it may even be heard in under-
tone among some of the lower animals.[3] Yet, notwithstanding

---

[1] Quoted by Hamilton, *loc cit.*, p. 514.

[2] On the evolution of the sentiment, see an interesting essay by Grant Allen,
Mind, iii. p. 324 ff.

[3] For a full account of the movements of smiling and laughter, and the range
of its manifestation in man and animals, see Darwin, *The Expression of the Emotions*,
chap. viii., also chap. v. p. 132 f.

the lowness of its origin, the vulgarity alike of its excitant and of its full muscular manifestation, the feeling is so rich an enjoyment as to be taken up into art, and made one of its chief constituents.

The causes of laughter are but very imperfectly understood. In its crudest forms it appears as a reflex movement on the application of certain well-known sense-stimuli, as violent cold.[1] A familiar case of *quasi*-reflex laughter is that excited by tickling. This appears, according to Preyer, as a reflex in the second month of life.[2] Laughter, visible and audible, occurs, however, earlier than this as the instinctive expression of joyful or glad emotion generally, *e.g.*, that excited by the sight of a beautiful curtain, the child's own image, and the like.[3] Among lower races, too, laughter appears as a prominent emotive expression of states of gladness generally, and especially the heightened sense of power, *e.g.*, after victory.[4] This is still more clearly manifested among idiots, who smile and laugh through sheer vainglory. Lastly, it may be observed that there seems a special tendency to carry out the movements of laughter as a relief from a prolonged, constrained attitude of mind involving forced self-control and inhibition of movement, as in listening to a serious discourse. The effect of very slight causes in provoking a smile on solemn occasions is well known.

In the feeling of the ludicrous, or comic laughter proper, the mode of excitation is restricted and specialised. Yet there is reason to think that this is an outgrowth from the more primitive form. The explanation of the effect of the ludicrous in nature and art shows the difficulties of psychological analysis at their greatest. The incident, action, or form of speech

---

[1] Certain pathological conditions appear to develop such purely reflex laughter. (See Höffding, *op. cit.*, vi. 9.)

[2] *Op. cit.*, p. 92. Tickling seems to produce the nearest approach to human laughter in monkeys. (See Darwin, *op. cit.*, pp. 132, 133.) Horwicz points out that tickling excites laughter only when prolonged, and infers from this that this last is a secondary effect immediately dependent on a primary emotive result (affect). (See his *Psychol. Analysen*, 2er theil, 2e hälfte, pp. 196, 197.)

[3] See Preyer, *op. cit.*, p. 185 ff. The gods are, according to Homer, like children in this respect, laughing out of mere "exuberance of their celestial joy," after their daily banquet.

[4] Savages laugh, like children, at trifles. (See Spencer, *Principles of Sociology*, p. 67.)

which provokes the reaction seems to do this immediately, in all truly laughter-loving men, so that all attempt to interpose subtle processes of reflexion, as is apt to be done, looks absurd. When, too, we come afterwards to reflect on the cause of our laughter, in order to spy out the exact aspect of the action which provokes the feeling, we find the analytical separation of the phenomenon into its elementary parts well-nigh impossible. All that can be done, then, is to bring out one or two main features which induction shows to be common, if not universal, conditions of the effect of the ludicrous.

(b) *Constituents of Feeling of the Ludicrous.* (1) To begin with, then, it seems certain that the feeling of glory, or of superiority, is a common ingredient in comic laughter. We laugh at all sorts of littlenesses, discomfitures, unworthinesses, and so forth, provided they are not serious enough to excite compassion, to offend our sense of decency, or evoke other incongruous feeling.[1] These exhibitions may be supposed to awaken a vague consciousness of superiority in the spectator. Such an effect must, however, be sharply discriminated from the primitive laughter of triumph over a discomfited foe or object of dislike. In this last there is a distinct apprehension of a special relation between the demeaned object and the spectator; whereas in the enjoyment of ludicrous spectacle, just because it is an æsthetic feeling, this reference to a special personal relation is wanting. The sense of superiority, so far as it enters into the effect of the ludicrous at all, is essentially an indirect and sub-conscious apprehension, the vague consciousness of elevation which grows out of the attitude of contemplation itself.[2]

The defects here referred to are particularly ludicrous when they appear in the great, that is to say, in those whom we have to look up to. Evidences of this are seen in the jokes of school-boys at the expense of their masters, and in the laughter of ordinary men at the whimsicalities of genius. From this it appears that we have in the feeling of the ludicrous a trans-ferred and refined form of the primitive brutal laughter of

---

[1] This limitation is recognised by Aristotle in the definition of the effect of the ludicrous, or of comedy, which he gives in the *Poetics.*

[2] This effect is analogous to that of contagious or sympathetic fear, which, according to Aristotle, constitutes along with pity an essential ingredient in the impression of tragedy.

triumph, intensified by contrast to, and relief from, a somewhat constrained and not too pleasant attitude of respect or reverence.

(2) In the second place, we have the more purely intellectual form of laughter, *viz.*, at the incongruous or the grotesque, in human action, speech, etc., as in the contradiction of our common notions of correctness and propriety in personal eccentricity and paradoxical extravagance. In this form the effect of the ludicrous is a staple ingredient in all varieties of wit, from the pun and its close kinsman, the conundrum, upwards. Here we have, as the most manifest ingredient, an effect of surprise, of contrast, and of contradiction of what is either customary or commonly accepted. This element of surprise is intensified in cases where, as in the conundrum, a serious idea, *e.g.*, a relation between two unrelated things, is first suggested, and then rudely dissipated, or where, as certain thinkers put it, the appearance of the ' idea ' is resolved into nothingness. A good deal of what is called wit will be found to owe its effect to a sudden disillusionary dissipating of a half-formed ideational combination, as when the form of words strongly suggests a similarity, while inspection of the sense shows the real point conveyed to be a contrast. The laughter in such a case may be regarded as analogous, to some extent, to that early type which comes through relief from a difficult and constrained attitude. By its element of unreality, its elfish make-believe, wit is an excellent means of compelling us to throw off for the nonce the shackles of serious thought. In enjoying a delicious incongruity of ideas, our intellect is exchanging the severe attitude of work for the relaxation of play. At the same time, the feeling of power plays a part in this intellectual laughter too, partly because the incongruity laughed at may always be viewed as a possible intellectual slip, and partly because wit is itself a kind of masterfulness, a daring setting at nought for the moment of the limitations of our serious thought.

(3) Lastly, in what the modern world calls humour we have a mixed feeling, in which the "crowing" laughter over discomfiture or degradation is still further softened and disguised by admixture of other feelings which give worth to the object, and more especially affection and sympathy. Thus the laughter at the foibles of genius is tempered by the feeling

of admiration which we have for the great man. The most characteristic form of modern humour occurs where there is a touch of kindly or humane feeling, as in the impression produced by those defects which appeal at once by aspects inextricably interwoven to our sense of fun and to our tenderness. This applies to much of our amusement at the naïve errors of children, the whimsicalities of old age. Shakespeare's daringly conjoint appeals to laughter and tears, *e.g.*, in following the self-caused calamity of Lear, are the highest achievement in this domain of humorous effect.[1]

In this way it seems possible to regard the feeling of the ludicrous as continuous with, and a development out of, primitive laughter. In order to account for this evolution of what is seemingly one of our most refined sentiments out of so humble an origin, we have to assume all that is meant by the processes of culture. Thus it is only as the higher intellectual development is attained, with its rapid and varied play of imagination, that the subtle perceptions of incongruity entering into the effect of the ludicrous become possible, and that the primitive, coarse feeling of glory is able to take on its highly-derivative, refined and disguised forms. The growth of knowledge and experience discovers many a new fountlet of laughter, as of tears. Men may laugh less boisterously than boys, but they are able to laugh at a vastly larger *variety* of objects.[2] The very growing demands of the serious affairs of life supply new sources of ludicrous effect through the workings of the irrepressible, and eminently salutary impulse to seek a momentary relief from oppressive constraint. Nay more, the larger and profounder sense of the misery of the world, though it sometimes overpowers our laughter, may also be forced now and again into its service. The healthy nature finds recruitment in a momentary escape from its burdens, in a half-illusory shifting of the habitual serious point of view, by which the heap of cares appears to melt away into unreality. The deepening and widening of our social feelings render

---

[1] Höffding thinks humour rests on a basis of sympathy, and so is analogous to the laughter at our own shortcomings. (*Op. cit.*, pp. 294, 295.)

[2] Not necessarily at *more* objects. Children, small or grown up, laugh at much that matured minds do not laugh at, but the range of ludicrous aspect in their case is surprisingly narrow.

possible those mixed feelings of humour in which we laugh at that which all the while we esteem and love. Lastly, through the imaginative extension of feelings primarily excited by man to animals and to insentient nature, we develop new directions of humorous gratification.

Whether these sources of ludicrous effect can be reduced to one is uncertain. According to the theory of Hobbes, as developed by Bain, all laughter springs out of a sense of glory or power, or, to use Bain's words, is connected with "the degradation of some person or interest possessing dignity".[1] According to this view, every effect of oddity or incongruity is secondary and derivative, the ultimate though disguised provocative being the loss of dignity in something. It would be exceedingly difficult to prove this, and no serious attempt has, so far as I know, been made to establish the proposition on an adequate inductive base. We laugh at old-fashionedness in children, that is, at the simulation of something *too dignified* for their years, just as we laugh at the *un*dignified childishness of the grown-up. This alone might suggest that degradation is not the sole cause, but that the presentation of the incongruous as such, of the foreign and unfitting, is sufficient to provoke mirth. The laughter of children at all make-believe or acting, *i.e.*, semblance of a foreign character, can only be brought under the head of crowing over loss of dignity by a good deal of forcing, and it is far better explained as merriment at the oddity or "funniness," and the unreality, or, more correctly, the rebound of disillusion from the half-acceptance of reality, of the mask or simulacrum itself. In this connexion it is worth noting that Preyer, one of the most careful observers of children, says that, while he saw roguish laughter towards the end of the second year, he has never observed jeering (höhnisches) laughter before the completion of the fourth year.[2]

The phenomena of reflex laughter, the quick, primitive laughter of sudden joy, as well as that sort which comes as 'reaction' from a constrained attitude, alike suggest that the group of movements involved are readily excited, and are peculiarly liable to be excited through a kind of escape-valve arrangement after a state of central tension. If our knowledge of the nervous system and of its mode of action were more complete we might be able to bring all laughter regarded as a neuro-muscular process under the head of such relief-current.[3]

[1] See *op. cit.*, chap. xiv., especially p. 257 ff. The germ of this theory is to be found in Aristotle, who connected the effect of comedy with the presentation of meanness or deformity (provided that this does not excite other and painful feelings). Bain's theory differs from Aristotle's and also from Hobbes's by accentuating the loss of dignity, or the transition from a dignified to an undignified state.

[2] *Op. cit.*, p. 187.

[3] According to Herbert Spencer, laughter may be explained physiologically as an escape of nervous power after a state of tension. (*Essays*, vol. i. essay iv., "Physiology of Laughter".) The teleological view of laughter here suggested is developed more distinctly in the theory of E. Hecker, that the laughter in the case of tickling at least is a purposive movement, causing a rhythmic series of inhibitions of the blood-supply in blood-vessels generally, and so in those of the brain. (*Die Physiologie und Psychologie des Lachens und des Komischen; cf.* Höffding, *op. cit.*, vi. 9.)

§ 16. *Æsthetic Feeling and Æsthetic Judgment: Faculty of Taste.* The effect of the beautiful is primarily an emotive phenomenon. Strictly speaking, the mind knows nothing of beauty save so far as it can appreciate, that is, enjoy. At the same time, an intellectual element, contemplation, is, as we saw just now, always involved in this appreciation. In order to enjoy a beautiful, natural scene, we must observe, compare, and imagine. Not only so, what is called taste implies that its possessor is able to analyse objects so as to note the precise sources of æsthetic enjoyment, *e.g.*, harmony of colour, nobility of idea. To discriminate real beauty from its counterfeit is to judge of the æsthetic worth of an object by reference to a standard-idea, which is the product of previous æsthetic experience and education.

The development of this power of æsthetic judgment illustrates the continual action of intellectual on emotive culture. Our æsthetic experience begins with a vague feeling of admiration. Thus a child enjoys a colour-spectacle, and most persons enjoy music, without any distinct apprehension of the special attributes or constituent features of the object which give it its charm or delightfulness. By successive processes of reflexion, *i.e.*, analysis and comparison, this vague feeling gives place to a " clear " or "illumined " feeling, that is, one accompanied by an intellectual recognition of its source or object. All æsthetic development tends to render this discrimination and recognition more precise and just. Hence the man of taste can give the grounds of his admiration, and can criticise, that is, distinctly point out, the sources of æsthetic value in objects.

In the normal course of development this æsthetic judgment grows out of, and bases itself upon, a feeling of æsthetic delight, or its opposite. Yet this relation is apt to be modified in the case of the individual by the action in the social environment, æsthetic tradition, and education. As we come to know what is considered æsthetically right, or in good taste, by others, we tend to make the authoritative utterances of their opinion our guide, in addition to, and in place of, the inner voice echoing our own past æsthetic experience. In this way there arises an artificial mode of æsthetic judgment, which is independent of feeling. We pronounce a thing beautiful, not because we individually feel its beauty, but because we follow the objective rule which external authority supplies. Hence the " cold head-knowledge that is divorced from enjoyment," of which Mr. R. L. Stevenson speaks.

The object of æsthetic education should be to reconcile the claims of individual feeling and of authority. Its first aim is to make the subject emotively responsive, to bring him under the varied spell of beautiful things, to widen and enlarge his æsthetic experience, and *pari passu* with this to stimulate him to reflect on his experiences so as to organise his impressions into a standard-idea. This primary individual judgment should then be set in juxtaposition with the external or objective standard : first of all, that of the particular age and community to which the learner belongs, and then the more comprehensive and philosophical standard reached by a comparison of the tastes of different nations and of different periods. Such extension of view should, however, in every case be supplemented by the verifying base of individual feeling. That is to say, the wider and more objective standard is to be adopted only as it is re-cognised as representing and formulating the growing indi-vidual experience. In this way the rights of the individual, which, as well-known maxims tell us, are particularly strong in the æsthetic domain, will be respected, while all the advan-tages of enlarged experience, breadth of view, and sympathetic solidarity with our kind will be secured.

(C) The Ethical or Moral Sentiment.

§ 17. *Characters of Moral Sentiment.* By the moral or ethical sentiment is meant the feeling which attaches itself to the idea of right or duty, and which is commonly spoken of as the sense of duty, as the feeling of moral approbation and dis-approbation ; and, in one of its most important manifestations, as conscience. Here, it is at once evident, we have a feeling which, while on the same developmental level as the other two sentiments just examined, presents marked differences from these. In its common form, approval of what is right, disap-proval of what is wrong, the sentiment has something in com-mon with the æsthetic feeling. Like this, it is excited by a contemplation, purely disinterested (*i.e.*, free from all reference to self and its interests), of certain attributes or relations in given objects. We are immediately pleased when observing or imagining a morally good action, just as when we are observing or imagining a beautiful object. Yet the exciting quality and the resulting feeling are widely different. Moral approval or disapproval differs from æsthetic in that it always fastens on a human action, whether another's or our own, and on that particular aspect or relation of the action which we call its rightness or wrongness.

We may now examine a little more closely into the characteristic differences of the moral sentiment.

(a) The most obvious peculiarity of the moral feeling is connected with the nature of its excitant or object, viz., a conscious and willed action, and the intention or disposition which precedes and conditions the overt act. As an approval or disapproval of somebody's doing or conduct, the feeling of duty is differentiated from a purely contemplative one (like the æsthetic), and becomes pre-eminently a *practical*, i.e., action-controlling, feeling. This is plain enough when the conduct contemplated is our own. To recognise and feel my projected action to be right is to be impelled towards it ; wrong, to be deterred from it. But even where we are considering another's conduct this active element is present. Here, however, the feeling prompts to a secondary and *external* form of control, e.g., prohibition of the action, punishment of it, commendation of it.

(b) In the second place, the moral sentiment is *regulative* or magisterial. By this is meant that, as a feeling for what is right, it is a recognition of claim, and a consciousness of what we call " oughtness " or obligation. Other practical feelings, e.g., the egoistic feeling of ambition, are wanting in this distinctive feature. It is only so far as the moral feeling comes in that we experience this peculiar sense of authoritative claim. In other words, the moral sentiment is, as common language expresses, a feeling of duty, that is, of what we " owe " or " ought".

This characteristic, again, implies that the emotion is a supreme and governmental force. It is not called forth towards actions to which we are impelled by primitive inclination, e.g., the satisfying of appetite. It comes in to modify and control these primitive impulses. Thus it prompts to lines of effort which are originally difficult and painful, it restrains or inhibits us from indulging our natural inclinations. In this way, at least during the process of moral development and education, the feeling for duty involves a consciousness of a certain antagonism between a higher and a lower nature.

(c) Another characteristic of the moral sentiment seems to follow from this judicial function. It is an unimpassioned pleasure, of a pale ascetic hue, as contrasted with the rosy tint of the pleasure-loving æsthetic sentiment. Like the emotion which attends the pursuit of knowledge, the feeling for duty has in it a considerable ingredient of dissonance or opposition, and of disagreeable effort. It is commonly held that the painful half of the feeling, indignation at wrong, remorse, is much more intense than the pleasurable half, moral approval. As Mr. Meredith pungently expresses it, conscience is " like the spleen whose uses are only to be understood in its derangement ". It is only so far as we outgrow the early sense of compulsion, and learn to follow the lead of duty heartily and enthusiastically, that the pleasures of a good conscience become a considerable quantity. In such cases, however, self-feeling (self-complacency) becomes a prominent factor. The man who does right with no sense of struggle or friction, and, at the same time, without the luscious accompaniment of self-gratulation, may feel a gentle stir of properly ethical gratification, i.e., pleasure in the pure consciousness of doing right ; yet such perfect moral adaptation, if ever realised, would seem to bring its subject somewhat near to a condition of sub-conscious automatic response to duty, which has no room for feeling. For the rest, morality only becomes the excitant of a strong and massive pleasurable emotion where it is exhibited in some rare and transcendent form, as preternatural honesty, self-denial, and the like. This admiration, however, of moral excellence or virtue, as distinguished from ordinary compliance with duty, is largely an æsthetic feeling—that is, a delight in what is rare, fine or noble, and great.[1]

The precise nature of {the consciousness of obligation and the relation of it to moral feeling are points of dispute in ethics. I have here assumed that the feeling of obligation, of the claim of the moral law, is an integral element in all ethical approval. Moral development tends, no doubt, as Kant shows, to raise us out of a state of bondage to external rule, into a state of comparatively free acceptance of moral commands. Yet, even when thus made internal and imposed by the ego on itself, they do not lose their character as commands, and the peculiar

[1] The same applies to Kant's feeling of reverence for the majesty of the moral law. Kant takes this along with the starry heavens as his illustration of the sublime.

feeling of "oughtness" which differentiates the ethical sentiment from other emotions includes in its most highly evolved form a more or less distinct recognition
of claim and obligation. That this is so seems borne out by the curious fact that
whenever we feel urged to do a thing which in the mood of the moment is not at
once attractive we tend insensibly to fall into a *quasi*-ethical mode of regarding
the action. Thus a man will say ' I ought,' ' I must,' with respect to some purely
personal enjoyment, such as going to the theatre, if at the moment indolence leads
him to stay where he is. It would seem to follow that the ideal state of things
foreshadowed by Herbert Spencer in his *Data of Ethics*, in which man is so perfectly adjusted to his social environment as to take to duty as easily and unerringly
as a duck takes to water, would mean a disappearance of the moral sentiment as
imperfect mortals experience it.[1]

(*d*) The last distinctive characteristic of the ethical feeling
is its *sociality*. In our appreciation of knowledge and of beauty
we are showing ourselves social beings, members of a community with common interests. But the social consciousness,
the feeling of solidarity of the self and the community, is not
always distinctly present here. In the case of the moral
feeling, however, there is always a measure of this consciousness. This is obvious enough when we approve or disapprove
another's action; for in this case, in the very exercise of the
judicial function, we are distinctly asserting our social relations.
It is true also of our moral self-approval or disapproval. And
here the element of sociality appears in a double way. On
the one hand, the action which we commend or blame has
reference to others. Our duties *are* our social relations, that
is to say, include actions and abstentions from action just so
far as these bear on the interests of others, the happiness of
the community to which we belong. On the other hand—and
this is only the obverse side of the same fact—our approval or
condemnation of our own action is the reflexion of external,
that is, of others' judgment of us (whether actually carried out
or only anticipated by us). Thus, whenever we experience the
moral feeling, we are realising our social nature and our social
condition. To approve a thing as right is to speak in the
name and with the support of the community. As we shall
see presently, the moral feeling grows out of, and is indeed
the highest and most enlightened form of, social feeling.

[1] I take no account here of the element of practical reason said to be present in
all cases of enlightened moral sentiment, since this is plainly not a differentiating
circumstance. We may rationalise our conduct without moralising it, as when we
recognise the superiority of Norway to Switzerland as a summer resort.

It follows that no man can exercise the moral function as an isolated individual. The private subjective working out of what is right has no meaning save with reference to a community which would benefit by the line of action, and (supposing it sufficiently enlightened) would consequently approve it. The approving community may happen not to be that with which the person has most to do in his daily life, and may even be an ideal, as yet unrealised conception. Yet in every case the consciousness of a community and of our relation to it is present.

§ 18. *Sources of the Moral Feeling.* The origin of the moral sentiment may be explained by a reference (1) to a number of simpler feelings or impulses in the child, together with (2) the influence of certain external agencies which are always at work in fully-developed social structures or communities.

(1) (*a*) Taking the constituent feelings, we have, in the first place, to recognise that what we call morality, though essentially a social feeling, has one of its roots in the *egoistic* feelings. The individual's regard for others, his desire to do his duty by others, presupposes the instinctive, self-preserving impulses dealt with above. Thus the peculiar feeling of condemnation of a wrong action can be traced down to the instinctive reaction of a purely individual or egoistic resentment. As was pointed out in connexion with sympathy, a feeling for others' good is an enlargement of the primary feeling for our own good.

(*b*) Next to this instinctive base in the egoistic feelings we have as an important contributing element the semi-social (ego-altruistic) feelings, *viz.*, the regard for others' opinion, the dislike of blame, and the love of praise. This is a powerful aid to morality, especially in the early stages of moral development, alike in the race and in the individual.

(*c*) The highest element in the moral sentiment is sympathy, that is, regard for others' welfare for its own sake. Since morality is concerned with others, with the needs, interests, claims, of our fellow-creatures, it is evident that a disinterested regard for it implies the existence of this purely social feeling of sympathy. Hence we never find the moral sentiment developing when the social feelings are wanting. In the case of the race and of the individual alike the sense of duty develops *pari passu* and in close organic connexion with the feeling of sympathy.[1]

[1] An integral factor in all the higher developments of moral feeling is intellectual discrimination, and generally the power of comparing and measuring. But this factor will be best considered by-and-by.

(2) Coming now to external agencies, we find that a true development of moral feeling is reached only where there is at work some form of external authority. The feelings just enumerated would not of themselves constitute a sense of *duty*. As already remarked, the peculiar shade of sentiment indicated by this expression involves the recognition of an external or objective claim. Such a claim asserts itself and makes itself felt, in the first instance, through what we call authority, or the imposition of commands by a superior will-power (*e.g.*, parental, tribal as solidified in custom).

With this brief account of the sources of moral feeling we may proceed to trace out the more important stages of its development.

§ 19. *Development of Moral Sentiment : the Animal Conscience.* In seeking to explain the growth of the moral sentiment we find ourselves at once confronted by two problems which have to be distinguished one from another. We may, in the first place, address ourselves to the task of retracing the historical development of the moral sentiment during the progress of the human race, or, to take a still wider view, during the evolution of animal life (zoological evolution) ; or, in the second place, we may content ourselves with following the growth of the sentiment in the human individual under the pre-existing condition of a moral community. The two problems will, it is evident, differ, in the case of the moral sentiment, very much as in that of language, in important respects, and can hardly be dealt with as one. We shall do well, therefore, to begin with a brief glance at the historical or racial problem, and then take up the problem of individual development.

Beginning with the zoological evolution of the feeling, we might at first suppose that a rudiment of the moral sentiment is discernible in those gregarious animals which, through the working of their social interests, unite to defend themselves against common enemies, to help one another in distress, and so forth. Yet there is no reason to suppose that the creatures in question have any feeling analogous to our regard for duty. Even if they had intelligence enough to grasp the idea of a collective tribal interest, they would still lack the recognition of its superior obligatory character. So far as we know, these species have not anything corresponding to the human love of approval, still less any analogue to human authority (commands, etc.). A faint adumbration of a true feeling of duty may be conjectured to exist among animals having a form of leadership. At the same time, it may be said with some confidence that the conditions of a feeling of duty only begin to be realised in the case of the dog, the elephant, and other intelligent animals which come under a careful system of human training. All good observers seem agreed that here we come upon a real germ of a conscience. In the case of the dog, more particularly, we see all the signs of submission, of self-control in deference to recognised authority, of complacency in winning approval, and even of remorse at wrong-doing.[1]

It is argued by Darwin that the superior strength and persistence of the social instinct in gregarious animals would in itself constitute a sufficient basis for a true moral consciousness, the animal realising through the greater strength and persistence of this social instinct that an omitted course of correspondent action "would have been better," and "ought to have been followed ".[2] This, however, seems to me to overlook the true *differentia* of moral feeling. As suggested above, there may be a struggle of impulses, and even a recognition by "reason" of the relation of a greater to a less without the peculiar sentiment of "oughtness ". This seems, in every case where it is distinctly recognisable, to be developed by help of authority, commands, and the correlative sanctions.[3]

§ 20. *Historical Development of Moral Feeling in the Human Race.* Let us now turn for a moment to the historical evolution of the moral sentiment in our own species. Here, it is evident, we have new generating conditions. Of these it is sufficient to point out the higher intelligence aided by the possession of a system of language. This of itself would render possible a distinct expression and understanding of approval and disapproval, and so favour the development of the correlative feelings. It would, moreover, by reacting on the social feelings, tend to bring about a more distinct appreciation of the relation of solidarity, of the effects of individual action on the common or

[1] One of the clearest examples of canine conscience I have met with was given me by a friend, the owner of the dog, and the witness of the action. The animal, a variety of terrier, was left in the dining-room, where were the remains of a cold supper. He got on the table, and secured a piece of tongue. But, without eating a morsel of it, he carried it into the drawing-room, deposited it at the feet of his mistress, and then crawled out of sight, looking the picture of abject misery.

[2] *Descent of Man*, pt. i. chap. iv.

[3] For a fuller examination of Darwin's theory of conscience, see my *Sensation and Intuition*, pp. 17, 18.

tribal weal, and so forth.   In this way a distinct consciousness of a " tribal self," as it has been called, could emerge.   Not only so, by the aid of language and tradition, particular modes of action, found to be necessary to the tribal welfare, would be generalised, or formulated in the shape of general rules.   There is every reason to think that a large part of our " common-sense," *i.e.*, popularly accepted, morality, has originated in this way through the gradual fixing by custom and tradition of lines of action found with growing experience to be socially useful or beneficial.[1]

Lastly, these same distinctively human conditions would supply the needed factor of authority.   The growth of intelligence would in itself lead the members of a community to distinguish between the promptings of their individual interests and desires, and the utterances of common interests, as recorded in custom and tradition.   Yet, as already hinted, this purely intellectual distinction would not of itself explain the origin of that peculiar shade of sentiment which we call sense of duty.   In order to the development of this we have to suppose the evolution of the social organism carried to the point of a differentiation of headship, authority, government, on the one hand and of subjection on the other.   All that we know or can conjecture of the early stages of human culture tells us that these conditions have been present from the beginning.   In the family organisation with its head, in the clan-like arrangement growing out of this, and so on, we find the seat of such an authority, a recognised power in certain hands to enforce by penalties respect for common interests.   Later on, we have as an important added element the priestly caste armed with a system of supernatural sanctions.   By such solidification of the utterances of the tribal impulse of self-preservation into commands, imposed and enforced by a definite authority, the social feeling may be supposed to have acquired the characteristic traits of a sense of obligation.[2]

---

[1] See, for an illustration of this, Darwin, *op. cit.*, pt. i. chap. iv. p. 116 ff. ; *cf.* H. Sidgwick, *The Methods of Ethics*, bk. iv. chap. iii.

[2] For a fuller account of the morality of savages, and the early phases of the evolution of the moral sentiment in man, see Darwin, *The Descent of Man*, pt. i. chap. iv. ; Lubbock, *Origin of Civilisation*, chap. vii. ; and E. B. Tylor, *Anthropology*, chap. xvi.

§ 21. *Development of Moral Sentiment in the Individual :* (a) *Instinctive Roots of Moral Feeling.* We may now turn to the question of more practical moment, *viz.*, the development of the moral sentiment in the individual member of a civilised and moralised community. How, it may be asked, does the child come to enter into the system of moral rules, to take on the common feeling of duty, the sense of justice, regard for truth, and so forth, which he finds in force around him ?

Here the first question that presents itself is that of the innateness of the feeling. It has been contended by certain moralists that the distinction between right and wrong and the correlative feelings are instinctive. If by this is meant that the peculiar type of feeling described above appears independently of the moral control, training, education which is supplied by the community, the idea must be rejected. At the same time, there is, undoubtedly, a real and considerable instinctive basis of the moral feeling. Thus, as we have seen, the child has an instinctive tendency not only to seek others' society, but, what is far more important in the present connexion, to desire the good opinion of others. These constitute in themselves a natural bias towards morality. Such instinctive feeders, how-ever, would be quite inadequate to generate a distinct feeling for rightness or duty, apart from the direct action of the community in the imposition of commands, and all that we mean by moral training.

The question how much of the moral feeling is instinctive is one of peculiar difficulty. Observation of infants suggests that something analogous to deference or submission to others, *e.g.*, in the effect of frowning, of an emphatic " Hush ! " appears very early. Yet it seems impossible to be sure that we have to do in this case with a properly moral feeling. According to the Lamarckian theory, as de-veloped by Herbert Spencer, ages of human experience of authority, commands, etc., might be expected to transmit to the individual an instinctive disposition to fall in with the demands of the community. It is possible that this is so. Yet there seems as yet no certain way of separating out and quantitatively estimating the hereditary factor, if such there be.[1]

(b) *Educative Action of Authority.* The normal course of

[1] On the question of the instinctiveness of morality in the individual, see Bain, *Mental and Moral Science, Ethics*, pt. i. chap. iii. On the bearings of heredity, consult H. Spencer, as quoted by Bain, *ibid.* p. 722 ; and Darwin, *Descent of Man*, pt. i. chap. iv. (p. 123).

development of the moral sentiment in the individual appears to proceed on this wise. The child endowed with an outfit of instinctive feelings and tendencies, egoistic and social, finds himself acted upon by a system of government or authority with definite commands, backed by punishments and rewards. These operate, in the first instance, on his egoistic feelings. He does what he is told in order to avoid the pain of punishment, or to earn the promised reward. At the same time, this apparatus of authority gives a definite direction to the instinctive workings of his social or semi-social impulses. Thus the innate impulse to win others' approval becomes fixed and in a measure moralised, as a desire to carry out lines of action uniformly approved by those whose good opinion is sought. In this way a sentiment of reverence for command or law is developed, which is in part a transformed egoistic fear of the stronger will above the child, or of the power it possesses of enforcing itself by penalties, and in part the semi-social instinct to stand well with others, fixed and concentrated on a particular person or persons, and on the expressed wishes of these.

This view is opposed to the common doctrine of English Associationists—Hartley, James Mill, and others—that the childish feeling for duty is wholly a product of associative transference, that is, a secondary repugnance to particular lines of action because of a primary egoistic shrinking from their disagreeable consequences. That this transference of feeling by associative integration is a large concurring factor is undoubted, and the fact, as some of these writers have clearly shown, is fraught with important educational consequences. At the same time, another and a potent influence co-operates from the first. The force of a command on a child cannot be wholly attributed to experience and prevision of consequences. It shows itself too early, and is out of proportion to the range and intensity of the experiences of punishment. Here, then, we have, as it seems, to do with a "residual phenomenon" which we must regard as instinctive. This instinctive deference to an uttered command is in part referrible to the superior power of external stimuli or sense-presentations generally in our mental life. A command given with emphasis (special loudness and distinctness of tone, accompanied by intent look) is the most powerful way of initiating or bringing on the corresponding movement (or inhibition of movement). In this respect it stands on a level with the actual presentation of an action by another, which, as we shall see, has a powerful tendency to call forth an imitative response. This force of external verbal suggestion, the effect of which we have already seen in the domain of normal belief, is illustrated further in the phenomena of hypnotic suggestion, which Guyau has recently brought into an instructive analogy with the moral influence of education.[1]   The natural impulse to comply with commands is, however, more than this, and involves a rudiment of regard for what others think and say of us as intrinsically valuable, that is to say, what we have dealt with under the head, love of approbation.

(c) *Higher Stage of Disinterested Morality.*   The crude, indistinct feeling of reverence for law thus developed becomes clearer as experience widens, the social feelings proper, *i.e.*, affection and sympathy, expand, and individual reflexion is added.   Thus the growth of a feeling of affection for and of trust in the parental governor will lead the child to take his commands as something acceptable or good.   In the early stages of moral growth, when obedience is very much respect for a particular person rather than for an abstract law, this force of affection counts for much.   Hence the importance of early home-training in morality, when the source of commands is also the person fitted by his or her other relations with the child to call forth his first warm affection.   In order, however, that this feeling may become a true respect for morality as such, the general vaildity of the commands upon the child must be recognised.   He will learn to view a command as intrinsically right in proportion as he discovers that it is not the enactment of a particular governor (the parent) for a particular subject (himself), but that it is imposed on, and in general submitted to by, others.   General custom is a mighty force in moulding alike our belief in what is true, and our ideas of what is beautiful, and morally right.   Nothing contributes so much to make clear to a child the objective character of moral distinctions and their independence of any particular personal source as this recognition of their universal imposition.[2]

The last stage in this development is reached when the

---

[1] Guyau considers that suggestion sets up in the hypnotised subject a sense of 'must,' or of obligation closely analogous to a moral feeling.   (See his volume, *Education and Heredity* (Eng. transl.), chap. i.)

[2] This involves a double process of cognition.   The child first notes that other children are subjected to a personal control like that under which he himself lives. He then observes that grown-up persons, though not subjected to a governor as he is, yet do most of the things he is bidden to do.   Here custom, working through the imitative impulses of the child, comes in to reinforce the other influences making for moral obedience.

grounds of such uniform subjection to law begin to be understood.   Here the growth of sympathy and of rational reflexion
is all-important.   It is when the child enters into others' feelings that he sees *why* he has to do this action as right, and to
abstain from that action as wrong.   As was shown above, it is
sympathy which brings home or makes real to each of us the
existence of our fellow-creatures with feelings, interests, and
aims like our own.   Hence it is by a growth and expansion of
sympathy that the child comes to grasp the social bearings of
his actions, as the injury he does another by an explosion of
anger, by an underhand trick, and so forth.   It is by a like expansion of sympathy that he comes to feel hurt when somebody else does an injury to another.   Such expansions of a
vividly realising sympathy with others' happiness and misery
are virtually new experiences for the child, which, when added
to the early everyday experience that his own happiness is
affected in various ways by others' actions, lead him on step by
step to discern or cognise the social grounds of moral commands
and of duty.   Certain classes of actions, he now begins to reflect, are enforced as right or forbidden as wrong, just because
they cause a benefit or an injury to others.   The moral law is
thus seen to be a social institution, a means of securing
that reciprocal good behaviour, which is the vital bond of a
community.

It is evident, from this brief sketch of the development of
the moral feeling, that it presupposes a considerable growth of
intelligence.   Mere feeling, apart from thought, would not
yield the moral sentiment or moral consciousness as we know
it.   The " sense " of duty is a product of egoistic and social
feeling with processes of reflexion, comparison, measurement
added.   Thus the sense of justice which leads a child to regard
another's happiness as equal in value to his own, which dis-
poses him to accept equal division of good things, including
affection, as also the proportioning of praise and blame to good
and ill desert, implies not only that he can sympathetically realise
a *tuum* as well as a *meum*, but that he can, to some extent, compare and measure his own happiness and his own claims with
those of others.

§ 22. *Differentiation of Moral Feelings.*   We have thus far considered the moral sentiment in its most general form as a feel-

ing of obligation, or of reverence for duty. This feeling, however, takes on a number of modifications according to the nature of the particular action, and its relation to the subject of the feeling. As an illustration of the influence of the former circumstance, we may take the peculiar shade of moral dislike attaching to all meanness of conduct. Here, it is evident, the moral sentiment is reinforced by a more primitive feeling of contempt. A man affected by a moral disgust shows the characteristic " tone " of his feeling in his very expression. As another illustration, we may take the moral horror excited by acts of extraordinary brutality, as the unnatural cruelties of the Emperors Caligula and Nero, and the barbarity of the " Black Hole of Calcutta ". On the other hand, the pleasurable feeling of moral approval takes on, as already hinted, special and distinctive features. As instances, we may take the glow of affectionate commendation with which we respond to all manifestations of generosity, the complex ethical and æsthetic emotion of moral admiration which is excited by the presentation of a rare and transcendent deed of virtue, as the heroic self-devotion of a Horatius or a Grace Darling.

With respect to the second circumstance, the relation of the action to the subject of the feeling, there is, as already hinted, a distinct and profound difference in the tinge of the emotion where we are approving or disapproving our own action. The gratified or wounded self-feeling, which is here present, gives its characteristic colouring, and even its expressive reaction, to the whole emotive state. Thus the feeling of remorse, even in its purest or most disinterested form, as pain at wrong-doing, is a feeling of self-humiliation, and displays itself as such.

22a. *Connexion of Moral and Æsthetic Feeling.* The consideration of the difference in the several forms of the ethical sentiment leads up to a question much discussed in ancient and modern times, viz., the relation of the moral to the æsthetic feeling. In the calm objective contemplation of another's moral action there is undoubtedly something analogous to the attitude of æsthetic contemplation. Both æsthetic and ethical approval and disapproval are modes of agreeable and disagreeable feeling called forth by the observation of what conforms or does not conform to a pre-existing standard. Hence the tendency in Greek, and certain systems of modern, thought to identify the æsthetic and the ethical on their subjective side as feelings, as also on their objective side (beauty and moral goodness). There are, however, as pointed out above, even in the case of the moral approval or disapproval of the impartial and dispassionate spectator, certain *differ-*

entiæ marking it off from æsthetic approval.   Whenever we are the subject of moral feeling our practical instincts or dispositions are engaged, and the affective result is pre-eminently that mild form of agreeable feeling which we call the satisfaction of a practical requirement or demand.   Where the pleasure of contemplating morality becomes intenser or fuller, there, as we have seen, we can detect the co-operation of a true æsthetic feeling.

It may be added that the feeling of moral approval takes on a more distinctly æsthetic form when it is *ideally* satisfied by the presentation to the imagination of a historical or fictitious moral character.   Here, it is evident, the working of the practical instincts is checked.   In imaginatively surveying the perfidy of an Iago or the fidelity of a Cordelia, our mental state is largely reduced to one of pure intellection and emotion.   The supreme interest of the moral side of character has always led the historian and the poet to gratify the imagination by a vigorous presentment of this aspect ; and it may confidently be predicated that true literary art will always seek a chief part of its agreeable effect in the ideal gratification of the moral sentiment.[1]

§ 23. *Moral Sentiment and Judgment.*   As in the case of the æsthetic sentiment, reference must here be made to the relation between the two factors, the emotive and the intellectual.   We have seen that the moral feeling is developed into a clear distinct sentiment through certain processes of reflexion.   It is only as the child examines carefully into actions, the dispositions from which they flow, and the effects of these, that he comes to a clear appreciation of the moral quality which we call rightness and wrongness.   As we all know, there is a blind and so-called 'instinctive' feeling of duty which is precedent to, and often stands in the way of, this rationalised sentiment.   A man feels a thing to be right, and that is enough for him.   To propose to examine the grounds of its rightness may, in such a case, savour of profanity.   Nevertheless, there can be no complete moral *faculty* where this investigation is wanting.   It is the object of moral education to lead on the learner to a clear discernment of the relations on which morality is based, to construct for himself a rational standard or general principle of right conduct, by the help of which he can confidently judge of the moral quality of actions in particular cases.

---

[1] On the relations of the moral to the æsthetic feeling, see Leslie Stephen's *Science of Ethics*, chap. viii. § 3.   Volkmann regards the moral sentiment as a species of æsthetic feeling, differenced from its other varieties by its direct reference to the ego or subject. (See *Lehrbuch der Psychol.* vol. ii. § 134.)

The adjustment of the rival claims of feeling and of intelligence or reason in the domain of morals is one of the difficult problems of ethics. The tendency to elevate into special prominence now the one factor, now the other, is curiously illustrated in the nomenclature employed by different moralists, *e.g.*, the "Moral Sense" of Hutcheson, the "Moral Sentiments" of Adam Smith and others, the "Practical Reason" of Kant. Fully considered, a moral consciousness is always a complex phenomenon with an emotive and an intellective factor, though the proportion of these varies greatly in different cases. In the crude, early stages of moral development, feeling (fear, respect for a person) is uppermost, though even here a subordinate intellectual element is present in the discrimination of a right and a wrong class of actions, *e.g.*, lie from truthful statement. Hence the early stages of moral education are chiefly concerned with fixing for life a habit of swift and unfailing emotive response, *e.g.*, repugnance at, and shrinking from, a lie. A person is not morally developed who has not this irresistible "instinct" to greet and to reject, prior to any process of reflexion, particular sorts of conduct. At the same time, all the higher stadia of moral development depend, to a large extent, on prolonged, and nice processes of reflexion. Only in this way can we reach, for example, a fine and true feeling even for such homely duties as honesty and truth. As modern literature is ever reminding us, morality, instead of being the simple affair we took it to be in childhood, is immensely and perplexingly complex; and each of us has, to a considerable extent, to evolve a new code of moral maxims to meet the peculiar circumstances of his own individual life.[1]

§ 24. *Moral and Religious Feeling.* In close connexion with the moral sentiment, psychologists are accustomed to treat of another and related emotion, *viz.*, the religious sentiment. To the modern man the relation of the two feelings appears as a particularly close one, and it may be as well to follow the custom and to touch on the highest and most complex of the emotions in this connexion.

The religious sentiment is one particularly difficult to define owing to the protean variety of its changes. We may, however, roughly mark it off as that mingled emotion of awe and delight which is specifically excited by the idea of the unseen world, and more especially of the mysterious Power that is supposed to preside over human life. It is easy to see that this feeling is a highly composite one, gathering up into itself the products of our various emotive sensibilities. Thus, as a feeling of personal dependence on, and resignation to, a divine order, the emotion includes a sublimated form of self-feeling. Again, as a sense of a common human destiny, it is eminently a social feeling. On another side, as a recognition of the world

---

[1] For a brief account of different views of the moral faculty, and of the bearing of this part of psychology on properly ethical thought, see below, Appendix L.

as an intelligible whole, it is an intellectual sentiment. In yet another of its aspects, as an expansion of the feeling for nature and the beauty of the cosmos, it is an æsthetic emotion.

Modern research has busily occupied itself with the early forms and course of development of the " religious conscious-ness ". The problem is one of the most intricate in the science of anthropology, and cannot yet be said to be finally solved. One or two conclusions, however, appear to be pretty firmly established.

The supposition of the classical world that religion was born of fear ("timor Deos fecit ") is now regarded as crude and inadequate. Even in its early and coarser forms we can see the complexity of its sources. No doubt, as is attested by the various known forms of primitive superstition (including even the mythology of the Greeks and Romans), the religious im-pulse was excited by that which was extraordinary, and especi-ally that which threatened human life. The basest forms of religion are embodiments of what to the civilised man is an abject terror. The practical instincts, however, co-operated from the beginning, and there is some reason to conjecture that the objects of primitive human ' worship ' were beings who could be worked upon and propitiated. The detachment of the feeling at this early stage of its evolution from the moral feeling is sufficiently shown by the character attributed to these first deities. In addition to these impulses springing out of the egoistic nature, higher feelings probably began to take part at an early period. Thus, in the worship of ancestors, which, according to some, is the primitive form of religious feeling, we see, along with the practical impulse to propitiate a power that can injure, the germ of a feeling of family piety. This social constituent in the religious consciousness would become more marked through the development of a fixed common ritual. Once more, it is probable that a germ of intellectual or rational feeling entered into, and helped to sustain, the religious emotion from the first. Although it would be an error to regard primitive man as a philosopher inquiring into the causes of things, there is reason to suppose that his crude ideas of the supernatural were suggested by certain ill-understood pheno-mena, such as dreams, and supplied to his mind an explanation of these. It has been said that every mythology is, at the same time, a cosmogony. The fact that, in the earliest times, art was bound up with religious institutions and observances suggests that a crude æsthetic feeling combined at an early stage with the religious. The modern developments of the religious feeling are the outcome of that intellectual and emotional culture which characterises a highly civilised community. Thus the growth of knowledge has substituted monotheism for polytheism. The higher development of the social, æsthetic, and moral feelings has led the Christian world to conceive of the Deity as a being who combines in himself all intellectual and moral perfections, as the sublime omnipresence and omniscience whom finite man can reverence, as the great father to whom he can confide all his interests, and as the perfectly good and righteous ruler who will readjust the seeming anomalies of life. The growing connexion of the religious with the moral feeling is seen in the fact that the development of the latest phases of Christian belief and sentiment has been due quite as much to the action of our moral feelings as to that of our modern scientific conceptions.[1]

REFERENCES FOR READING.

On the nature and development of the abstract sentiments, the following may be referred to : Bain, *The Emotions and the Will*, pt. i. chaps. xii., xiv., and xv. ; H. Spencer, *Principles of Psychology*, vol. ii. pt. viii. chap. vii. and following ; Nahlowsky, *Das Gefühlsleben*, 2 buch, 2 abschnitt ; Volkmann, *Lehrbuch der Psychol.* §§ 133, 134 ; Wundt, *Physiol. Psychologie*, ii. cap. xviii. 3 ; Horwicz, *Psychol. Analysen*, 2er theil, 2e hälfte, buch 2 and 3 ; and Höffding, *Outlines of Psychology*, vi. c. 8, 9.

[1] On the nature of the religious sentiment and its relation to the moral, see Volkmann, *op. cit.*, vol. ii. § 134 ; Wundt, *Physiol. Psychol.* vol. ii. cap. xviii. ; and Höffding, *Outlines of Psychology*, vi. c. 8. The early forms of the religious emotion among lower races are dealt with (among others) by Tylor, *Anthropology*, chap. xiv. ; and, more fully, by Spencer, *Principles of Sociology*, pt. i.

PART V.

CONATION OR VOLITION.

–

CHAPTER XVII.

VOLUNTARY MOVEMENT.

We have now surveyed the principal stages in two out of the three directions of psychical development. It remains to carry out a similar process in the case of the third direction, that of conation or volition.

§ 1. *Range of Conative Phenomena.* Here it becomes important, at the outset, to indicate, with some precision, what processes are to be included under the head of conation or volition. These terms are used in a variety of meanings, some wider and more comprehensive, some narrower and more special. Popular thought recognises a phenomenon of will where there is action with full consciousness of activity, of an antecedent purpose. But there are active phenomena having a psychical aspect which fall short of this, and which must be taken into account if the more complex and highly developed forms are to be understood.

One thing is certain. No actions of the organism which are carried out *unconsciously* can fall under the head of conation, seeing that, according to our definition of mind, they lie outside the properly psychological domain altogether. But between these and the highest manifestations, deliberative and selective actions, there are intermediate grades, *e.g.*, the actions described as instinctive. It seems best, on the whole, to make

*

(172)

the terms conation, volition, comprehensive enough to include all actions which have a conscious accompaniment, and which we will henceforth mark off as *psychical* actions.

Besides the (psychical) movements of the bodily organs which we generally think of, first of all, in connexion with volition, there are the processes which fall under the head of attention. Here, again, as pointed out above, we have a lower (reflex) and a higher (volitional) form. Being, however, both modes of psychical or conscious action, they may be embraced under the head of conative or active phenomena.

§ 2. *Characteristics of Conation : Active Consciousness.* The most obvious common characteristic in this variety of actions or conative processes is, as already suggested, that peculiar element which is best marked off as active consciousness. To consciously move the limb, to direct attention on a difficult point, is to have a particular and unique sort of experience, the *differentia* of which we can only describe by help of the term active, or some equivalent expression, as sense of exertion, or of effort.

As suggested above, the peculiar colouring of these active psychoses is in most, if not all, cases connected with the working of the motor side of the nervous system. The active phase in all bodily movement depends on the generic peculiarities of muscular sensation (and its ideational substitute) in its contrast to passive sensation. So far as bodily movement is concerned, one may safely say that were it not for this muscular consciousness we should not think of marking off volition as a distinct function of mind.

With respect to the processes of attention or " mental activity," the presence of the muscular consciousness is less obvious. Yet we saw reason to hold that here, too, the distinctive active colouring is determined mainly by a psycho-physical muscular process.

It would thus seem to follow that the most obvious general differentiating circumstance in all conative phenomena is the presence of the psychical correlative of muscular action. Our consciousness of activity is based upon the common peculiarities of our muscular sensibility.

§ 2a. *Psychological and Physiological View of Muscular Action.* In thus connecting the peculiarity of conative phenomena with a motor or muscular element we must be careful to distinguish between the psychological and physiological points of view. The terms "movement," "muscular action," and the like, are clearly names of *physical* processes, central innervation of motor nerve, conduction of excitation to muscle, muscular contraction, etc. In certain cases the physical process is all that is meant. This applies to those movements of the vital organs which are effected by the unstriped or 'involuntary' species of muscle. Even in the case of psychical movements the psychical and the physiological factor must be carefully distinguished. All our expressions for describing active phenomena, *e.g.*, "muscular action," "consciousness of energy," are apt to suggest to the unwary that the psychical element involves at least a *cognition* of the material process. This misapprehension must be strenuously guarded against. As psychologists, we employ the expressions "muscular action," and the like, merely in order to mark off a certain group of psychical phenomena by a reference to their well-known physiological concomitants. No doubt, so far as we take a psycho-physical view of mental processes, we shall include a reference to the nervous conditions involved. But, here, again, as psychologists, we shall only be concerned with the immediate correlatives of the psychical process. Thus we shall leave out of account what takes place in the motor nerve, and in the fibres of the muscle, since these have no immediate psychical concomitants.

A corollary of this rigorous demarcation of the psychical from the physiological aspect of action is that active consciousness is differentiated from other kinds, whatever the precise physiological base of muscular sensation turns out to be. As pointed out above, the psychological distinction between active and other consciousness does supply a certain presumption in favour of such a strongly marked difference of physiological correlative as the theory of an *efferent* psycho-physical process lays down. On the other hand, should the neural base of muscular consciousness turn out to be wholly afferent, this fact would not make the peculiar consciousness of active exertion one whit less peculiar. The physiological question only seems to affect the psychological, because of the common confusion of thought just referred to. Thus the active consciousness is described as "consciousness of outgoing energy," and this way of speaking strongly suggests that it contains some knowledge of the process of central innervation ;[1] whereas, in truth, there may be a distinct active consciousness in a baby that knows nothing of nerve-centre or nerve, innervation, or muscular contraction.[2]

§ 3. *Conation as Psychically Initiated Action.* While, however, the terms conation and volition may thus be stretched so as to cover all psychical processes into which the "active

[1] Ferrier goes so far as to say that the hypothesis of an innervational psychosis (*i.e.*, of course, a mode of sensational consciousness *concomitant with* central innervation) is absurd, because it involves the idea of a consciousness " *of the molecular processes of our brains* ". (*Functions of the Brain*, p. 382.)

[2] The idea that a reference of muscular sensation to an afferent base would destroy its distinctively active character has, no doubt, been suggested by the fact that psychologists who most emphasise the contrast of active and passive consciousness have naturally connected this fundamental distinction in our experience with a strongly marked neural contrast. (See Bain, *The Senses and the Intellect*, p. 77.)

consciousness " enters, they connote in all their higher and more specialised forms other ingredients as well. A *voluntary* process, whether a phase of what we call bodily movement, or of intellectual activity, always carries with it, not merely a psychical concomitant (active consciousness), but a psychical antecedent. This antecedent may be roughly described as a forecasting or prevision of the action itself and of some at least of its results. The precise nature of this initiative process varies, as we shall presently see, considerably in different cases. In some instances there is hardly anything discernible but the idea or representation of the active experience itself, the psychical aspect of the movement. In other cases, the prominent psychical antecedent is what we variously call a consciousness of purpose, a foregrasping of a desired end, or an inducement or motive. The fully-developed conative process includes each of these two initiative elements. Since, however, the presence of a conscious purpose is, as we shall see, the factor which most distinctly differentiates the higher and more specialised forms of volition, it is customary in defining these to make special reference to this factor. Accordingly we may say that, along with the element of active consciousness, a second main differentiating ingredient in the conative process is consciousness of purpose, or forecasting of end.

§ 4. *Conation in its Relation to Feeling and Cognition.* The *differentiæ* of conative phenomena now reached, *viz.*, active consciousness and psychical initiation through representation of an end, serve to mark off with some distinctness the domain of volition from that of intellection and of feeling. A word or two may serve to make this plain.

Taking feeling first of all, we see that conation contrasts with this in respect of its passivity. Pleasure and pain are non-active. It is true, as has been shown above, that all feeling has motor concomitants which contribute psychical elements to the emotive state. In certain emotions, moreover, *e.g.*, anger, the active phase of consciousness becomes more prominent, and here we see the domains of feeling and conation overlapping, and becoming, in a measure, confused. In the lower stage of development, where feeling and conation are less completely differentiated and specialised, this want of a sharp boundary between the two is particularly apparent. Yet this circumstance does not affect the validity of the distinction here drawn with respect to all the more specialised forms.

The difference between feeling and conation is further seen in the peculiarities of the psychical initiation of voluntary action. The representation of end when examined will be found to involve the peculiar phenomenon known as *desire*, and this again has to do with the gratification of feeling. We desire, in general at least, that which is viewed as pleasurable, *e.g.*, the quenching of one's thirst. Yet here, it is evident, feeling takes on a new form. As a stimulus or motive to action, it no longer remains mere feeling, but undergoes a special modification by entering into the peculiar complex state of desire.

In like manner our *differentiæ* serve, in general, to demarcate conation from the region of intellection. The order of presentations and ideas has its passive side, as determined by external conditions and the laws of suggestion. From this we can mark off that active or conative factor which appears as soon as we begin to modify and control these passive sequences by attention. Our higher intellectual products involve, as we have seen, the co-operation of this conative factor.

A special difficulty in marking off conation from intellection arises from the circumstance that every psychical action has an intellective phase. Thus the active consciousness can be discriminated as that answering to a particular kind of movement, and retained and reproduced for future use. In this way, as we shall see, the development of action proceeds by help of intellectual processes carried out on a particular class of sensations or presentations (muscular or " motor "). Again, the psychical initiation of action or forecasting of end is clearly a mode of expectation which has been dealt with under intellect. All this, however, while showing that volition is only one phase of a larger organically conditioned process, does not serve to efface the distinctions here laid down. The muscular sensations, so far as these are discriminated and retained, come, no doubt, under the rubric of presentative elements, that is, of intellectual phenomena, along with the passive sensations. Yet the characteristic of active consciousness remains, and, so far as this is realised, the phenomenon may claim a place among the processes of conation.

*

§ 4a. *Different Views of Conation.* In the various theories of volition proposed by psychologists we find great differences in respect of the range of phenomena included. It has been the tendency, on the whole, to restrict the term too much, to define as essential characteristics features which are present only at a comparatively high stage of volitional development. Thus writers have maintained that there is no volition where there is not conscious choice. Popular usage, however, justifies us in saying that there is volition where there is action initiated by desire for end. It is common, too, to make self-consciousness, under the form of an apprehension of personal agency or causative power, an integral element in conation. No doubt, this is a common feature in all mature fully conscious types of action. Yet, as we shall presently see, it is a variable concomitant, which does not appear in a distinct form at least in the lower stages of development, and which may be absent, too, in certain absorbing modes of adult action.[1]

Recent psychological theories show a tendency to deny to conation the rank of a fundamentally distinct psychical function. The way in which this is done is by viewing active experiences *merely* as a group of presentative elements or sensations, and by regarding the typical psychical process in volition as motor idea (representation), followed by motor sensation (presentation). This way of conceiving the conative process meets us in the scheme of Ward already referred to, in which the conative phase is marked off from the cognitive as voluntary feeling-initiated attention to motor presentations in contradistinction to non-voluntary attention to presentations in general.[2] This view, shorn, however, of its important qualification, initiation by feeling, appears in the ingenious psycho-physical study on volition by Münsterberg, who tells us that will is in its psychical aspect merely a particular grouping of sensations, or, as it is otherwise expressed, an idea of movement followed by the sensation.[3] James expresses a similar view in his general delimitation of the volitional process.[4] On the other hand, Wundt, following Bain in England, contends strongly for the fundamental peculiarity of the active consciousness.[5]

§ 5. *Attention and Psychical Movement.* It has been pointed out that the conative process takes one of two roughly dis-

---

[1] The reader must be careful to note that in speaking of conation as containing the ingredient of *active* consciousness we do not imply active *efficiency* or power to produce effects. In our mature consciousness the two elements are closely conjoined, but an effort of analysis suffices for distinguishing them.

[2] *Loc. cit.*, p. 44. Ward, indeed, adds the production of change in the motor continuum, but this hardly differentiates conation in his theory, seeing that this makes attention to all classes of presentations a means of intensifying these.

[3] *Die Willenshandlung*, pp. 62 and 67 ff.

[4] *Principles of Psychology*, ii. p. 492 ff.

[5] Wundt, *Physiol. Psychol.* ii. cap. xx. § 1. It is to be noted that while Bain and Wundt hold that the active consciousness is correlated with the initial process of central innervation, Münsterberg and James take the opposite view.

tinguishable directions, *viz.*, conscious movement and attention. They are sometimes contrasted as bodily and mental actions, and also as external and internal directions of activity. These distinctions are, however, very far from being exact. As we have seen, bodily movement only constitutes a conative process, so far as it has the *mental* or conscious concomitant. Again, as was pointed out above, attention in all its earlier forms is directed externally to presentations of sense, and involves motor adjustments of the organ concerned.

If we take the later complex mental life of the adult, we, no doubt, find a certain demarcation of two fields of conation, *viz.*, intellectual activity and voluntary movement. In the one there is a fixing, intensifying, and, on the other hand, an inhibiting of presentations generally under the stimulus of a properly intellectual feeling, the desire for knowledge. In the other there is the conscious initiation and direction of (muscular) movement under the stimulus of practical needs. Yet the common separation of these two domains does not involve a fundamental distinction in the conative process itself. Intellectual and motor activity are alike forms of the active consciousness, and are commonly described by the same language, *e.g.*, exertion, effort, strain. As has been pointed out above, not only is the active phase of intellectual attention determined in all cases, in part at least, by a psycho-muscular ingredient, but the psychical process which constitutes the initiation of movement involves a special direction of attention.[1] As we shall see, in all fully developed volitional processes, in which the forecasting is distinct, there is a fixing of the attention on certain ideas, whether that of the on-coming action itself (motor representations) or of some of its results.

It is thus probable that the two branches of conation spring from a common stem, and that the primitive form of activity is at once, according to the aspect in which we view it, both attention and conscious muscular action. These two aspects become differentiated, to some degree, through the growing complexity of the psychical life, and more particularly through the differentiation of the life of thought with its bodily motionlessness, and the practical life with its system of movements.

---

[1] *Cf.* above, I. p. 67 and p. 147 ff.

As just hinted, while intellectual attention and the volitional direction of movement have each an ingredient of muscular consciousness, this varies, to some extent, in the two cases. Thought (save where accompanied by special movements,
as those of the eyes in reading) is characterised by a more or less prolonged state
of muscular tension, involving a balancing action of the antagonist muscles.
Voluntary action or movement, on the other hand, is pre-eminently a *changing*
muscular consciousness involving the particular mode of muscular sensation
marked off above as that of (free) movement. Whether, in addition to this circumstance, there is given in the nervous process which intensifies the action of the
sensory (or ideational) centre during a state of attention a further differentiating
psychical accompaniment it is, in the present state of our knowledge, impossible
to say.[1]

Voluntary Movement.

We may now proceed to trace the origin and growth of the
volitional process. And in doing this we shall be concerned
mainly with the development of the process in connexion with
movement. Our problem is, first of all, to determine the
factors which combine to generate what we understand by a
conscious voluntary movement, as when we stretch out the
hand in order to take an object ; and, secondly, to trace out the
successive stages in the development of this movement.[2]

§ 6. *Roots of Voluntary Action : Instinct and Experience.* A
glance at what we mean by a voluntary action shows us that it
presupposes two factors. The first of these may be marked off
as the Original or Instinctive root of volition. Such is the
impulse to seek that which is agreeable and beneficial, and to
avoid what is painful and harmful. It is evident that this disposition is primordial, and has to be presupposed in any attempt
to account for the growth of the volitional process. It may be
added that this impulse or disposition shows itself, first of
all, in a sub-conscious form, in what is sometimes specially
marked off as *Impulse* (Trieb), that is, a rudimentary and
essentially vague process of craving, or striving.[3] This phe-

---

[1] *Cf.* above, Vol. I. p. 147 ff.

[2] As we shall see by-and-by, this branch of the volitional process involves not
merely positive movement, but the arrest of muscular action, as when we voluntarily
inhibit a movement of the limb.

[3] The student should pay particular heed to this word "impulse," which has
brought much confusion into psychology. Here, again, we see the tendency to
indicate by one and the same term a purely physiological, and also a psychological
or psycho-physical process. That is to say, 'impulse' means now a mere central
nervous tension involving a tendency to motor discharge, now a more or less con-

nomenon will meet us in the region of what is specially marked off as instinctive action. In its later and clearly conscious form it becomes what we know as *Desire*. This element of impulse or desire is, as we shall see, organically related to motor activity, and always involves some amount of the active consciousness. It is thus essentially an *active* phenomenon.

In the second place, the volitional process, say in willing a movement of the arm for the purpose of plucking fruit, presupposes experience. A child brings with it into the world no prophetic prevision of its doings and their results. Before he can carry out a conscious intelligent direction of a movement to a particular result, there must have been some experience (or series of experiences) by which he has learnt first of all the particular result which he now aims at ; secondly, the particular conscious movement which he now wills to carry out ; and thirdly, the (causal) connexion between these two. It follows from this that clearly volitional movement is preceded by earlier forms of movement. It is in the experiences obtained by means of these Primitive Movements that each of us learns how to execute what we call our voluntary movements.

We see from this that the development of a clearly conscious volitional action is a gradual process. A distinct representation of objects of desire as ends of action, and the selection of appropriate, *i.e.*, effectual, movements, arise little by little out of indistinct processes in which intelligence or reason plays hardly a greater part than in the instinctive actions of the lower animals.

One other aspect of this process of volitional development may be just referred to. It was pointed out above that the development both of intelligence and of feeling proceeds from the outer life of sensation to the inner life of ideation. The course of volitional development is similar. In the earliest stages of this development we shall find movements called forth in immediate response to sensations. Little by little this crude form of movement will be seen to be complicated by ideational

---

scious process of striving away from the actual present to a represented non-present. (*Cf.* Höffding, *Psychology*, vii. B. 1, *a*.) The term will be used in this work for the latter process only.

processes, suggestion of desirable object, and of appropriate action, till in the highest type of volition this internal ideational factor becomes the main determinant under the form of delibera-tion and rational choice.

We may now proceed to trace this movement of volitional development, beginning with an account of those primitive movements which precede the distinctly volitional type.

PRIMITIVE MOVEMENTS.

§ 7. (a) *Movements not Psychically Initiated : Random, Auto-matic Movements.* To begin with, then, there is reason to suppose that in the case of the child, as well as of other young animals, certain movements arise independently of sen-sory stimulation, and through the "automatic" excitation of the central cellular substance spoken of above.[1] They have been variously called Spontaneous (Bain), Automatic (Wundt), and Impulsive (Preyer). They may also be marked off as random movements. Illustrations of this class may be found in the movements of the chick in the egg, and in some of the earliest movements of the infant, as stretching out the arms, the legs, rolling the eyes on waking while the lids are still closed, and so forth.

The most striking psychical characteristic of these random or automatic movements is the absence of all psychical initia-tion. They are preceded by no anticipatory consciousness, either of the movement itself, or of anything resulting there-from. Accordingly they only claim a place in a psychological account of movement by reason of the active consciousness or motor experience which they yield.

Biologically considered, these movements are purposeless or useless to the organism. That is to say, not only are they wanting in *conscious* purpose, but they cannot be shown to subserve unconsciously any useful result. This considera-tion would suggest that at most they play a very subordinate *rôle* in the life of the child.

The question of the existence of strictly automatic movements, and, granting this, of their number or extent, is not easy to answer. By those who regard all movement as of the reflex or sensori-motor type such unconsciously (though cen-trally) initiated movements are ignored. Since the sensorial mechanism is continu-

[1] See p. 46.

ally being played on by its several orders of stimuli, it is, of course, exceedingly difficult in the case of any movement to show that it is a *pure* central product, that is, wholly the result of automatic action of the brain-cells. The most careful demarcation of the group of automatic movements is that of Preyer, who argues with considerable force that the movements specified by him cannot be brought under the sensori-motor form.[1]

§ 8. (b) *Sensori-Motor Movements : Conscious Reflexes.* As pointed out above, most, at least, of the lower class of move-ments take on the reflex or sensori-motor form. With such reflexes as involve only the lower centres, and have no conscious concomitant, we are, of course, not here concerned. The type of reflex which we have to consider is that in which both the sensory and the motor stage yield a conscious element. In other words, we are interested in those sensori-motor actions in which sensation constitutes the psychical initiative. These may be called *conscious reflexes.* Such are the movement of closing the eye-lid when an object is brought near the eye, and of starting at a sound. These conscious reflexes differ, as already implied, from automatic movements in the important circumstance that they have a conscious antecedent, *viz.*, a sensation, more or less distinct. In this respect they bear a certain resemblance to *voluntary* movements, for, as we shall see, these last commonly have a sensation or its ideational representative as an initiative factor. For the rest, these con-scious reflexes are distinctly marked off from true volitional actions, first of all, by the absence of all idea of purpose or end :

---

[1] *Die Seele des Kindes*, p. 127 ff. Preyer's term "impulsive" may be objected to on the ground already urged that it is properly applied to simple psychically initiated movements. Spontaneous movement is made a central feature in Bain's theory of Will (*op. cit.*, p. 303 ff.). He seems to include under this head all the diffused playful motor activity of vigorous young creatures, the result, as he conceives it, of an "overflow" of energy from the motor centres. The importance of such automatic movements is also upheld by Höffding, *Psychology*, vii. A.; but, on the other hand, is disputed by Wundt, *op. cit.*, ii. p. 488 f. A physiological difficulty in the way of admitting them is the observation of Fleichsig and others that the new-born child is wanting in the motor nerve paths (pyramidic path) which subserve central (motor) innervation.[2] The theory of spontaneous movement has recently been severely criticised by A. Herzen (*Psycho-physiologie*, p. 66 ff.), who, however, appears to suppose that since such movements are wanting in a sensory stimulus they must be wanting in *any* organic conditions.

[2] Quoted by Raehlmann, *Zeitschrift für Psych. und Physiol. der Sinnesorgane*, ii. p. 58.

and, secondly (what is closely connected with this), by their unvarying mechanical character—the same motor response always occurring when the particular sensory stimulus recurs.

Some of these movements are perfect, or approximately so, at birth. This applies to the necessary actions of inspiration and expiration, swallowing, and the like, which, as we have seen, are commonly carried on by means of sub-cortical centres, and do not involve distinct consciousness at all. It applies also to more distinctly conscious reflexes, as, for example, closing the fingers over a small object, as a pencil, brought in contact with their anterior surface. Others first occur later. This applies to many movements of the eyes.[1] These conscious reflexes constitute a considerable group of infantile movements. Thus Preyer instances six different reflexes among ocular movements. As already observed, all these primitive reflexes are determined, mainly at least, by inherited organic arrangements.

It is to be added that, in addition to such original reflexes, acquired voluntary movements themselves tend by repetition and the lapsing of the element of conscious purpose to take on a reflex character. Many of our demonstrably acquired movements, e.g., brushing away a fly from the face, putting out a hand to stop an object approaching us, offering our hand in response to the invitation of another's outstretched hand, have this reflex or sensori-motor character. In addition to these restricted and specialised reflex reactions there is a more diffused form of motor reaction of the same reflex type. Thus it has been proved by recent experiments of Féré that every sensorial stimulus tends, according to the degree of its strength, to innervate the muscles generally. This diffused form of reflex motor reaction is, as we shall presently see, important as supplying unformed material for volitional selection.[2]

As pointed out above, specialised reflex movements, just because they are definite responses to particular external sense-stimuli, have in general a recognisable biological character as useful functional actions. In many cases this element of

[1] See below, Appendix B.

[2] Féré is able thus to measure sensation indirectly, by the dynamometer or instrument for measuring muscular energy. (See his *Sensation et Mouvement. Cf. Revue Philosophique*, xxiv. p. 572 ff. ; also James's account of the experiments, *Principles of Psychology*, ii. p. 379 ff.)

utility (unconscious) purposiveness is patent, as in the actions of swallowing, etc., already referred to. And in all cases where the reflex movement is restricted and specialised, and not merely a diffused motor result of sensory stimulation, the character of the movement suggests a purposive adaptation to the particular stimulus. Thus in reflex action, unconscious and conscious alike, we see the simplest form of a special motor reaction of the organism in response to some particular mode of environmental action.[1]

Some writers use the term reflex only with reference to unconscious movements carried out by means of the lower centres, *c.g.*, the movements of the legs of a decapitated frog. Carpenter, again, marks off conscious reflexes from unconscious by help of the terms sensori-motor (or consensual) and excito-motor. Since, however, all the movements included above have on the physiological or neural side the reflex or sensori-motor form, it seems best to distinguish them by the addition of the words conscious and unconscious.[2]

§ 9. (c) *Instinctive Movements.* Closely allied to conscious reflex movements, and not easily distinguished from these, are a third group of primitive reactions, *viz.*, instinctive movements. In a sense all original *unacquired* movements determined by congenital organic arrangements are instinctive,[3] and the word instinct is often used in psychology with this wide reference. In the narrow and stricter sense, however, " instinctive movement " stands for one particular variety of the primitive sensorimotor type of reaction. These instinctive movements are, physiologically considered, distinguished from properly simple reflexes by their complexity. Many of the instinctive actions of the lower animals, *c.g.*, the building instinct of the beaver, are complicated series of movements. This applies even to such an apparently simple instinct as sucking.

Psychologically considered, instinctive actions are characterised not merely by the richer active consciousness which this motor complication implies, but also by a fuller and more im-

---

[1] On the biological conception of reflex movement as evolved and fixed on account of its utility, see Münsterberg, *Die Willenshandlung*, i.; and Ziehen, *Leitfaden der physiol. Psychologie*, p. 6 ff. Ziehen brings out the fact that unconscious reflexes are, through their very rigidity and insusceptibility of modification with modification of the stimulus, useful only in a *general* way, *i.e.*, in the majority of cases.

[2] The nature of reflex action is dealt with by Carpenter, *Mental Physiology*, chap. ii. par. 47, 66 ff. ; G. H. Lewes, *Physical Basis of Mind*, prob. iv. ; Wundt, *op. cit.*, ii. cap. xxi. p. 489 ff. ; G. H. Schneider, *Der Menschliche Wille*, cap. ii.

[3] The reader must carefully note that acquired actions which become automatic by practice and habit, though popularly described as instinctive, are not included under instinct as just defined.

portant psychical initiative. In the case of many conscious reflexes the psychical concomitant is, as was remarked, indistinct and fugitive. It is otherwise with instinctive movements. Many of these at least are preceded by sensations of considerable intensity. Moreover, and this is a capital distinction, the sensational element in the initiation of instinctive movements has a marked *affective* concomitant. Thus the instincts of birds, *e.g.*, incubation, migration, appear to be determined by sensations having a strong accompaniment of painful feeling, *viz.*, one of discomfort or distress, which element comes distinctly into view whenever the appropriate movements are not at once forthcoming. A striking illustration of this is seen in all the instinctive, that is, original, appetites. As is well known, the infantile instinctive movements of sucking are preceded and prompted by the distinctly painful sensation of hunger.

Instinct is marked off from purely affective phenomena by its active element. By this is meant that peculiar and energetic stir of muscular activity during the state of discomfort which, unlike a merely expressive movement, *e.g.*, the smile, appears to indicate a state of vague craving or striving after something not realised at the moment. The migratory bird overtaken by its peculiar sensation, and the child seized by its hunger, are restless, and the motor activity excited appears in its more characteristic part to be related to the satisfaction of the instinct.

This leads us to consider the most important aspect of instinctive movement, its purposiveness. It is commonly allowed that instinctive movements show much more distinctly than reflex movements the purposive character, the specialised adjustment of functional activity to a particular biological utility. Hence the common way of grouping instincts according to the biological end they subserve, *e.g.*, self-preserving, race-preserving instincts.[1] Being thus preceded by a feeling of uneasiness which always manifests itself in movement of a more general or special character, and having, too, a definitely purposive value, it seems natural to refer them to a *quasi*-volitional

---

[1] *e.g.*, by Schneider, *Der Thierische Wille*, p. 397 ff.; *cf.* Wundt, *Physiol. Psychologie*, ii. p. 419.

element, *viz.*, *impulse ;* and this is commonly done.[1]　It is, however, exceedingly difficult to say how far the psychical antecedent here resembles desire for end, or a consciousness of purpose. That the bird which incubates under the stress of certain sensations has no distinct idea of an end to be gained seems to be sufficiently shown by the blind mechanical character of the response in many cases, the absence of a clear perception of the precise circumstances which only could make the action useful. Thus when the flesh-fly, misled by smell, deposits its eggs in the flowers of the " carrion plant," it shows that if there is intelligence in the case it is very imperfect.[2]　Indeed, it has not been seriously contended that instinctive actions are performed with a clear consciousness of end and of means, such as accompanies " rational " action in the case of man.　At the same time, it seems undeniable, as Wundt and others contend, that we have in the excitation of the instinctive prompting a truly active phenomenon in the shape of a semi-conscious impulse to act, which impulse involves the analogue of what we call a desire for an object, *viz.*, a vague, undefined craving for something, and a restless striving towards its realisation.[3]

It has been customary to assign instinct to the lower animals, and to attribute the actions of man to intelligence. Recent research, however, goes to show that though instinctive movement plays a smaller part in the life of the child than in that of the young animal, it is larger than has been generally supposed. The instinctive factor in human action includes but few specialised and perfected instincts, such as sucking.

[1] Here, again, the student must beware of the ambiguities of psychological language. Impulse is, as we have now seen, used in three senses : (1) the single prompting element or motive in simple *voluntary* action, (2) the vaguely conscious antecedent in the case of instinctive movement, and (3) the purely physiological antecedent in the case of "automatic" movement.

[2] For a full account of such " imperfect instincts," see Romanes, *Mental Evolution in Animals*, p. 167 ff.

[3] This condition of striving towards something not clearly represented may perhaps be understood by comparing it with more familiar psychical phenomena, *e.g.*, fear of an undefined evil, which is specially observable in young children, the teasing sense of having forgotten something (we know not what) in which all *distinct* representation is wanting. At the utmost, there seems to be involved in such a blind impulse a rudimentary differentiation of the consciousness of presence and of absence.

It appears for the most part under the form of an original organically conditioned impulse or tendency of a more or less vague
character, requiring the specialising influence of experience
and education. Such are the instinctive promptings to movements of the arm and hand in grasping, of the legs in walking,
of the vocal organ in the first infantile " la-la-ing," and so forth.
As we shall see presently, these instinctive promptings to particular groups of movements under the stimulus of particular
sensations and sense-feelings constitute a valuable aid to the
early growth of voluntary action.[1]

§ 9a. *Nature and Origin of Instinct.* The demarcation of the limits of instinctive movement, and the explanation of the nature of instinct, are problems of particular difficulty in the case alike of animal and of human psychology. As already
pointed out, instinctive action is of the same fundamental type as reflex, and shades
off into this. Hence the two are dealt with together by some writers, as when H.
Spencer calls instinct a compound reflex,[2] and Münsterberg defines instinctive
action merely by the *negative* criterion, the absence of preceding idea of end, thereby
apparently including conscious reflexes.[3] It is, however, important, from a psychophysical point of view, to mark off instinctive actions as a separate class, by reason
of the important differentiating *psychical* character, the presence of feeling, and of
that restlessness which, in its external manifestation, is analogous to a properly
conative phenomenon, *viz.,* the movement-prompting influence of desire.[4].

With respect to the precise psychical nature of instinct little is known. From
a metaphysical point of view, the instincts of animals have been described by Hartmann as unconscious prevision, or a kind of clairvoyance. That instinct stands in
close relation to intelligent or volitional action is suggested by a number of facts.
Thus instinctive actions appear, from physiological experiment, to involve the
cerebral hemispheres. They are, at least, greatly interfered with by the destruction of these. Again, they do not exhibit the rigidity of reflex movements,
but can be modified by experience and habit.[5] Once more, instincts are found
among many animals which admittedly have a high degree of true intelligence.
At the same time, it would be an error to assimilate instinctive movements to consciously purposive and intelligent actions. Their appearance in early life before
the higher centres are developed,[6] their liability to deception where the exciting

---

[1] On the range of instinct in man, see Preyer, *op. cit.,* cap. xi.; James, *Principles
of Psychology,* ii. chap. xxiv.

[2] *Principles of Psychology,* pt. iv. chap. v. *Cf.* Schneider, *Der Menschliche Wille,*
pt. ii.

[3] *Willenshandlung,* pp. 92, 93.

[4] On the way in which instinctive shades off into voluntary action, see Volkmann, *Psychologie,* i. p. 315 ff.

[5] This is well brought out by Romanes, under the head "Plasticity of Instinct," *Mental Evolution in Animals,* p. 203 ff.; *cf.* James, *op. cit.,* ii. 394 ff.

[6] This is, of course, not true of all instincts, *e.g.,* the sexual, which presupposes
certain processes of organic development.

presentation resembles the normal excitant, these and other circumstances point to the conclusion that instinctive actions stand on a lower psycho-physical plane, and involve only a dim and rudimentary sort of consciousness.[1]

Closely connected with the question of the inmost psychical nature of instinct is that of its origin. That instinct has close analogies to habit, that is, to acquired action which has grown semi-conscious and automatic through long practice, is obvious, and is illustrated in the popular way of describing such habitual or secondarily automatic action as "instinctive". This affinity makes it appear natural to suppose that instinctive movement may have arisen in the course of racial and zoological evolution by the working of the law of habit aided by heredity. According to this view, instinct is "lapsed intelligence"; it represents actions which, in the case of remote ancestors, were found to be useful and intelligently carried out, though now they have become "organised," by help of definite nervous arrangements transmitted from parent to offspring, into sub-conscious and non-intelligent actions. This view obviously rests on the assumption that acquisitions ("acquired characters") can be transmitted, a proposition which, as we have seen, has been vigorously attacked by Weissmann and his followers. It is, moreover, beset with a special difficulty, *viz.*, that, if instinctive actions are to be looked on as originally intelligent, they involve a knowledge of remote ends, *e.g.*, providing considerably beforehand for a "rainy day," for the needs of offspring, and also of complicated adjustments in order to reach those ends, which seem in some cases to equal if not to surpass the ordinary powers of human intelligence. On the other hand, the theory that such highly complicated and beautifully adjusted actions can have arisen from "accidental" variations of nervous structure in particular individual members of a species has its own difficulties. Hence it is not surprising to find that certain writers, as Romanes, postulate a *double* origin for instinct, *viz.*, natural selection and "lapsed intelligence".[2]

§ 10. *Instinctive Movement and Emotive Reaction.* One other form of primitive or unacquired conscious movement has to be referred to, that already dealt with in part as Emotive Reaction, or Expressive Movement. In treating of feeling it was pointed out that all excitation of pleasure or of pain has for its direct psycho-physical result movement, including that of the "voluntary" organs. This movement is in part general or diffused, though, as development advances, it becomes specialised into a variety of well-defined *expressive* reactions, *e.g.*, the movements in anger, tender emotion.

Since these movements are determined from the first by

---

[1] Since instinctive movements are primitive and unacquired, the supposition that they are consciously intelligent is beset with all the difficulties attending the hypothesis of transmitted psychical elements and hereditary memory.

[2] On the difficult question of the kinship and genealogical relation of instinct and intelligence, see Romanes, *Mental Evolution in Animals*, chap. xi. and following, especially chaps. xiii. and xvii.; *cf.* also Schneider, *Der Menschliche Wille*, pt. ii.; James, *Principles of Psychology*, ii. chap. xxiv. and chap. xxviii. p. 678 ff.

congenital nervous arrangements, they obviously have a close affinity to the group of instinctive movements, which, as we have just seen, are also preceded by some element of feeling. Indeed, some writers deal with such connate expressive movements as a variety of instinct. At the beginning of life, more particularly, before the conative and the affective manifestations have become clearly differentiated, the regions of instinctive and expressive movement are not easily marked off one from another. Thus the crying of the hungry infant seems at once the expression of distress, and the effect of a *quasi*-conative impulse, the craving begotten of appetite. In proportion, however, as this differentiation progresses, we see that instinctive movement (in the narrow sense) comes to be distinguished from expressive movement by its clearly marked *purposive* aspect.[1]

§ 11. *Genesis of Voluntary Movement: Random Movement as Starting-point in Volitional Development.* The various groups of primitive or unacquired movement just described would suffice to bring into play the " voluntary " motor mechanism, and so supply the active consciousness or experience of active movement. And this, aided by the reflective or reproductive power, would, it is evident, contribute one important factor to the production of voluntary movement, *viz.*, that motor idea or representation which, as we saw, is one ingredient in its psychical initiation.

We have seen, however, that before this experience of movement could issue in a volitional process something else is necessary, *viz.*, the experience of certain *results* of movement, as favourable or of benefit to the child, that is to say, as bringing him satisfaction or pleasure in some way, and the association of these results with particular varieties of movement. We have now to inquire how this more complicated experience, *viz.*, that of pleasure-producing movement, may arise.

---

[1] It is to be added that this differentiation tends to be counteracted by the transformation of once conative, purposive actions into emotive expressions. Thus a girl's sudden act of turning and looking round in the midst of writing a letter when overtaken with a difficulty, though appearing to be a purposive appeal for help, may be merely the expressive survival of early purposive actions. (*Cf.* above, p. 69.) For a good account of the several classes of early movement of the human being, see Preyer, *op. cit.*, cap. viii. *et seq.*; *cf.* Lotze, *Medicinische Psychologie*, buch ii. kap. iii. § 24.

It is evident that this experience will be forthcoming when by an accidental coincidence a movement involuntarily carried out in one of the ways described above brings about a favourable change in the child's condition, whether this be to remove or lessen discomfort or to introduce a positive element of pleasure of a sufficient amount to excite the child's attention to the sequence. It becomes important, then, to ask how far such coincidences are likely to occur. With respect to Random Movement, it has been supposed by some, especially Bain, that owing to its plenitude, and the diversity of directions it takes, it would again and again lead to such noticeable additions of pleasure. It is to be remarked, however, that random movement, since it does not presuppose an antecedent state of discomfort, would have to depend altogether upon the production of positive pleasure. The later view of this group of movements, which regards them as few in number and restricted in their character, does not favour the supposition that they constitute a considerable factor in the development of voluntary movement. At the same time the organic fact made prominent in Bain's theory of spontaneous movement, *viz.*, that the fulness and variety of our muscular action are determined throughout by the condition of the motor apparatus, central and peripheral, is a fact of considerable importance in tracing the development of the voluntary powers.

Bain's theory of spontaneous movement might perhaps be modified by regarding such movement as preceded by a vague form of feeling, *viz.*, pleasurable consciousness in the shape of a feeling of vitality or vigour, and in this way his doctrine might be assimilated to that other theory (to be considered presently) which derives voluntary from feeling-prompted movement. The other part of his doctrine, *viz.*, that such spontaneous movement as happened to bring pleasure would thereby be sustained and prolonged (according to the law of self-conservation, which says that pleasure is attended by a heightening of vital functions generally, and so would tend to further any motor activity going on at the moment), seems of less psychological consequence. This action of a heightened vitality would involve also the excitation of other non-useful, including some *obstructive*, movements, so that the loss from a conative point of view would presumably compensate the gain. The persistence or following up of the happy hit would be effected by a *psychical* cause, *viz.*, a true though crude germ of true conative selection. But of this more presently.[1]

[1] For an account of Bain's theory of the genesis of voluntary movement, see *The Emotions and the Will* (The Will), chap. i.

§ 12. *Reflex Movement as Precursor of Volitional.* Let us now turn to the second class, reflex or sensori-motor movements. These are marked off by the important circumstance that they are initiated by sensations, and that they thus disclose a particular state of things in the environment to which the responsive movement may be adapted. They would seem, then, to offer a more promising field for the development of ' happy hits ' than the vague possibilities of random movement.

Here, however, a distinction must be drawn. So far as the reflex is from the first a specialised reaction organically connected with the particular psycho-physical process of the sensation, it remains what it was originally, at best but a sub-conscious psychical phenomenon. These movements, though, as Preyer has shown, they may be taken up into voluntary ones by a purposive on-bringing of the sensational stimulus, as in the complex movements of grasping, do not rise to the level of distinctly conscious phenomena.

On the other hand, the more diffused or scattered type of reflex movement offers a better starting-point. Here there seems to be some room for the occurrence of a happy coincidence. Thus the movements caused by a sudden sound may lead to a turning of the head in the direction of the sound, and so bring about a new pleasure. At the same time, the very fugitiveness of reflex movements would prevent their being of much use in this way. It is only when the excitant sensation becomes distinct and persistent in consciousness, and when it is accompanied by an appreciable element of feeling, *e.g.*, when the child is confronted with and mentally roused by a bright light, that the resulting movements are likely to issue in favourable changes of the child's feeling. This leads us on to consider the conative aspect of the third group of movements.

The theory that reflex action is the starting-point in the development of voluntary movement is based mainly on physiological grounds. Thus, from a physiological point of view, the movements of the lowest organisms, as well as those carried out by means of the lower nerve-centres in the case of higher organisms, and constituting their earliest movements, are of a reflex type. Not only so, conscious intelligent actions themselves can be assimilated on the neural side to this type by supposing the central process intervening between sensory stimulation and motor reaction expanded and complicated by the addition of the higher circuitous arc. Hence we find that writers on physiological psychology are apt to take reflex movement as their starting-point. This applies (with certain qualifications at least) to Lotze, and more manifestly to Herbert Spencer, and to some later writers, as Münsterberg. That there is a certain amount of *psychological* soundness also in this view will appear presently. Only it must be borne in mind that, when the psychologist takes reflex action as his fundamental type, he does not mean specialised or unconscious reflexes, but movements of a sensori-motor form with an intervening element of feeling.[1]

§ 13. *Transition from Instinctive to Voluntary Movement.* It is to the group of instinctive movements, with which we may take the early undifferentiated expressive movements, that we must look for a true starting-point in the development of voluntary action in the human individual. Here we have among the psychical antecedents of the movement a feeling of pleasure or of pain, a circumstance of great importance in this connexion, and one which in itself brings the type of movement much nearer to that of the voluntary action. Not only so, we find in the element of impulse which is so marked a constituent of instinctive action a distinctly active element which forms the analogue, and in a sense the true genetical antecedent, of the conscious pursuit of an end.

It has been pointed out above, in connexion alike with expressive and with instinctive movement, that all feeling originally tends to excite the "voluntary" muscles to action, the range of this effect varying with the intensity of the feeling. We can easily observe that during states of feeling the infant carries out a number of movements of the limbs, the head, etc. It is this type of wide-ranging, unspecialised, feeling-prompted movement which supplies the nucleus of a truly volitional action. Out of the variety of movement thus arising certain elements will modify the pre-existing feeling. Thus, if the child is expressing a feeling of pleasure, the movement may react by prolonging and intensifying the pleasure, as in producing agreeable sound from a toy. If, on the other hand, it is giving vent to a feeling of discomfort or distress, some of the move-

---

[1] On this point consult Lotze, *Medicin. Psychologie*, p. 289 ff.; Spencer, *op. cit.*, vol. ii. pt. iv. chap. ix.; Münsterberg, *Willenshandlung*, iii.; and Ziehen, *Leitfaden der physiol. Psychologie*, p. 14 ff. It is, however, one thing to say that reflex movements do not supply a starting-point in the genesis of voluntary movement, another to argue, with Wundt and Ward, that all reflexes are to be viewed by the psychologist as a product of volition, that is, as analogous to the "secondarily automatic" actions of man. (See Wundt, *op. cit.*, ii. 496; Ward, art. "Psychology" (*En. Brit.*), p. 43, col. a.)

ments called forth may tend to relieve the pain. In this way, for example, it might hit on the movements which relieve cramp in the limb, which banish the feeling of cold by bringing it nearer the mother's body, and so forth. Attention to these changes in their connexion with the particular movements bringing them about, which would be secured by the deeply interesting nature of the changes, would serve to fix them in the memory. In this way an opening would be supplied for the instinctive prompting referred to above, viz., the seeking of the beneficial or pleasurable and the avoidance of the hurtful or painful. Hereafter, through the recalling of the sequence, the child would be able to represent beforehand the relief or on-coming of pleasure, and also to consciously initiate the appropriate movement by an idea of the same.

From this brief sketch we can see that, given the instinctive impulse to seek pleasure and avoid pain, and a sufficient amount and variety of the experience of feeling-prompted movement with its reflex effects on the prompting feeling, voluntary movements might arise by a process closely analogous to that of natural selection. That is to say, particular movements among a miscellaneous group would be consciously selected and preserved because they were found to be beneficial or useful to the agent.

We have here assumed that the field of selection is as wide as the "voluntary" muscular system, that whatever the nature of the original feeling, it is just as likely to prompt to one kind of movement as another. But, as we have seen, this is not the case. The child is endowed at the first with more or less specialised impulses or instinctive promptings which appear to stand in an organic relation to particular groups of movements. Thus hunger, which in its active form is known as the nutritive impulse, begets along with a weaker and more diffused excitation of movement a specially energetic excitation of movements of particular organs, viz., those fitted to bring about the satisfaction of the impulse.[1] Still more clearly is this original restriction and specialisation of the motor excitation seen in the case of those instinctive locomotor movements (alter-

---

[1] At least this is true of very early periods. Of course it is difficult to exclude in this case the effect of experience.

native movements of the legs) which are produced by bringing
the soles of a baby's feet in contact with one's lap.

So far as such special impulses or instincts are at work in
the case of the child, they must, it is evident, influence the result.
To begin with, they would serve greatly to expedite the process
of selective adjustment. By supplying through congenital
organic arrangements particular preferential lines of motor discharge they would spare the child the need of wide-reaching
tentative. But this is not all. The instinctive prompting has
in it, as we have seen, an element of vague craving, which
shows itself when not immediately satisfied in characteristic and
energetic muscular effects. Thus the child when hungry is
specially predisposed to carry out the movements appropriate
to this craving, and to find a peculiarly intense gratification in
satisfying its active impulse.[1] Hence the eagerness with which
the movements are seized, and the energy with which they are
maintained during the prompting of the appetite. All this would,
it is evident, greatly favour the selection and retention of the
appropriate movements. In other words, in proportion as the
element of instinctive prompting (Trieb) is powerful, and its
direction specialised, the process of volitional acquisition is
shortened.

§ 13a. *Genetic Relation of Instinctive Impulse to Volition.* It is a question of
dispute how far special instinctive impulses, other than the general impulse to seek
pleasure and avoid pain, enter into voluntary action. The English associational
psychologists have been wont to overlook the co-operation of specialised active
impulses, and motor dispositions. This, however, is to take too abstract a view
of volition, and to ignore the genetic continuity of human and animal action.
On the other hand, recent writers, under the influence of the idea of evolutional
continuity, incline to the view that *all* voluntary movement arises out of such
special instinctive prompting. Thus it is said that the instinctive impulse forms
the psychical root of a large part of human action, *viz.*, all that which has to do
directly or indirectly with the satisfaction of the bodily wants. This view naturally
connects itself with the doctrine already touched, on that pleasure is a secondary
phenomenon based on conation. According to this way of envisaging the matter,
the child gets pleasure as a result of satisfying his active impulses ; he does not first
experience pleasure, and then begin to act with a view to its fruition.

That there is some truth in this contention has been admitted. Instinctive
promptings of special kinds exist antecedently to experiences of particular pleasures, *e.g.*, those of eating. The conscious desire for the pleasures of the table is a
secondary phenomenon superinduced and largely based upon the active prompting.
Yet to make such special instinctive prompting the *sole* starting-point in volitional

---

[1] *Cf.* what was said above (p. 17) in connexion with the pleasure of activity.

development is to take a totally inadequate estimate of the complexity, the many-rootedness of human action; to overlook, for example, the whole field of the higher sense-feelings, as well as the wide group of emotions, such as the love of novelty and beauty, which has, from the first, as much to do with the prompting of human action as the desire to keep the wolf from the door. The attempt to derive all kinds of voluntary action from special instinctive roots leads to an almost grotesque endowment of the child with such problematic impulses as secretiveness.[1] This tracing back of all human action to special instinctive promptings is, moreover, opposed to the biological conception of the human brain, and its indefinite modifiability or educability. That we should come into the world endowed with instinctive tendencies to perform the functional activities necessary to life is what we should expect on biological grounds. But to suppose that a child is born with a special instinctive prompting to pull things to pieces, to run after another child that starts running hard by, and so forth, is to try to rob nature of her reputation, well-earned in the main, for economy or thrift. The child is perfectly well able to find his way to the delights of destruction through the rich prompting of muscular activity, and the impulse of curiosity with which nature has undoubtedly supplied him. In like manner, his love of bodily exercise, his imitative impulse, and other known feelings of childhood, are quite sufficient to conduct him to the sportive pleasures of the mimic chase.[2]

We may now pass from the sub-volitional domain and confine our attention to the process of voluntary action itself. We have to inquire a little more closely into the volitional process, and to explain the nature and development of each of its constituent factors.

§ 14. *The Factors of Voluntary Action.* The process involved in a voluntary action may be best seen by means of a simple example. The child has tasted an orange. You offer him another, and he puts out his hand and takes it. The psychical event in this case seems to consist of the following stages. The complex of visual sensations supplied by the orange suggests, according to the law of contiguous association, the representation of the taste and the pleasure accompanying this. This representation of a pleasurable experience closely connected in time with an actual presentation excites the state of desire. That is, the child craves a renewed enjoyment of the orange-sucking.

---

[1] This grotesqueness seems to me to be illustrated in James's list of human instincts. (*Op. cit.*, ii. p. 432 f.)

[2] On the genetic relation of instinct to voluntary action, see Wundt, *op. cit.*, cap. xx. i. Wundt criticises (on grounds which, I confess, are not clear to me) the hypothesis adopted above, that the child finds his way to a large part of his voluntary pursuits by what we must call accidental coincidences. The reference of all voluntary action to original promptings is implied in the scheme of impulses (Triebe) drawn up by Schneider and his successors.

The idea of the succulent pleasure-giving orange, fixed and sustained in the state of desire, suggests in its turn (also by associative reproduction) a particular action or series of movements by means of which the pleasure may be realised. Here we have the representation of certain movements, the last psychical antecedent of the actual (objective) movement. The ensuing physiological process, the innervation of the muscles, lies outside the psychical domain. Lastly, there is the stage of realisation of end or accomplishment, *viz.*, the active consciousness correlated with the motor or muscular process, and the substitution of the real experience of sucking for the representation of the same. This, however, is not so much a phase of the volitional process as its psychical consequence under normal and favourable conditions.

We may now consider more fully each of the main factors in the volitional process, *viz.*, the highly complex state of desire and foregrasping of end ; and, secondly, the factor of motor representation regarded as a conative phenomenon.

DESIRE.

The phenomenon known as desire has, as we have seen, its dim prototype in instinctive impulse. It constitutes in a greater or less degree of intensity an integral element in all true conative processes. In a sense it is prior to and independent of all definite forms of conscious action itself. There is a large region of desire which does not issue in a complete volitional process. It stands, moreover, in a peculiarly close relation to feeling. Hence certain writers deal with it before coming on to volitional action. Since, however, desire only assumes a clear and definite form as experience, including active experience, advances, it is best on the whole to discuss it after the earlier and sub-conscious stages of action.

§ 15. *The Analysis of Desire.* (1) Since all definite desire is of some object or perceptible result, one obvious element in the psychical state is an idea or *representation*. When a child desires an object, say an orange, or a playmate's society, he is imagining this object as actually present or realised. In this way all desire is related to the *intellectual* side of mind. Where there is no knowledge there can be no desire. We must have had experiences and be able to recall these with some degree of clearness before we can have a desire for new and similar ones. Our desires multiply as our experience widens and grows more varied.

The representation involved in desire may be either a mere reproduction of a past experience, as when the child desires the orange, or may involve in addition a constructive process. We are able to desire things of which as yet we have had no fruition, provided that they resemble actual gratifications closely enough to allow of our forming the necessary images. Hence desire is apt to accompany not only the recallings of past personal experience, but the imagination of untried situations, as in listening to others' recitals, in reading, in weaving images of possible futures, and so forth.[1]

It is to be noted that the representative element in desire differs from that in intellectual imagination. In desiring a thing, even an objective sense-presentation, *e.g.*, in longing to visit a country in order to see its scenery or its art, there is a reference, more or less distinct, to the self. That is to say, we desire things not as mere objects (as we *know* things) but in their relation to ourselves, and *as affecting our condition*. This ingredient of self-consciousness becomes particularly distinct and prominent in certain forms of desire, as where we wish to possess a thing, to rise in the social scale, and so forth. We may thus say that while the objects of intellectual imagination are believed in as *present existences*, the objects of desire are always thought of as non-existent, that is, as unrealised situations or experiences of the self.

(2) A closer inspection shows us that all representations do not excite desire. Many images, *e.g.*, those of familiar objects in our surroundings, other people's doings, and the like, may arise without any appreciable accompaniment of mental craving or desire. This peculiar psychical state is only aroused by the representation of objects so far as they excite our *feeling*, and more particularly are thought of as fitted to benefit us or bring us pleasure. In desiring a succulent fruit a child represents the delight of eating it : in desiring a good social position

---

[1] Whether desire may be excited by general or abstract ideas is a point that had better be reserved for the present.

*

or a high reputation a man represents the coveted situation on its pleasurable side.

Now we have seen that the representation of something pleasurable has itself a pleasurable tone. In mentally fore-casting the incidents of a coming tour abroad we have an ideal 'sip' of the actual pleasure. But in ordinary cases this ideal element is greatly inferior to the reality, and is recognised as such. And this consciousness of inferiority lies at the very root of the state of desire or craving; for to desire a thing is to experience or feel the absence or *want* of this thing. As soon as this sense of discrepancy between the actual and the ideal or imagined state of things disappears desire expires. In intense expectation, in the vivid imagination of unattained delights, as in reading a work of fiction, and in absorbing states of moral and religious aspiration, desire succumbs, giving place to a momentary illusory sense of actual fruition. This sufficing pleasure of vivid imagination is the prerogative of those gifted with high imaginative powers. Children often experience it. Thus George Sand has a delightful account of the perfect satisfaction, free from all desire, which she got as a child from listening to the nursery ditty :—

> Allons dans la grange
> Voir la poule blanche
> Qui pond un bel oeuf d'argent
> Pour ce cher petit enfant.[1]

It seems paradoxical at first to speak of the representation of a pleasurable experience which is aware of its own shortcoming. It might appear as if we necessarily realise our object in the measure of completeness in which we represent it. But this sense of non-realisation in desire is by no means a solitary mental phenomenon. In memory, for example, we are aware of the inferiority of the present representative image to the past presentation for which it stands, and this recognition is an essential element in our belief in a past.[2] In other words, the mental image is attended by a peculiar mental state or feeling, namely, the assur-ance that there *was* something more, unrealised at the moment. It has, moreover, been pointed out above that in representing a class of objects by means of a "general idea" there is a similar mental concomitant, *viz.*, an awareness of the existence of something only very imperfectly realised at the moment.[3]

---

[1] *Histoire de ma vie*, ii. pp. 154, 155.

[2] *Cf.* above, I. p. 280 f.

[3] See, on the whole subject of such vague accompaniments of our more distinct mental states, James, *Principles of Psychology*, i. chap. ix.

The relation between feeling and desire here brought out is a particularly close one, and deserves a word or two of further explanation. We have seen that, though in its abstract purity a passive phenomenon, feeling vents itself in movement, and to this extent has an active accompaniment. It may now be added that all our feelings, whether in the actual or ideal form, tend to excite the psychical state, desire. Not only does the representation of a pleasure wholly unrealised at the moment arouse this craving, the actual experience of pleasure appears in most cases to beget something like a desire for its prolongation, if not also for its intensification. This excitation of desire by feeling is, however, much more apparent in the case of our pains. As hinted above in the account of primitive movement, actual suffering produces a restlessness, an appearance of a craving for, and a striving after, relief. Actual pains are, indeed, quite as much prompters of desire (that is, of relief) as imagined ones.

The excitation of desire in connexion with an actual pleasure is probably a similar process to that involved in its excitation by a representation of pleasure. Enjoyment so far as complete, that is, considered simply in itself, is not desire-provoking but satisfying or quieting.[1] But no enjoyment remains long at one and the same level of intensity. As we saw above, the prolongation of any pleasurable stimulation tends to diminish its effect. And it is probably *the sense of a falling-off* which is the real excitant of the ever-renewed desire which we commonly find in these circumstances.

While feeling is thus an antecedent and main condition of desire, the latter state contributes in its turn new elements of feeling. As pointed out above, one great class of pains are those of want or craving. The root-element of desire, the sense of the inferiority of the ideal to the actual, is distinctly painful, and when the state is fully developed, that is to say, is not immediately replaced by its satisfaction, the painful ingredient may become intense. We thus see that desire, viewed as an affective state, is a complex phenomenon, in which a pleasurable element, the accompaniment of the representation, is opposed to, and in conflict with, a painful element, the sense of deficiency or shortcoming, which last grows more intense, and may ' quench ' the

---

[1] Mr. Stephen expresses this by saying that pleasure is a state of equilibrium, or one in which there is a tendency to persist. (*Science of Ethic*, chap. ii. § 12.)

pleasurable element if the state of non-realisation is unduly prolonged.[1]

(3) While desire thus stands in relation to each of the two other phases of mind, it is sufficiently marked off as an *active* phenomenon.   It is in virtue of this characteristic that it constitutes the connecting link between knowing and feeling on the one side, and willing on the other.   In desiring a thing, say an approaching holiday, we are in a state of active tension, as if striving to aid the realisation of that which is only represented at the moment, and recognised as such.   This innermost core of desire has been variously described as a movement of the mind (*e.g.*, by Aristotle), and more commonly as a striving towards the fruition or realisation of the object.

This element of active prompting in desire appears under each of the two phases which, as we have seen, are always present in our active states, *viz.*, attention, and muscular consciousness.

It is evident, in the first place, that in desiring a thing, as a position, a prize, our attention is closely fixed and concentrated on the idea.   In the degree in which the idea is interesting and exciting, so will it tend to persist and monopolise consciousness.

This calling forth of a strong reaction will in itself, conformably to what has been said above, give the colouring of active consciousness to the state of intent desire.   But there is more than this.   The direction of the attention to an idea tends, as we have seen, to develop and intensify this idea.   Now, so far as this becomes a conscious process, we have, it is obvious, a new and very important ingredient of activity.   In fixating the agreeable idea we tend to pass insensibly into a state of *striving towards an end*, *viz.*, an intensification, or fuller degree of realisation of that which is desired, and so recognised as not yet fully realised.   We thus see that there is in the very process of mental concentration, as soon as this becomes consciously directed to the representation of something agreeable and desirable,[2] the germ of a purposive activity, the striving towards an end.[3]

---

[1] The relation of desire to feeling is carefully discussed by Volkmann, *Lehrbuch der Psychologie*, vol. ii. § 143.

[2] The student of ethics need not be reminded that I am not here employing the term desirable in its ethical sense, *viz.*, *worthy* of desire, but merely as a convenient psychological expression for that which is, as a matter of fact, desired.

[3] This process is variously conceived by different writers.   The Herbartians

In the second place, as we have already seen in connexion with the phenomena of sub-conscious desire (impulse or organic appetite), desire involves an accompaniment of muscular activity and the correlated " active consciousness ". To desire is to be incipiently active, to be stirred to muscular exertion. Desire and complete muscular inaction are incompatible, and the motor agitation in this case seems differentiated from that which accompanies states of passive feeling.   It has the look, at least, of incipient, tentative reaching out towards attainment. This appearance of striving is most distinct, of course, in cases where experience has taught the mind that a certain mode of action leads to the realisation of desire.   But, antecedently to the teachings of experience, we see desire prompting to that mode of incipient, ill-directed movement which is the indication of the restlessness or dissatisfaction inherent in the state.

§ 16. *Desire and Aversion.*   The great contrast in the region of feeling between pleasure and pain has its counterpart in the domain of activity.   While the representation of what is pleasurable excites the positive form of desire, that is, longing to realise, the representation of what is painful awakens the negative form of aversion, or the longing to be rid of.   We strive *towards* what gives us pleasure, and *away from* what gives us pain.   If the pain be an actual experience of the moment, aversion takes the form of craving for relief, a form of desire which, as has been pointed out, seems to be the most primitive. If, on the other hand, a pain be merely imagined, the aversion assumes the aspect of a mental recoil or shrinking back.

Here, again, we may connect the active phase of desire with the process of attention.   Just as positive desire for what is grateful involves an exertion of the attention, with a more or less distinct purpose to fix, intensify and fully realise the agreeable presentation, so the recoil of aversion appears to involve a *withdrawal* of attention from the ungrateful presentation.   Thus the longing for relief from a present actual suffering is a casting

speak of this tendency of the idea to persist and grow distinct and vivid as if it were the result of the idea's own activity (self-assertion of the *Vorstellung* against limitation or hindrance). Wundt, in his somewhat peculiar doctrine of Apperception, finds the root of all volition in this ' striving ' aspect of attention. (*Cf.* Volkmann, *op. cit.*, § 139 ff., and Wundt, *op. cit.*, cap. xviii. i.)

about for other ideas which shall displace the painful presentation. Still more plainly is the shrinking back from an anticipated pain a *rejection*, that is, an extrusion from consciousness by an act of attention, of the painful idea. This withdrawal of attention, as we shall presently see, directly favours the mental fixation of any ideas of movement which promise to contribute to the desired relief or avoidance.

There seems, at first sight, a contradiction between the idea that pain evokes an expulsive movement of attention, and the fact that attention is frequently detained by painful ideas in what is known as fascination. Such fascination, however, as in watching something horrible, always seems to depend upon the presence of an element of *pleasurable* excitement. To this extent the action conforms to the general psychological rule. At the same time, in morbid states of mind, in which there is an unwholesome bent towards what is painful, a phenomenon not unknown among the " cultured " classes, we have to recognise an interference with what we understand by the normal volitional process. The entertaining of distinctly painful ideas *for their own sake*, that is, without any desire to get a secondary result of pleasurable consciousness from so doing, is seen by all to be non-volitional and abnormal.

It will be noted that while positive desire always points directly to the absent, or non-realised, negative desire may point primarily to that which is present and actual. In the primordial impulse to get away from pain the representation is that of putting an end to what is actual. It may be added that even when shrinking from a possible suffering we appear first to represent it as an actual state, and then to direct our thoughts to getting rid of it.[1] Again, the difference between positive desire and aversion carries with it as a further consequence a dissimilarity in the mode of arousing action, a point which deserves a moment's attention.

Pain, as that which we dislike and strive to get away from, is a particularly powerful stimulus to action. The child when in pain moves its limbs, twists its body and head, cries, and so forth. There are evident biological grounds for this phenomenon: if the painful, *i.e.*, the injurious, did not rouse us to exertion we should perish.[2] Hence the general belief in the urgent efficacy of pain as a motive, a belief plainly attested by our systems of scholastic and criminal discipline.

While, however, the desire to get rid of pain thus evokes, in a special manner, all the resources of active energy, its effect is definitely limited in its range or extent. As soon as the pain is got rid of the impulse to action ceases. Where, on the other hand, the motive is a positive desire for the pleasurable, its realisation cannot end in this abrupt manner ; for, however great the fruition, we can go on desiring an increase, or at least a prolongation of the enjoyment.

In thus contrasting positive desire and aversion, we must not, however, regard the distinction as an absolute one. It was pointed out above that relief

---

[1] Perhaps this is what Waitz means when he says that aversion involves belief in the reality of the pain, whereas desire involves no corresponding belief in the reality of the pleasure. (*Lehrbuch der Psychologie*, § 42, p. 443.)

[2] This superior biological significance or utility of pain was discerned by Descartes in his *Traité des Passions de l'Âme.*

from pain is itself a condition of pleasure, and in craving for exemption from suffering we are virtually, and more or less consciously, desiring the *pleasure* of relief. Thus in physical suffering we often distinctly represent to ourselves the delicious sense of cessation. This fusion of desire and aversion is strikingly illustrated in the case of "relative" pleasures, as liberty and health. In desiring these we are at once shrinking from pain, and desiring the pleasure of the transition to the opposite condition; and generally, since most of our pleasures and pains stand in certain relations one to another as opposite modes of one and the same kind of experience, *e.g.*, muscular activity and fatigue, good and bad combinations of tones and colours, pride and humiliation, and so forth, in desiring relief from the painful experience, we tend to represent, with some degree of distinctness, the correlative pleasurable experience, and conversely, in desiring the latter, to represent deliverance from the former.

Lastly, a bare reference may be made to the fact that since all desire, when prolonged and fully developed, involves a painful element, the longing for anything as good or pleasure-bringing tends, where the non-realisation is delayed, to be accompanied by aversion in the form of a struggling away from *the torment of desire itself.* The impatient longing for a thing after prolonged expectation, *e.g.*, the rise of the curtain in the theatre, the arrival of a long-absent friend, involves the conscious wish to get rid of the pain of craving. Such aversion constitutes, it is evident, a secondary form of desire which at times eclipses the primary desire for a positive gratification.[1]

§ 16a. *Desire and Pleasure.* We are now at a stage at which we can profitably refer to the much-disputed point whether in desiring a thing we are consciously representing the attainment of a pleasure.

That we do actually desire many things, situations, activities, so far as they appear pleasurable, seems to be generally accepted. In desiring, for example, fine weather, success in our work, the approval and sympathy of others, and so forth, we are certainly representing things which, if realised, will bring pleasure, and we appear moreover to be desiring them *as pleasurable*.

That desire, which is incipient action, should direct itself to pleasure and away from pain is, further, what we should expect on biological grounds. It has been shown that our pleasures tend to coincide with what is beneficial to us, pain with what is hurtful.[2] This consilience acquires a biological significance on the supposition that in conscious volitional processes we act for the sake of pleasure or freedom from pain. That is to say, it is assumed that, owing to our deep-rooted and unalterable impulse to seek the pleasurable, the correspondence between the pleasurable and the beneficial will tend to be brought about by the processes of individual variation and natural selection. From this point of view we should regard as normal and healthy, desires for things at once pleasurable and beneficial; as abnormal or morbid, desires for things pleasurable but injurious, and still more (if such a thing is possible) desires for what is painful as well as hurtful, *e.g.*, ghastly horrors.

While, however, common observation and scientific theory alike suggest the

---

[1] It might be thought that a desire *to escape desire* would inevitably lead not to a strengthening of the primary craving, but to the extinguishing of this by putting away the idea of the coveted object. In many cases, however, this may be impossible, owing to the fixity of the exciting idea. In other cases, as in delayed though still confident expectation, it would be unreasonable.

[2] See pp. 18, 20, and 36 ff.

consilience of the desirable and the pleasurable, the general truth of the proposition has been denied by psychologists and by moralists. Thus it is said that we desire many things, *e.g.*, knowledge, virtue, others' happiness, without at the moment thinking of the attainment of the objects as pleasure-yielding. We shall have to deal with the matter again at the end of our whole treatment of the volitional process. Here it will be enough to point out one or two considerations which appear to throw light on the difficulties touched on and yet to save the general truth of the proposition that what we represent in desire is something pleasure-bringing.

The first and most obvious consideration here is that, as we saw above, feeling is the comcomitant (or, as some would say, the affective aspect or tone) of certain presentative elements. We only attend to a feeling by attending, in a manner at least, to its presentative base.[1] It follows that there is no such thing as a representation of *mere* pleasure in desire. In desiring the pleasure of a friend's companionship or the success of a business venture my attention is directed to an ideational complex, corresponding to certain objective facts or circumstances in their relation to myself. This, however, does not affect the truth of the contention that the desire is awakened by such representations *qua* pleasurable, and in the measure in which they appear to us at the moment to promise pleasure.[2]

Closely connected with this circumstance is another, *viz.*, that since most of our pleasures are found to follow from the realisation of certain previous conditions, as, for example, health, a sufficiency of wealth, friendship, we come with growing intelligence to make the attainment of these conditions more and more the object of representation in desire. Hence in this case the idea of a realisation of pleasurable experience falls back into indistinct consciousness, save so far as the attainment of these conditions itself comes, according to the principle of associational transference spoken of above, to be regarded as pleasurable. This fact will be more fully illustrated presently, when we discuss the nature of the higher motives to action.

The general truth of the proposition here insisted on, that we desire the pleasurable, and in proportion to the amount of pleasurableness with which we invest the object at the moment of representation, is modified by other features of desire. It has already been pointed out that this phenomenon, as active, owes its intensity in part to organic conditions. These may be sai to consist, first of all, of a strong general disposition to activity, the outcome of a healthy, vigorous motor system ; and, secondly, of specialised instinctive or acquired tendencies to particular actions under particular circumstances. The vigorous youth eager for activity is wont to invest the " object of desire " in all kinds of active pursuits, as games, with an exaggerated hedonic value. The desire for the pleasure of activity appears in this case to beget a *quasi*-illusory belief in the greater pleasurableness *of the result.* Similarly in the case where special instinctive impulses are presumably at work, as in that of the hunter, the soldier, and possibly the explorer and the *savant.* Here a powerful organically-conditioned impulse to a particular line of action leads to the heightening of the apparent pleasurableness of the object of desire (success of victory, fame, etc.). The effect on desire and impulse of acquired dispositions to particular actions is seen in what we call habit. Here we have a prolonged pursuit

---

[1] See what was said above, I. p. 77, on the difference between intellectual and feeling-serving attention to presentations.

[2] Of course this anticipation may be illusory in almost any degree. We may desire as pleasurable what will turn out pleasureless, and even positively disagreeable.

of, and apparently also a prolonged desire for, certain objects after the original pleasurableness has sensibly diminished. Thus a man may go on pursuing knowledge after the first youthful delight of study has passed, and the habitual drunkard is said to continue to crave the cup after he knows that indulgence means bitterness.[1]

Even in these cases, however, the general correspondence between strength of desire and represented pleasure is not disproven. We must remember that desire has its other aspect, *viz.*, aversion to the disagreeable, and that when we talk of desiring pleasure we must include with this the absence of the opposite state of pain. It is probable, as we saw above, that this aversion frequently comes in as a concomitant of the desire for pleasure. Since, as was previously pointed out also, instinctive impulses when unsatisfied are painful states, it is reasonable to infer that they commonly excite a secondary impulse of aversion or longing to escape from a painful condition. The same remark seems to apply to the phenomenon of habit-fixed desire, for habit, as we saw, when unfulfilled, tends to produce a painful sense of deprivation. So far as this urgency of pain underlies our impulses the view of Plato and the modern pessimists that action is an attempt to get away from a painful condition is justified. How far, however, desire is thus resolvable into a craving to escape from pain is doubtful. It has been assumed above that the pleasurableness of activity is increased by the fact of there being a strong organic disposition at work; and, so far as this consideration is valid, the desires of the sportsman, the soldier, and so forth, may be said to be hedonically rational in the sense that their intensity is proportionate to the depth of the *whole* pleasurable satisfaction ensuing.[2]

---

[1] The stubborn survival of this desire is finely suggested by Tennyson in *The Northern Cobbler*, where the self-cured drunkard, who keeps his enemy, a bottle of gin, before his eye as a warning, cannot help remembering that

" Fine an' meller 'e mun be by this, if I cared to taäste ".

[2] English psychologists have, on the whole, maintained that desire is for pleasure and the absence of pain. This view is ably set forth by Prof. Bain, *The Emotion and the Will*, pt. i. chap. viii. § 7. The question has played a prominent part in English ethical writings, owing to the natural connexion between it and the utilitarian theory of morals. On this side of the subject, see Prof. Sidgwick, *The Methods of Ethics*, chap. iv.; L. Stephen, *Science of Ethics*, chap. ii. § 11 and following; F. H. Bradley, *Ethical Studies*, iii.; and S. Alexander, *Moral Order and Progress*, bk. ii. chap. v. ii. One of the most recent psychological discussions of the subject goes astonishingly far in denying apparently all special connexion between pleasure and pain and volition. The volitional process, according to Prof. James, may be introduced by any idea which has a motor character and. so a movement-prompting influence, whether or no there be an accompaniment of feeling. (*Principles of Psychology*, ii. p. 549 ff.) This writer appears to ignore the problem of desire. It may be added that in Germany, as in England, psychological opinion is divided on the question. *Cf.*, for example, Waitz, *Lehrbuch der Psychologie*, § 40, who holds that what is desired is pleasure, and Volkmann (*op. cit.*, §§ 139-143), who opposes the old dictum that in desiring a thing we necessarily represent it as good or pleasurable ("nihil appetimus nisi sub specie boni"). *

§ 17. *Conditions of the Strength of Desire.* The state of desire admits of different degrees of strength or energy. Our desires range through all degrees of intensity and persistence, from vague, fugitive wishes, up to intense and. absorbing longings. These differences show themselves in various ways. Thus, a strong desire prompts to great and prolonged activity or exertion, whereas a weak one fails to do so. Again, strength of desire may be measured by the amount of pain incurred if the craving remains unsatisfied.

The most important circumstance determining the strength of desire or active prompting is the magnitude of the pleasure represented. In general it may be said that the greater the pleasure represented the stronger will be the desire, and the more energetic the current of active impulse. Thus a schoolboy's activity (mental and bodily) is roused to a much greater extent by the prospect of a whole holiday than by that of going home half-an-hour earlier than usual. At the same time, it is to be borne in mind that the representation of the desired object may not accurately correspond with the degree of the actual enjoyment. As philosophers, ancient and modern, have been wont to remind us, that which is near influences us, by way both of attraction and repulsion, more powerfully than that which is remote. The strength of a desire may thus be said to turn on the magnitude of the pleasure *as represented at the present moment.*

It follows from what was said above that by adequate representation in states of desire is not meant an intensity of the affective element approximating to that of the actual realisation. We may represent a pleasure, say that of a visit to a new country, as great without realising its full intensity. By combining what has just been said with what was said before, we may see that a strong desire involves, first, a sense of the *relative* magnitude of the represented pleasure, *i.e.*, the superiority of the reality to the representation; and, secondly, an appreciation of the *absolute* magnitude of the enjoyment desired, or, more correctly, its intensity as compared with that of other actual pleasures.

In addition to this effect of distance on the representation of pleasure there is another modifying circumstance, *viz.*, the superior force of pleasurable over that of painful ideas. There seems no doubt that, in the case of most men, the prospect of a pleasure counts for more than that of a pain. This is curiously illustrated in all games of chance, the readiness of people to wager, to take a stake in lotteries, and so forth. Here the pleasurable idea of gain extrudes all thought of *the more probable result of loss.* Similarly, when people continue to betake themselves to certain forms of social "entertainment" *on the chance* of finding enjoyment, after experience must have taught them that weariness is the much more probable result.

STRENGTH OF DESIRE. 207

This general principle that we desire things in the ratio of their imagined pleasurableness must be further qualified by one or two considerations. In the first place, it is to be remembered that a man is not at all times equally disposed to activity. A more powerful inducement is needed to stir active impulse when we are inactive and indolent than when we are strongly inclined to activity. The same thing shows itself in the case of different men. In indolent persons desire does not rise to the intensity of a real craving, or at least the craving does not take on the full active phase of restlessness or impulse. Such differences in the excitability of desire are probably connected with differences in the degree of vigour, and the consequent instability or readiness to discharge, of the central motor organs.

This disposition to muscular action seems to be more specially connected with well-recruited motor centres than with well-nourished peripheral organs, though of course the condition of the muscles themselves may react on the central structures. This disposition is an antecedent condition of a wide range of pleasurable activity: the more vigorous the motor organs and the more ready for work, the higher can the exercises be carried without becoming excessive and painful. The opposite state of active lethargy or indolence, on the other hand, corresponds with a restricted range of pleasurable activity, or, in other words, a wide range of excessive and effort-attended action. Hence, as we shall see by-and-by, the inclination to activity is commonly attended with a more or less distinct representation of the pleasure of the activity itself, as distinguished from that which constitutes the object of the primary desire. Similarly, indolence commonly implies a shrinking from a represented pain, that of excessive and fatiguing effort. It may be added that, though this readiness to act would directly strengthen merely the active outcome of the desire, it tends indirectly to strengthen the desire as a whole. A vigorous child, strongly disposed to act *somehow*, will, through the very working of this impulse, be more keen and persistent in entertaining any ideas of desirable objects which happen to present themselves. It follows from what has been said above that this same active disposition will secure an energetic carrying out of those processes of attention on which the stability and the effectiveness of desire depend.

One other modifying circumstance may be touched on, viz., that of volitional inertia or habit. The very fact of having erected an idea into an object of desire, and striven towards its realisation, generates a tendency to go on desiring and striving in this direction. This is seen in the persistence of prolonged and patient action, a valuable quality in the higher volition. It is seen also in the *recurring* pursuit of customary objects of desire, a pursuit which, as has been pointed out, may outlive the first intense experiences of pleasure.

§ 17. *Desire and Motive.* Hitherto we have dealt with desire merely as a state of craving without any reference to the nature of the desire as realisable or non-realisable. It is evident that we have many desires which do not go beyond this stage. There is the passing *wish* for this, that, and the other impossible thing, including even nascent longing for past enjoyments. Since unsatisfied desire is painful, an educated will seeks, among other things, to check all futile and unreasonable desires by *reflexion on their unattainability.* In this way there arises a tendency, never perfectly realised, but realised more and more as volitional development proceeds, to shut out all cravings that remain mere cravings. In other words, desires become restricted to *effectual desires*, that is, such as prompt to definite lines of action.

A desire when thus transformed into a practical incentive, or excitant to action, is what we call a motive.

A motive is thus a desire viewed in its relation to a particular represented action, to the carrying out of which it urges or prompts. The desire in this case ceases to be a vague, fluctuating state of longing, and becomes fixed and defined as an impulse to realise a definite concrete experience, *viz.*, the known and anticipated result of a particular action; or, since the object of desire is now foregrasped as the certain result of a particular active exertion, it assumes the form of the *end* of this action.

The terminology of desire is not quite fixed. Thus some writers distinguish wish from desire, regarding the latter as always involving some belief in the attainability of the object. Thus Münsterberg writes : "We can well *wish* to be near the stars, though we cannot *desire* them ".[1] It is true, no doubt, that what we commonly understand by 'desire' is a more or less definite and fixed state of longing, whereas 'wish' marks off the *nascent* desires which are only momentary, being instantly dismissed *as futile.* Whether, however, in all persistent states of desire we do consciously believe in the attainability of the object is exceedingly doubtful. No doubt, as explained above, the vivid imagination, which forms a central element of desire, carries with it a measure of vague belief. In desiring strongly we are, it is obvious, foregrasping an unrealised situation, and thus far thinking of this as a possi-

[1] *Die Willenshandlung*, p. 94. *Cf.* Höffding, *Psychology*, vii. B. 1*b.* It is to be noted that Aristotle distinguished desire from volition by means of the same differentia. He held that we may desire the impossible, though we cannot will it. (See Rabier, *Leçons de Philosophie*, p. 524 ff.)

bility. Also, it is evident that any suggestion of (absolute) unattainability tends to cool the ardour of desire.[1] We may thus say that all stable states of desire imply a vague rudiment of belief in the possibility of attainment, though there need be no representation of the way or ways by which attainment may be brought about.

§ 18. *Motor Ideas as Constituents in Volition.* We may now pass to the other factor in conation, *viz.*, Motor Representations. We have already, to some extent, examined into the nature of these motor ideas. We have now to view them under a new aspect as a factor in the volitional process.

In representing a movement, say that of throwing in a cricket-ball to the wicket, we have, as we have seen, to do with a complex experience. A conscious movement is a complex of sensations, passive and active, and a complete idea of the movement would, of course, involve all these constituents. Not only so, a movement is not only a sum of muscle, surface, joint, and other sensations connected with the particular moving organ, it is also a visual percept. When we move our arm we see the movement and its immediate result as a change in our visual field. In fully representing the act of throwing in the cricket-ball we should include the visual image of the moving arm, and of the ball flying into space.[2]

A little reflexion shows, however, that though we can thus represent a movement in this complete way before initiating it, we rarely, if ever, do so. The forecasting which suffices to bring about our ordinary movements is too rapid and fugitive to allow all of its aspects to be imaginatively realised. Moreover, even when the idea of movement becomes more fixed and is definitely attended to, the restrictive and selective action of attention tends to bring about a special imaginative summoning of a particular phase. Thus a cricketer throwing in a ball from the field would attend specially to the amount of exertion (as

---

[1] An apparent exception to this generalisation is that after desiring an object as realisable we may go on desiring it *more* keenly when we begin to discern its unattainability. Thus, on a recent visit to Rouen, on perceiving that my time did not allow me to see one of the smaller churches, I found myself setting a fictitious value on this comparatively unimportant feature in my plan. This is analogous to the sudden development of " hunger " in a child for a cake when you propose to take it from him. Here the withdrawal of the object seems to intensify desire by exciting a more vivid consciousness of non-realisation.

[2] In the case of vocal and other sound-producing actions, as clapping the hands, the idea of the *auditory* effect would take the place of the visual image.

also the direction of movement) needed by the particular position of the wicket, whereas a boy throwing a stone into a pond
would have his attention mainly directed to the idea of the
coming splash.   It will be found presently that in this way the
precise character of the motor antecedent in volitional action
varies considerably.

One thing, however, seems clear.   There is, in all cases of
properly voluntary action, a more or less definite foregrasping
of the action as conscious exertion or active consciousness.   To
will to move the arm is thus to enter already, in a measure, upon
an active experience.   It is this fact which makes the motor
antecedent so important and characteristic an element in the
whole conative process.

It was pointed out just now that the state of desire becomes
modified by attachment to the idea of a definite movement.
We have now to see how this attachment modifies the motor
representation.

As soon as experience and association suggest that a desire
is realisable by a particular action, the idea of this action becomes of interest and is fixed by the attention.   This is merely
an illustration of the general law that interest produces a selective transfixing of particular ideas.   But more than this is observable in the case we are dealing with.   As soon as an action
is thought of as a means to a desired object, the action is itself
desired.   That is to say, the object of desire is now envisaged as
the final stage of a larger experience in which the action itself
occupies a place.   Thus when a boy, feeling uncomfortably hot
and desiring a bath, thinks of the river hard by, he foreacts
more or less distinctly, and desires, the whole experience of
going to the bank, undressing, and plunging in.   We have here
an illustration of the effect of a *transference* of feeling or interest
to psychical concomitants.   Though it is primarily and mainly
the pleasure of the cooling plunge that he desires, the vivid
imaginative realisation of the dependence of this *desideratum* on
certain antecedent actions suffices to invest these two with
some of the interest of the result.

Since movement is itself pleasurable under certain conditions
the idea of it may of course be felt to be desirable on its own
account.   There is little doubt, as we shall see by-and-by, that
in a good deal of voluntary action this intrinsic interest of motor

activity combines with and reinforces the borrowed or transferred interest in the desired object or end.

It is assumed here that the desire arises before the idea of the movement and suggests this last by way of association. Agreeably with this the order of psychical events in volitional action would be desire for object followed by idea of related action. Here it is evident there is an inversion of the order of actual experience; for the realisation of the desired object *follows* that of the action which is its causal condition. Since the idea of end is here the antecedent and psychical cause of the whole process, the conative order supplies the typical example of that particular mode of causation which has been marked off as final, or as teleological determination, *i.e.*, determination by end. It must be remembered, however, that while there is thus in the preliminary stage of voluntary action a transition from the idea of desired object to that of the appropriate action, as soon as this latter presents itself the anticipation of the desired object as the *end* of the action takes its place as the final stage of the anticipation of the whole experience, *viz.*, action followed by certain results. We have thus two 'moments' or phases in the full development of the conative process, (*a*) preliminary phase, desire inducing idea of movement, (*b*) completed phase, desire of movement thought of as leading on to realisation of end. Throughout this somewhat complicated process the interest continues to come from the idea of the desired object, the fixation of which is the great sustaining force throughout.

§ 19. *Initiation and Actual Performance.* When the force of a desire is thus concentrated upon, and made the psychical support of, a particular motor idea, the initiative stage of action is completed. As soon as a desire prompts us with sufficient intensity or strength and a suitable action is suggested with the requisite distinctness and stability, the actual performance follows, provided that there is nothing to counteract this prompting. No additional psychical initiative in the shape of a *fiat*, or a conscious process of resolve " I will to move," is required in this case. We may thus say that the volitional process which initiates the actual movement itself, together with the conscious experience of this, as also of the attainment of end when the action is successful, is resolvable into the idea of a definite movement backed and sustained by the force of desire.

As already observed, the actual movement itself is only of psychological account so far as it is conscious process, that is, an active experience. We know, however, that the processes involved in the execution of a movement are in part purely physiological. Thus the efferent transmission of the cortical excitation to the muscles lies outside the limits of mind, and no more needs to be considered by the psychologist than the corresponding *afferent* process in the sensory nerve which precedes a sensation.

At the same time psychologists customarily make a point of dwelling on the relation of the conative process ("willing") to the actual performance. The great metaphysical interest of this relation, as appearing to illustrate the causal action of mind on matter, would in itself appear to justify this excursion beyond the strict limits of psychology.

From a scientific point of view, the process here referred to must be regarded as a succession of a purely physical upon a psycho-physical process. The action of certain regions of the nerve-centres, correlated with the state of desire and motor representation, is followed by the out-going current of motor innervation. How, it may be asked, is this effected ?

The psychologist is only interested in this properly physiological question in so far as it involves the fixing of the central nervous correlatives of the *psychical* process itself. And here we have two possible explanations. Those who hold to the out-going current theory of "active consciousness" (*e.g.*, Bain and Wundt) would say that the representation of a movement already engages those motor centres of the cortex from which the process of innervation issues. According to this view, we have first a stage of sub-excitation of central motor elements passing into a full excitation leading to an out-going discharge. The whole series of nervous events here is similar to what takes place when the sub-excitation of sensory tracts, giving rise to an idea, grows into a full excitation producing a 'subjective' sensation or hallucination. On the other hand, those who adopt the hypothesis that the whole conscious experience in muscular action is a product of peripherally induced sensations must of course connect all ideas of movement with *sensory* nerve-centres. It remains to show how this excitation of sensory centres leads on to the efferent process of motor discharge, and ingenious attempts have recently been made by James and others to explain how the needed nervous arrangements can be brought about.[1]

---

[1] See Münsterberg, *Die Willenshandlung*, iii., especially p. 141 ff.; James, *op. cit.*, ii. p. 580 ff.

Since, according to this view of Münsterberg, James, and others, all active consciousness is peripherally initiated sensation which succeeds the process of innervation and muscular contraction, it follows that all ideas, including the forecastings, of movement *have to do with its results*. Thus, in desiring or willing to move my arm, I am representing the complex of sensations which will be produced by those reflex centripetal nerve-currents from tendon, joint, skin, and possibly muscle itself which have been started by the physical process of moving the part. It follows from what has been said above that, even if this conclusion were established much more firmly than it is, the result would not alter the *psychical* nature of the idea of movement as a mode of the active consciousness.

§ 20. *Variations in Type of Voluntary Movement.* We have here studied the common type of the volitional process, *viz.*, the initiation of a movement by a desire coupled with an idea of that movement. This general form of the process admits of certain variations. A word or two on these will complete our analysis of voluntary action.

As already pointed out, the conative process is in many cases set going by a sensation, so that the whole action takes on the appearance of a complicated reflex movement. That this should be so is evident. Our voluntary actions are adjustments to particular external circumstances. Thus the movement of stretching out the hand straight in front is, primarily at least, definitely related to the fact of there being an object present which the agent can touch. In truth, a large part of the simpler human action, as observable in the first years of life, comes under the two heads of approaching or getting possession of agreeable objects, and receding from or getting rid of disagreeable ones. A special form of such reflex-like voluntary movement is illustrated in the movements called forth in the measurements of reaction-time (responses to signals), where, as we have seen, under special conditions, the customary order of the volitional process is apt to be modified by special central preadjustments and a consequent overlapping of factors which are otherwise successive.[1]

While, however, this is the fundamental type of movement, it is not the only one. As we shall see, the growth of experience leading to more complicated ideational processes brings about what is, in appearance at least, a new form of voluntary movement, *viz.*, internally or ideationally initiated. Thus we

---

[1] *Cf.* above, I. p. 155, footnote [1].

think of a needed book in another room or in a friend's house and go in quest of it. Here the process of desiring and acting is not started by any sense-stimulus of the moment. Even in such cases, however, the form of reflex movement is not wholly lost. For the processes of ideational suggestion, even when prolonged, depend, as we have seen, on sense-presentations as their starting-point. Thus the idea of the needed book will have been called up directly or indirectly by the perception of something in our present surroundings, *e.g.*, a reference in another book. Not only so, it is evident that since all muscular action is determined by the actual surroundings—for we cannot even move a limb save when certain external conditions are present, as room for movement—it follows that all volitional movements preserve something of the character of reflex or sensory-motor reactions.

One other variety of voluntary movement demands a moment's attention. It is said that actions are brought about by vivid and persistent ideas of movement apart from desire. These *ideo-motor* actions, as they have been called, are illustrated in imitative movements, as when a man reproduces some of the balancing movements of a rope walker. We shall have occasion to refer to these later on. Here it is enough to point out that, properly speaking, these only come into the category of volitional phenomena so far as they are initiated by some analogue of desire.

That this ingredient is present in many so-called ideo-motor movements is certain. Thus imitative movements are often prompted by a desire to try our hand at what we see another doing, an impulse to rival or excel, and so forth. Similarly the carrying out of a movement vividly suggested by a present situation, as a boy's leap at the mere sight of a gate, a brook, and so forth, is always (in a perfectly normal condition of mind) due to a special pleasurable interest in the motor idea, *e.g.*, as new and untried, as difficult, and so promising κῦδος, which interest, the sustaining force of the process, involves at least a nascent desire in the shape of *a wish to realise a pleasurable idea.*

Where, on the other hand, a movement follows quite independently of desire, it must of course be regarded as non-voluntary. This applies, for example, to those slight unintentional movements towards the object intently fixated in thought by help of which what is sometimes erroneously called thought-reading, but is more correctly styled muscle-reading, is carried out. Here the involuntary character of the movements is clearly shown in the fact that the person who thus gives the clue is, as he thinks, trying to inhibit all such indications.[1] A like execution of motor impulse in opposition to a true conative or desiderative process is observable in the working out of an *idée fixe*, *e.g.*, as when a man leaps from a precipice; for in this case we have, it is evident, to do with the effect of morbid fascination, in which a painful idea of what is harmful, instead of being shrunk from, persists, and masters the attention.[2]

Writers do not always make it clear how far they regard the mere idea of movement as constituting a volitional process. Thus when Wundt writes: " External voluntary action consists in the apperception (*i.e.*, attentive fixation) of an idea of movement " (*loc. cit.*, p. 470), it is not quite evident whether he excludes the effect of painful fascination. James, so far as I understand him (*op. cit.*, ii. chap. xxvi., especially p. 549 ff.), treats *all* movements initiated by motor ideas as on a level with, if not actually included among, volitional phenomena. This is of course to set down pleasure and desire to the rank of mere *accidental* features of the volitional process, a conclusion which James appears to accept. This seems not only bad psychology but bad biology. The ends of self-preservation require that the mere idea of movement should *not* produce movement which may be useless or even harmful; that movement should only take place when something beneficial is to be secured or something harmful avoided; a result which (on the supposition of the approximate correspondence of the pleasurable and the beneficial, the painful and the harmful) is secured by the volitional process as above set forth.

[1] The probable explanation of these slight involuntary movements is that, according to the theory of space-consciousness unfolded above, a vivid and steady representation of a particular spatial point or situation *includes* that of movement towards the spot. The new dynamo-metrical experiments might well occupy themselves with the problem whether steadily thinking of particular regions of space, as in front, behind, to the right, does not involve nascent forms of those movements of head, or eye, or of limb, which would bring the visual or tactual sense to bear on these localities. A somewhat similar case of involuntary movements following on a vivid idea of the same, even when the person thinks himself checking the tendency, is that of table-turning. (See Carpenter, *Mental Physiology*, p. 292 ff.) These phenomena illustrate further a tendency to that detachment of a sub-conscious from the fully conscious group of psycho-physical processes which meets us in the more striking manifestation of a split-up or disorganised personality in certain abnormal conditions (hysteria). (See W. James, *op. cit.*, i. p. 202 ff.)

[2] The nature of such ideo-motor action is well brought out by Bain, *The Senses and the Intellect*, p. 336 ff. Carpenter gives numerous illustrations of ideo-motor actions in his *Mental Physiology*, chap. vi. 3.

§ 21. *Development of Voluntary Movement : Growth in Pre-cision.* With this analysis of the process of voluntary action to guide us, we may proceed to trace the chief phases in its de-velopment.

This progress will, it is evident, depend partly on the growth of the feelings and desires, partly on the extension of motor experience, that is, experience of movement and its results, subserving the acquisition of a stock of motor ideas. The first factor will be best considered in the following chapter in con-nexion with a full account of motives. In the present chapter we shall trace the progress of voluntary action as determined by the second factor, the extension of motor experience. Here we shall assume that certain simple forms of desire growing out of the organic wants, the elementary experience of sense-feeling, and the instinctive emotions, are present and opera-tive.

The mastery of a particular movement for volitional purposes is a gradual process.   A succession of tentatives is necessary before the precise form of movement is consciously differentiated or isolated, and retained in the form of an idea. As we saw above, the volitional adoption of movement is a process of selection. The primordial tendency to general diffused movement with any on-coming of feeling has to be re-stricted or inhibited. A child, when learning to write, to play the piano, and so forth, has to separate out a particular group of manual movements from among a miscellaneous throng of useless ones.   Also the particular connexions of movement which are primordial and organically determined and known as " concomitant movements " (*Mitbewegungen*), as in the case of finger-movements, must be broken through.[1]

By such repeated performances, aided by the laws of reten-tion and reproduction, the learner acquires definite motor ideas, and attaches these to ideas of their results, both those which are constant, depending on organic connexions, *e.g.*, the visual (or auditory) results of movement, and those which are variable, depending on special circumstances, organic or environmental, as the state of hunger, and the proximity of food.

---

[1] On the early ataxic condition of childish movement, see Münsterberg, *Die Willenshandlung*, p. 21 and p. 143.

As already implied, the repetition of movement, or what we call practice, tends to facilitate the process of initiation. The psycho-physical association or co-ordination of a particular movement with a particular result becomes fixed, so that less preliminary attention to the idea of movement is required. More particularly such recurring performance serves to repress as no longer needed the distinct idea of the motor experience itself. The bare idea of the wished-for result, e.g., the look of the door opened by one's arm, the sound produced by an act of articulation, suffices now to reinstate the appropriate action.[1]

§ 22. *Complication of Movement: Construction.* The acquisition or mastery of particular movements leads on to the attainment of new and more complex forms by the help of these as elements.   In other words, every acquired movement supplies material for further and more difficult acquirement.[2]

The process which here takes place has already been indicated in the general account of construction. There presents itself a set of circumstances, with a correlative need, similar to previous ones, the motor actions appropriate to which have already been learned. This leads by an assimilative process to motor tentatives of a like character. Little by little the old type of action is modified in the required direction. That is to say, the preliminary guiding idea becomes clearer and more precise by successive modifications due to the experiences supplied by the series of tentatives.

This acquisition of new movement, by help of previously acquired material, takes on its simplest form in a mere *combination of elements.*[3] Such combination may be simultaneous, as in joining movements of the arms and speech-organ in reciting; or successive, as in the actions of feeding, dressing oneself, and so forth. The vast majority of useful actions are made up of such successive movements with more or less of simultaneous combination also. As shown above, these succes-

---

[1] See James, *op. cit.*, ii. pp. 518, 519.

[2] As already pointed out, the acquisition of voluntary movement proceeds to some extent by taking up and embodying as elements of more complex and partly voluntary movements reflex and instinctive movements.

[3] By ' element' is here meant relatively simple combination of muscular actions. As pointed out above, in connexion with the presentative aspect of the subject, every movement involves the co-operation of a number of muscles.

sions are held together in consciousness by the force of contiguous association.[1]

The process of modifying old motor acquisitions, though often spoken of as merely combination, is in reality combination and separation. It is obvious that each integration of movements into an organic whole tends to keep these combined in this particular way. Hence when we require to re-combine the movements in new arrangements there is the force of the old associations to work against. A striking example of this is the boy's tendency to move his tongue when he learns to write, a practice probably due to the fact that in writing a word the learner must at the beginning distinctly represent the sound and the corresponding articulate movement.[2]

§ 23. *Imitative Movement.* A considerable factor in the early development of voluntary movement is what is known as imitation. By an imitative movement is meant one which is called forth directly by the sight of that movement as performed by another. Thus it is an imitative action when a child pouts, shakes his head, and so on, merely in response to another's like movement.[3]

Imitation implies a connexion between the visual percept of a movement and the idea of the movement itself as a complex of active and passive sensations. Hence it only begins to appear when the correlated centres have attained a certain development. It is often said that imitation is instinctive, that is, unacquired. But since it only begins to appear about the fourth month, when simple *voluntary* action directed towards an end is also first recognisable, it is probable that imitation is acquired.[4]

Imitation presupposes a certain experience of movement and

---

[1] See Vol. I. p. 308 ff.

[2] This is well explained by Féré, *Revue Philosophique*, 1886, p. 546. The part played by isolation in the combining of movements is illustrated by Preyer, *Die Seele des Kindes*, p. 214.

[3] The reproduction of another's movement through the auditory effect, as in imitating singing, etc., follows the same line of development.

[4] On the first manifestations of the imitative impulse, see Preyer, *op. cit.,* p. 176 ff. Preyer regards imitation as distinctly acquired. Stricker, however, maintains that there is an early form of instinctive (unlearnt) imitation. (*Ueber die Bewegungsvorstellungen*, pp. 20, 21.)

a resulting stock of motor acquisitions. It presupposes, further, special attention to the movement in its connexion with the visual (or auditory) percept when the movement was carried out by the child itself. To wave the hand in response to another's action implies that the child recognises the similarity of the seen movement to the appearance of its own moving hand. It is a further question whether imitative movement should be regarded as truly volitional. As a rapid reaction of a sensori-motor form, it has the look of a mechanical process. As already pointed out, imitative movement is of the ideo-motor type, *viz.*, that in which movement follows upon an idea of the same. In many cases, as in the instance given by Lotze,[1] moving the arm in sympathetic concomitance with the movements of a billiard player, there appears to be no conscious purpose. In some cases, indeed, the imitative movement seems specially useless, as in imitative coughing.[2] When to this is added that imitation is at its strongest among children, savages, certain animals that just fall short of the higher human intelligence, and in states of idiocy, there is much to favour the view that it is purely ideo-motor, and so sub-volitional.[3]

While it is undoubted that a vividly suggested idea of movement is the main element in the initiation of imitative movement, it is rarely, if ever, under normal conditions, the only factor. As suggested above, imitation follows on the persistence of motor ideas *having a pleasurable interest*. The child does not by any means imitate *all* the actions it sees, but only certain ones which specially impress it. In some cases it recognises them as significant, and adopts them as expressive signs, as in waving the hand in adieu. In other instances, as in imitating the actions of persons at table, it assimilates actions which it sees to be useful. In other cases, again, the movement is rendered æsthetically interesting, as in the case of odd or funny gestures,

---

[1] *Med. Psychologie*, p. 293.

[2] In my own case, imitation is here a conscious getting rid of the subjective sensation of irritation or tickling in the throat induced suggestively by the sound.

[3] On the prevalence of imitation among savages and certain animals, see Darwin's account of the Fuegians, *Naturalist's Voyage Round the World*, chap. x. (p. 251) ; *cf.* his *Descent of Man*, pt. i. chap. iii. ; also Romanes, *Mental Evolution in Animals*, p. 225. It was pointed out by Gurney that imitation plays a conspicuous part in the hypnotic state (hypnotic mimicry). (See *Mind*, xxxvi. p. 493.)

as also new and difficult movements which bring glory. Where these motives are lacking there is often an intellectual incentive at work, *viz.*, the *curiosity to see how a thing is done.* Hence we have good reason to suppose that in most, at least, of a child's imitation there is a rudiment of desire. For the rest, the abundant imitative activity of early life illustrates the strength of the *playful* impulse, of the disposition to indulge in motor activity for the sake of its intrinsic pleasurableness.

In this account of imitative movement we have supposed that it is a faithful reproduction of an action that has been previously acquired under the pressure of some special desire. Here the process of learning is confined to assimilating the seen movement of another to the child's own previously executed and seen movements. But imitation is rarely of this simple form. Its value as a factor in the development of voluntary movement depends on the circumstance that it leads to the acquisition of *new* combinations. A child learns to talk, for example, by striking out for the first time particular combinations of articulate movement under the stimulus and guidance of others' speech. Here it is evident the process is more complex. The new imitated action must be recognised as partially similar to the child's own past movements, and so suggest these. The motor ideas thus revived will then be modified and combined very much as in the constructive process of independent acquisition.

Imitation, as just remarked, is a characteristic of early years. The later processes of development tend to restrict its range. The growth of the higher controlling motives which have to be considered by-and-by serves to inhibit the naïve primitive impulse to imitate what others do. At the same time the mimetic impulse is not rendered inoperative : it is only narrowed and specialised by an intensification of the element of conscious purpose. Thus the boy imitates what he thinks to be fine, as a particular gait or form of speech ; what he sees to be useful, *e.g.*, manual dexterities ; what his moral and his æsthetic sense tell him to be worthy, as brave deeds, and so forth. The purposive character grows specially distinct in all comic imitation or mimicry, which aims at a *quasi*-artistic effect of ludicrous spectacle.

§ 24. *Movement and Verbal Suggestion : the Word of Command.* Very closely related to imitative movement is that form called

forth by the presentation of some arbitrarily attached sign of movement, such as the gesture signifying ' Come here ! ' or a verbal
symbol. Here, too, there is an associative connexion between particular sense-presentations (sights or sounds) and corresponding
motor ideas; only that in this case the connexion is not organically determined from the outset, but formed by the processes of
education. In training a dog or a child to respond to signs,
that is, to obey, the object is to induce such a close connexion
between the rousing sign and the motor reaction as that the
latter shall follow certainly and immediately.

According to the common view, the responsive movements
which we call obeying are acquired by a properly volitional process, *viz.*, the desire to avoid punishment, to earn reward, or to
please the commander. And there is no doubt that this properly
describes what takes place in all drill-like discipline. At the
same time it must be remembered that there is a wider influence
co-operating at all stages. This influence is the tendency of all
motor ideas, when sufficiently vivid and stable, to realise themselves in actual movement. Words are, as already pointed
out, potent suggestors. To name a movement, as in the
request ' Give me your hand,' is to call up a vivid representation
of this movement. The interest of being addressed suffices
for a direction of the attention in the required way. Hence
the response is swift, easy, and but half-volitional, from the
first, *i.e.*, as soon as the signs are understood.[1]

This power of calling forth sign-provoked movements in
others is a chief instrument in education. The disposition in
young creatures to move when movement is suggested, aided
by the strength of those social feelings which lead to respectful
attention to what older people say, serves to give the educator
a large direction of early activity. By such educative direction movement is indefinitely modified and expanded, and so

---

[1] That such responses to verbal (or other) signs do take place under the masterful
suggestiveness of the word is illustrated in the now familiar phenomena of hypnotic
suggestion. According to the most recent views, the motor responses in this condition are conscious. Yet their immediacy in cases where the action is grotesque
and even harmful, as when the subject is bidden drink something which is visibly
repulsive to the normal state, shows that there is no intelligent realisation and
adoption of the movements.

the development of voluntary command of the motor organs greatly furthered.[1]

§ 25. *Internal Origination of Movement.* In all the forms of movement considered so far action is of the reflex or sensori-motor pattern, occurring in response to certain sense-presentations. A higher stage is reached when movement becomes detached from such external sensuous provocatives, and follows on an internal process of ideation. In this way it becomes internally, or, to speak in physiological language, centrally, initiated. This internal origination is illustrated whenever the idea of a particular movement is suggested by processes of ideation, as when a child goes back on his forgotten promise to do something, and then sets about carrying out the action. Here, as already pointed out, we have merely a complication of the primordial reflex type of movement through the interposition of certain psycho-physical processes answering to the ideational train.

It is to be noted that after a certain amount of experience of bodily movement, and the acquisition of something like a complete visuo-tactual map of the moving organs, together with a vocabulary of these and their various movements, ideas, of this, that, and the other movement, tend to recur with greater and greater facility. Moreover, the extension of the process of repetition or practice greatly furthers the *readiness* to move as soon as an idea arises. Hence there emerges a new type of movement, *viz.*, one detached from the special impulses and desires which first called it into existence. This is illustrated whenever we move a limb *from the mere wish to do so.* Here, it is evident, there is shown a special facility in re-presenting particular movements with the required definiteness ; for the desire to move may be a motive of the weakest intensity which we can hardly define save by calling it the capricious wish to realise an idea simply because it happens to be present.[2]

It is evident that the acquisition of this ability to move instantly, and with the required precision through the mediacy

---

[1] This is but one side of the result of external sign-working control of move-ment. As we shall see by-and-by, it acts also, and largely, in the direction of restraining or inhibiting movement.

[2] The nature of the motive will have to be considered later on.

of the mere wish to move, serves greatly to further the whole process of motor acquisition. By thus gaining a ready command over every variety of ordinary bodily movement, apart from other movements, and also from the promptings of particular external circumstances, the learner is put into a much better position for carrying out those processes of re-combination of elements through which further advance is secured.

§ 25a. *Voluntary Movement and Consciousness of Power.* · The succession of stages in the growth of voluntary movement just described brings with it, in an imperfect or obscure form, a new conative element, *viz.*, the consciousness of self as agent and of its power of producing certain effects in the world of presentations. A word or two on this concomitant may suffice at this · stage, seeing that a fuller consideration of the subject will have to be made hereafter.

It was pointed out above that the conative process, strictly speaking, ends with the transition from the idea to the actual presentation of the movement.   What follows (process of innervation, contraction of muscle, etc.) depends on extra-psychical, *i.e.*, purely physiological, causes.   At the same time, these physiological processes contribute new psychical elements, *viz.*, not only the full sensational form of the active consciousness itself (muscular sensations), but also immediately attendant changes in the presentation-scheme, such as that of the visual picture due to the movement of an arm across the field.   Now these changes in the presentation-scheme constitute the most interesting factor in what we call a voluntary movement.   The mere conative process, or volition, when, as in the case of a paralysed limb, it fails to originate these changes, is, as we know, regarded as incomplete, and the patient is apt to fall into the illusion that his arm has been moved.   It is through attending to these changes, as uniformly following (under normal conditions) particular conative processes, that we come by our idea of power or causal agency.

There is reason to think that the child begins to note these sequences at an early date, and so to arrive at a dim consciousness of its power.   At first, this may be supposed to take the form of a cognition of the dependence of an outward effect, *e.g.*, approach or removal of an object, on the movement of " my " limb.   As we saw above, the child's first idea of self is a sensuous image, *viz.*, that of its corporeal frame.　The actions of its limbs are, as explained above, localised in, or referred to, these, and thus connected with the bodily self.　Indeed, this first crude idea of self is probably hardly less one of a moving or active frame, than of a sensitive organism, subject to pleasure and pain.

As ideation develops the line of cleavage between the self and not-self will be shifted.　The agent will now be thought of as the imaging and desiring subject, and the bodily movement will tend, on its visual side at least, to be taken up into the group of resulting changes in the presentation-world.　This stage will have been reached as soon as movement assumes . the form described in the preceding paragraph.

To note the dependence of a movement on the internal idea of, and wish for, that movement, is to realise the causal agency of the ego as the desirer or wisher.　Here, then, we see the dawn of that consciousness of power, agency, causal relation of the self to the world of not-self, on the model of which, as hinted above, all our ideas of physical action and physical force themselves are probably fashioned.[1]

HABIT.

§ 26. *General Nature of Habit.*　The development of move-ment here traced out illustrates in a specially clear manner the working of the law or principle of habit.　The general nature of this principle and its bearing on our psychical life as a whole have been touched on already.[2]　We have now to consider it more fully as specially affecting the processes of voluntary move-ment.

Habit is a product of acquisition.　In this respect it differs from instinct, with which otherwise it has much in common. We say we do a thing from habit, *e.g.*, nod back when a person not recognised nods to us, when as a consequence of long practice and frequent repetition the action has become in a measure or-ganised, and thus shorn of some of its original appanage of full

---

[1] *Cf.* what was said above, Vol. I. p. 264 f., on the first idea of self, and, p. 444 ff., on the genesis of the idea of cause.

[2] See above, Vol. I. p. 57 and p. 201.

consciousness or attention. The characteristic note of habit is mechanicality. In its most forcible manifestation habitual movement approaches to a sub-conscious reflex, as in the case just referred to. Hence Hartley spoke of habit-transformed movements as "*secondarily automatic*," to distinguish them from the primary or congenital reflexes.

It is thus evident that in habit we have in a particular way to do with that lapse of the intenser degrees of consciousness which accompanies an approximation of nervous structures to a state of perfect adjustment to the environment. The oft-repeated action becomes habitual and so automatic because the nervous centres engaged have taken on special modifications, have, according to the customary physiological figure, become 'seamed' by special lines of discharge. The perfect fixation of a habit appears to liberate the highest cortical centres from all but the slighest measure of co-operation in the process, the greater part of the central work (transmission of a definite kind of afferent excitation into a definite path of motor discharge) being now carried out by help of stably fixed arrangements in subordinate centres.[1]

It is to be noted that habit, in its wide sense, includes *all* the results of repetition and practice. Hence an action begins to come under the law of habit as soon as it is acquired. It is evident, indeed, that the volitional process in its complete, fully-conscious form is restricted to new, or at least infrequent, actions. It is only when I have to do something new and un-familiar that I need to realise, with the maximum distinctness, by a special concentration of attention, the idea of the object or end and the idea of the required movement.

The on-coming of habit is shown by two principal criteria. First of all, repetition of movement tends to remove all sense of effort and to render the movement *easy*. This growing facility in executing a movement is a complex psycho-physical pheno-menon, depending partly on processes of organic growth

---

[1] Whether such a secondary automatic action involves the formation of a sub-cortical " short-cut " is doubtful. It seems reasonable to suppose that the effect of repetition and habit may be to facilitate the transmission of the nervous dis-charge along definite cortical ' paths,' thereby shortening the process, and so suppressing, or at least reducing to a minimum duration, certain members of the psychical series. (See Ziehen, *op. cit.*, p. 161 f.)

which involve the peripheral organs, as in the strengthening effect of exercising particular groups of muscles, shown in a lessening of the *muscular* strain ; partly on the formation of the central connexions already alluded to, which manifests itself in a diminution of the strain of *attention*.

In the second place, habit involves and manifests itself in a consolidation of the processes of association involved. One of the most familiar characteristics of habit is prompt succession of a movement on the recurrence of the idea of a desired object. Here the intermediate idea of the movement itself is repressed or skipped. This swift succession of movement, upon the bare suggestion of something obtainable, depends on the laying down of special lines of nervous discharge by which the transmission of the current is expedited.

A further and more striking result of this fixing of associative connexion is the co-ordination of particular sense-presentations with appropriate motor responses. This is illustrated in the recurring movements of everyday life, as taking out a latch key on approaching one's door. Where this process is complete there lapses not only the initiative idea of the movement, *but even the idea of procurable object*. Thus when a man automatically winds up his watch on taking it out of his pocket during the operation of dressing for dinner, the action seems to be wanting *in all ideational initiation*. The movement has for its sole psychical initiator the complex presentation of the moment, *viz.*, the handling and seeing of the watch under these particular circumstances. Here then the habitual movement approximates to a pure conscious reflex, a movement following at once on a sensational stimulus without intervention of idea of end or of movement.[1]

§ 27. *Habit and Chains of Movement.* As we saw when dealing with the process of association, series of movements tend by repetition to grow consolidated, so that each step calls up the succeeding ones without a distinct intervention of consciousness. Simple examples of this are to be found in the

---

[1] It is not of course easy to say when all vestige of ideation disappears. James instances the putting out of the hand to take a cherry from the dessert plate when all inclination to eat has subsided. This is a pretty clear instance of a mechanical movement, yet my own experience suggests that even here there is present a shred of desire, the irrepressible wish to eat a dainty even after appetite is sated.

series of movements involved in walking, dressing and undressing, in playing a piece of music from memory, reciting a familiar
poem, and so forth.[1]

Such chains of movement approximate in their lack of clear
consciousness, their mechanical regularity, and promptness of
succession to the motor sequences in breathing, and other
primarily automatic movements. These characters imply that
particular central nervous arrangements have become completed, by which a stimulus supplied by the carrying out of
one member of the series of movements instantly evokes the
discharge required for the carrying out of the next member.
What differentiates such habitual chains from primarily automatic successions is the initial volitional impulse. I must consciously and voluntarily *start* the walking, the dressing, and so
forth. But the start is all, so far as volition is concerned. The
succession then takes care of itself, and, what is more, is
carried out better for the non-intervention of attention. That
is to say, the intrusion of volition in the mechanical part of the
arrangements tends, as we all know, to derange these.

Such chains of movement lack distinct volitional impulses other than the
original one, and also distinct motor representations. As we saw above, a
succession of movements consists of a chain of motor experiences and of corresponding sensations of touch, sight, or hearing. When often repeated the muscular
sensation, together with the passive sensation attending the execution of any step
in the movement, appears at once to bring on the next movement without the intervention of a distinct representation of this movement. In other words, such chains
of movements are a succession of (partially conscious) reflexes. The fact that the
intrusion of a volitional impulse in the shape of a special direction of attention to
one of the steps in the process distinctly deranges such a habitual train suggests
that the mechanical effect here brought about depends on a perfect co-ordination of
functional activities of certain lower centres, which co-ordination would only be
interfered with by the intrusion of any action from the highest centres answering
to the processes of conscious volition.[2]

§ 27a. *Habitual Concomitants of Movement.* Closely connected with such habitual
series are habitual simultaneous combinations of movements. Where we have been
in the way of doing a number of things together, as using certain gestures while
speaking, the movements tend to become co-ordinated into one group. This result
is illustrated in the indispensableness of habitual, though otherwise quite useless,
tricks of manner. As is well known, persons are apt when carrying out certain con-

[1] *Cf.* above, Vol. I. p. 308 ff.

[2] Whether these lower centres are ever altogether sub-*cortical* is, as hinted
before, uncertain. On the central conditions involved in these secondarily
automatic chains, see Münsterberg, *Die Willenshandlung,* p. 149 ff.

scious actions, more particularly those of speaking in public, to accompany these by trick-movements, such as playing with a waistcoat button, or a watch-chain. They were probably adopted at first as a vent for or relief to nervousness, and have become fixed by habitual association. The action of habit, as " second nature " in such cases, is seen in the fact that when the concomitant sub-conscious trick is cut off, as by the practical joke of removing the button or watch-chain, the progress of the conscious action itself is interfered with. The two nervous processes have here become organised into one process, so that the derangement of the seemingly useless concomitant throws out of gear the whole central apparatus involved.[1]

§ 27b. *Habit and Routine.* In the above account of habit the influence of the principle has been traced in the production of movement only. Yet, as was suggested above when dealing with the feelings and again in treating of the desires, habit has to do also with these important antecedents of voluntary movements.[2] Taken as including the whole effect of repetition in inducing subsequent recurrence, habit shows itself in the tendency of most men to be regular and periodic in their desires and impulses. The man of routine goes through the whole cycle of his daily avocations largely under the lead of habit. Thus certain desires recur at stated times, and in this way take on something of the character of those original bodily appetites which arise through periodically recurring organic feelings. In this way the whole series of daily pursuits, the rising and dressing, breakfasting, going to the place of business, and so forth, tends to get welded into a single chain of actions, with, at most, a partial development of desire for each succeeding pursuit when the proper hour arrives.

§ 28. *Degrees of Habitual Co-ordination.* It follows from our general definition of the principle, that habit shows itself in very unlike degrees of strength. The process of organic attachment is more or less complete in the case of different movements. We may now glance at these differences in the strength of habit, and seek to determine their conditions.

We may estimate the prompting force of habit in more ways than one. The most obvious index to its influence is lapse of psychical initiation as seen in the swiftness of the motor response. All the popular examples of habit, as the story of the victimised soldier who dropped his dinner at the word "Attention!" shouted by some practical jokers, illustrate this feature. The swifter the response to a particular

---

[1] *Cf.* what was said above (Vol. I. p. 150) on the motor concomitants of ideational attention. See, further, Carpenter, *Mental Physiology*, pp. 372, 373. The survival as habitual trick of what was presumably in its inception a useful action (as giving relief) is analogous to that of a once-useful action under the form of an expressional movement. On the whole subject of habitual association of movements, see James, *op. cit.*, i. p. 114 ff.

[2] See p. 32 ff. and p. 204 f.

sense-stimulus the more of force of habit is there indicated. Another criterion is *speciality* or precision of response. All habitual actions of the more pronounced type are definite varieties of movement specially co-ordinated with equally definite varieties of sensation-complex. When only partially formed a habit continues to display something of the uncertainty, the 'Weitläufigkeit,' of the half-acquired movement. The soldier's loss of his dinner was due to the unerring precision‾ of the habitual reaction, the swift dropping of the arms into the vertical line on the recurrence of the customary signal. The stronger the habit, the more definite or exact will be the response. Another measure of strength of habit closely connected with the preceding is uniformity, or *unfailingness* of response whenever the proper stimulus occurs. This criterion, together with speciality or definiteness, gives to habit its unvarying and monotonous character, its resemblance to the actions of a machine, and to those lower nervous reflexes which come nearest to mechanical actions. Lastly, the strength of a habit is directly measurable in terms of the difficulty of modifying it by special volitional effort. Half-formed habits can be easily altered: wholly formed, only by dint of extraordinary volitional effort.

Employing such criteria, we can draw up a scale of habitual movement. At the upper extremity we have the " secondarily-automatic " type of movement. This extreme variety approaches the primarily automatic in more ways than one. Thus it resembles instinct in its stubborn blindness and refusal to discriminate. Just as an animal under the force of instinct sometimes lays its egg in the wrong place through overlooking difference, so a ·man will now and again under the force of habit take out his latch key when approaching somebody else's door. The amusing stories of the quaint, useless things done by learned men and others in a " fit of abstraction " illustrate this absence of discriminating attention to the objects actually before them.[1]

---

[1] Students all the world over love to tell such stories of their instructors. At a certain Germany University a story used to circulate of a professor who, walking out one moonlight night along a tree-sentinelled road, leaped over a whole series of tree shadows under the impression that they were ditch-like hollows. Such want of discrimination, the necessary consequence of lapse of attention, serves to limit the speciality, or degree of exact adjustment of the habitual movement.

From this extreme variety downwards we have a series of
manifestations of habit with less and less of the characteristics
just dwelt upon.   Thus the kicking away a stone lying on one's
path is less habitual than the warding off a blow with one's
right arm.   It is less swift, less specialised, for it will be sometimes removed by the other leg or by a switch of our stick,
and less uniform, for we only thrust it aside at all in a certain
percentage of cases.

The main conditions on which these varying degrees of
habit depend appear to be the following: (1) The amount of
time and attention given to the particular movement or combination of movements so as to make it our own.   Since habit
is superinduced on a volitional process, it is evident that the
action must first be perfectly acquired through a conscious
process of acquisition.   (2) The frequency with which the
particular stimulus has been followed by the particular movement.   This condition, repetition, or frequency of performance,
is the great determinant of strength of habit.   (3) The unbroken
uniformity of past responses.   By this is meant that a particular stimulus S should have always been followed by a
particular motor reaction M, not sometimes followed at other
times not, or followed by another sort of movement, as M'.   This
condition evidently goes to determine the degree of unfailingness, as also of specialisation in the habit.   Thus, children
who are sometimes required to do a certain thing by their
parents, but now and again allowed to intermit the action,
never acquire perfect habits.   Similarly in all cases where a
choice of action is left over.

It is evident from this that habit and memory are, as psycho-physical
processes, closely related.   The operations of (intellectual) learning are an illustration of habit, inasmuch as they show the effect of repetition and practice in
facilitating a psycho-physical formation.   Since, moreover, all learning has a
motor factor, e.g., articulating a series of words, it comes in part under the head of
habit viewed as organisation of motor processes.   Conversely, as we have just seen,
habit is the result of a process of acquisition which is subject to very much the
same psycho-physical conditions (formation of physiological "traces," and neural
'paths') as intellectual acquisition.   The difference between conscious memory or
reproduction and habit is that the former (in its complete form) includes an extended
central process and its psychical concomitant, distinct ideation.   The reinstatement may be sudden and mechanical, like the on-bringing of movement by habit;
but it is the reinstatement of a *psychical* phenomenon, an idea.   It is only in the
reproduction of ideational series or trains, where (as we have seen) the constituent

members may receive the *minimum* of attention, that we see the true counterpart of motor habit in the region of memory. For the rest it is plain that if, with Hering, we extend the proper psychological meaning of 'memory,' making it cover unconscious, *i.e.*, purely physiological, modifications, habit will be viewed as one phase of such 'organic' memory.[1]

§ 29. *Habit and Plasticity of Movement.* It is evident from our account of habit that it is essentially a process of fixation, a restriction of movement to definite lines. Habitual actions, just because they become sub-conscious and largely non-voluntary, are rendered stable and unalterable. Habit thus presents one aspect which is opposed to all that we understand by development or progress. Itself the product of development, it tends in its turn to obstruct to some extent further development. We see this in the boy's manual feats, moving the hands or fingers in ways opposed to the habitual, which therefore as master-strokes excite admiration ; in the difficulty the tyro at the oars encounters in turning his boat, rowing with one arm and backwatering with the other ; and in the common failure of stout resolve to break through noxious habits.

While, however, in its narrower and more rigid form, habit diminishes the plasticity of the neuro-muscular apparatus, it would be an error to suppose that it is wholly an obstacle to progress. This would be to overlook the range of the principle, its influence in cases where action falls far short of the automatic stage, and also to misunderstand the nature of motor development. What we call new movements are never wholly new, and, as pointed out above, the perfect mastery of particular movements always helps us to the mastery of others. Thus the movements of equilibration and locomotion in skating are, as every learner knows, greatly furthered by previously acquired and habitual movements. The learning here consists in a few and comparatively slight modifications of old combinations in particular directions ; and though the modifications may be difficult through the obstructive force of the previous co-ordinations, they are a far less difficult operation than would be the learning of the whole group of movements *de novo*.

§ 29*a. Habit and Volition.* The characteristic of habit here touched on brings out the relation of habit-prompted to voluntary movements. Volition, as a

[1] *Cf.* above, Vol. I. p. 353.

conscious process, is selective, and voluntary action consequently modifiable.[1] Habit, though a product of volitional activity, tends to become sub-conscious and automatic, and so fixed and unalterable. So far as this is the case, the movement ceases to be our own in the sense that we consciously initiate it, and can inhibit or alter it as circumstances require. How far our ordinary habits are thus cut off from the psycho-physical process of volition it is difficult to say. In early life, at least, while the nerve-centres are as yet plastic, habits may be modified if only a sufficient strength of motive is forthcoming. This is seen in the possibility of learning new combinations of movement, as in dancing, and so forth. Even in later life long and stubborn habits may be broken through by men of exceptionally strong will. Yet there comes a time when the process of organisation resists all further modifying influence.

§ 29b. *Biological View of Habit.* As hinted above, habit, that is, facility, promptness, and certainty in carrying out specialised modes of acquired reaction, is useful to the organism, in so far as like circumstances recur, giving rise to like needs. The readiness with which we stretch out the hand to take what we want, and vary the direction of the movements according to the seen differences in the position of the object, illustrates the economising of organic force. Had we to concentrate our attention on the action as the baby has when first acquiring it, we should never get a thousandth part of our life-work done.[2]

At the same time, the organising process of habit is clearly harmful, so far as it hinders that ever new modification of movement which the changing character of our environment and of our own needs demands. Thus the soldier's hasty dropping of his dinner is an illustration of the biological harmfulness of habit when it becomes a specialisation of response fixed *in excess of the uniformity of outward circumstances.* Our tendency to go on dating our letters, etc., with a particular year after entering on its successor illustrates the obstructiveness of habit to further progress.

The ideal biological conception would be fixed, unvarying mode of response for all cases where the circumstances require and can permit it, want of perfect fixity in the degree in which variation is called for. Such an ideal conception is, however, never perfectly realisable. Nature, that is, the laws of our neuro-psychical organism, gives us a compromise which, though it has its drawbacks, works on the whole fairly well. What we find is a practically unalterable degree of habit for all cases where the response is useful *generally,* that is, in the large majority of cases. Thus, though the dropping of the dinner was a mistake, it followed on exceptional artificially induced circumstances, and so did not affect the validity of the general rule of military life that swift response to command is beneficial to the doer. In like manner it may be said that temporally useful habits, as that of dating letters, are beneficial (because in the main appropriate), provided that the effort of altering them when the time comes is inconsiderable.[3]

---

[1] The converse is not necessarily true. It is found that the actions of a frog which has been deprived of its cerebral hemispheres will vary adjustively within certain limits, though consciousness, and therefore volition, are presumably altogether absent in the case. (See Ziehen, *Leitfaden der physiol. Psychologie,* pp. 8, 9, who still further adds to the confusion of psychological nomenclature by calling such unconscious yet variable reactions " automatic ".)

[2] *Cf.* above, I. p. 78 ff. and 203 ff.

[3] A precisely similar line of remark may be applied to instinctive actions.

REFERENCES FOR READING.

On the general nature of conation or volition, the reader may consult the following : Ward, *loc. cit.*, pp. 42 f. and 72 ff. ; Höffding, *Psychology*, vii. A. ; Wundt, *Physiologische Psychol.* ii. caps. xviii. and xx.; two articles by O. Külpe on " Die Lehre vom Willen in der neucren Psychologie," in Wundt's *Phil. Studien*, bd. v. p. 179 ff. and p. 381 ff. For an account of the primitive forms of movement and the development of voluntary movement out of these, he may refer to Bain, *The Emotions and the Will*, bk. ii. chap. i. and following ; James, *Principles of Psychology*, ii. chap. xxvi. ; and Münsterberg, *Die Willenshandlung*, especially iii.

Though sometimes wrong or missing the mark, they are right in the vast majority of cases. As we saw above, Ziehen brings out the fact of a similar general or preponderant utility in the case of reflex actions.

CHAPTER XVIII.

COMPLEX ACTION: CONDUCT.

§ 1. *Simple and Complex Action.* In the previous chapter we have traced the process by which each of us acquires the command of his moving organs. It is in consciously bringing these into play that we first carry out conative processes, or, as we popularly express it, exercise our will. Such voluntary movements, moreover, are the necessary præ-condition of all higher and more complex conative processes. As already implied, the highest manifestations of will include conscious muscular action as their essential and differentiating factor.

This higher type of complex action is distinguished from that simple type of action which we have hitherto studied, *viz.*, voluntary movement, partly, by the greater complexity of the motor factor itself. That is to say, the later and more difficult actions consist of integrated series of mutually adjusted movements or combinations of movements. This is seen by a glance at the complex co-ordinations which make up the occupations of a man's business or profession. Here numerous movements and combinations of these corresponding to the writing of a letter, conversing with a client, and so forth, have to be integrated into a systematic course or plan of action.

Such complex co-ordinations are, it is evident, only possible by help of a good deal of preliminary forecasting of result or what we have called ideational initiation. It is the growth or expansion of this internal ideational factor which most plainly marks off the higher region of action from the lower region of voluntary movement. What we call a mature action, the full expression of a man's will, is one that is preceded by a good deal of reflexion, and is, moreover, the outcome of a rich

complex play of feelings or motives. Early action is characterised by impulsiveness, that is, simplicity of initiative, late
action by rationality or 'thoughtfulness,' that is, complexity
and fulness of initiative.

This is seen in the everyday use of language. We commonly call an action
a *movement* when little representativeness enters into it, and when consequently
the movement itself bulks as the chief part of the whole process. On the other
hand, we dignify it by the term *action* when the results of the movement, so far as
they are represented, become a prominent feature. Thus stretching out the arm
in imitation of another would generally be described as a movement: whereas the
same movement if performed in order to present something to a person would be
called an action. The relation is seen, too, in the fact that very different movements, if leading to the same kind of result, and having the same motive, would be
spoken of as one and the same kind of action, *e.g.*, greeting an acquaintance
(whether by head or arm movement, etc.).

§ 2. *Intellectual Factor in Growth of Will.* This transformation of simple into complex action is conditioned by the
progress of experience and the growth of the higher province
of intellectual activity, *viz.*, ideation or representation. The
effect of this expanding ideation will be to suggest a larger
variety of desirable objects. The child's cravings are prompted
by actual sensations, *e.g.*, hunger, or by percepts, *e.g.*, the
sight of something agreeable. As the inner ideational life
unfolds, ideas of pleasure-bringing objects are added through
processes of suggestion. Thus, as experience teaches the uses
and enjoyments of common objects about him, desires will be
awakened by a larger and larger number of percepts. Not
only so, the growth of ideation will enable him to imaginatively
move forward to a greater and greater distance from the actual
present. And while in this manner more conative material
is supplied by the ideational trains, there is a greater power
of steadily fixating this material. In this way ideas of
desirable things, realisable not only in the immediate but in
the remote future, acquire a footing in the mind, and so come
to be effective stimuli to conative exertion.

Not only does the growth of ideation thus amplify the range
of conative impulse by multiplying occasions of desire, but it
serves to introduce greater complexity into action. More
particularly the processes of associative integration tend to
enrich the initiative stage of voluntary action. Thus, as we

shall see more fully by-and-by, where our desired object is found by experience to be incompatible with others, the idea of the one will be apt to be accompanied with that of the other, and so the conative process complicated by the element of reflexion.

This process of associative integration will, it is evident, embrace the motor ideas which enter into voluntary action. It is by such associative integration that a child learns to co-ordinate a series of movements into a planful action directed to a distant result. In carrying out such a methodical course each stage of the action must be represented in its right place in relation to other stages, and the whole series associatively connected with the idea of the desired object.

Again, one and the same action may be found to be followed by a number of results, some of them not intended. Here experience introduces another kind of complexity. When the action is afterwards thought of, the idea of the undesired, as well as of the desired, results will be apt to recur. By the formation of such associations, as we shall see more clearly presently, action takes on a still higher degree of reflective consciousness.

§ 3. *Affective Factor in Growth of Will.* The growth of will is conditioned, in the second place, by the development of feeling. Since, as we saw above, feeling supplies the spring or impelling force in conation, it is evident that the expansion of the affective life must tend to advance the development of conation.

In studying the early growth of will we assumed that only the simpler feelings came into play. The command of the bodily organs is gained to a considerable extent under the stimulus of the sense-feelings. The first desires and aversions which rouse the muscular organs are connected with the plea-sures and pains of the bodily life and the senses. With these impulses there co-operate from an early period others derived from the primitive or instinctive emotions, such as the love of activity and of displaying one's power, rivalry, and the early crude form of curiosity, or the desire to inspect objects. The effect of the first awakening of social feeling is seen in the impulse of imitation, and still more clearly in that of obedience to commands, both of which, as we have found, contribute,

in an important measure, to the early development of movement.

As the feelings grow in number and the higher forms of
emotion begin to appear, the conative process is prompted by a
larger variety of desires. Thus the child begins to act for the
sake of earning praise, of giving pleasure to others, or of doing
what is right for its own sake. In this way each new advance
in emotional development tends to widen the range of desire in
a corresponding measure.

It is to be observed that the effect of this development of the
feelings on conation involves a further increase in the ideational
or representative factor in voluntary action. As we saw above,
the higher feelings are marked off from the lower by their
greater complexity and degree of representativeness. Accordingly, as action comes under the dominion of these higher
feelings it necessarily takes on more of a reflective or thoughtful
cast. Thus a child who aims at winning the commendation of
his parent or teacher is *ipso facto* representing a remote and not
easily imagined result of his action, *viz.*, the effect of it on
another's opinion, and through this on the relation between
himself and this person.

§ 4. *Motive-Ideas.* As this last illustration suggests, there
now appears as a result of this development of ideation and
feeling a new form of conative stimulus, which we can describe as Motive-Idea. The simplest form of a conscious pursuit of end involves, as we saw, merely the representation of a
single pleasurable experience realisable by a particular action.
Conation does not, however, remain at this level. The forecastings of end become fuller and richer, or, in other words, the
impelling motive grows more complex.

This increased complexity may be seen by examining any
common motive of action, as ambition or thrift. The impulse
to put things aside for the future grows out of complex experiences and processes of reflexion, as the evil of wanting them, the
difficulty in getting them the moment they are wanted, the
fact that when put aside they have the same utility to-morrow as
they would have now, and so forth. It is to be noticed further
that the motive-idea answering to thrift or providence is an
integration of a number of affective elements. Thus, while
starting from the painful experience of want and the correlated

impulse of aversion, it includes, as a secondary factor, pleasurable elements, making up the comfortable sense of security.

And here it may be well to point out a further characteristic of this higher class of motives. It was remarked above that all desire when fully developed becomes consciousness of want, or of an incomplete self. This consciousness of need accompanies the imaginative reaching out towards the pleasurable experience, the one or the other factor being preponderant according to circumstances. In the higher region of motive-ideas this constituent of self-consciousness becomes much more distinct. Thus in thrifty action a man is feeling his dependence on certain external conditions, the misery of a state of unthriftiness, and the comfort of the opposite state. Here, then, desire takes the form of an impulse to realise a new condition of the self.

In the case of the other motive named, ambition, there is a still further complication of the motive-structure. The boy's desire to get on, to rise to the top of his class, or be the captain of the cricket eleven, involves still more plainly than the desire to make provision for the future the idea of self. This self-consciousness, again, as implied in our previous account of it, is highly complex, comprehending a mass of indirect gratification through the acknowledgments and plaudits of others. Not only so, all ambition is an impulse of progress, of advance from a lower to a higher stage, and as such presupposes a way of viewing life as a whole or in its continuity.

§ 5. *Unification of Action: Permanent Ends.* This last feature of the higher conation, the viewing experience as a continuous whole, constitutes the most important of its distinctive characters. Early crude action is piecemeal action, that is, the isolated pursuit of this, that, and the other temporary gratification.[1] The development of reflexion and self-consciousness leads to an organisation or unification of action into a connected system. Thus ambition when fixed as a steady incentive means a recurring motive-idea, leading to a succession of progressive actions, the whole constituting the pursuit of a permanent end.

Such permanent ends arise as the product of a growing con-

---

[1] This applies alike to the workings of instinctive impulse, as hunger, and to the experience-taught pursuit of particular pleasures.

sciousness of the persistent self, its abiding relations to the environment, and the interests correlated with these. In this way we all of us learn to think of our health as a stable possession to be maintained day after day by a repetition of certain habitual actions. In like manner we come to erect knowledge, reputation, æsthetic culture and enjoyment into enduring interests and recurrent motive-ideas which lead to consistent *courses* of action.

Such unification of successive actions into a course of action involves, just like the integration of a series of movements into a methodical action, a reference not merely to the temporary, but to the permanent results of each member of the train. Thus a boy who is learning to aim at knowledge as a permanent interest will be prompted to pursue this and that piece of knowledge, not only for the sake of the particular pleasure which every intellectual acquisition brings at the moment, but for the sake of its subsequent value as an element in the permanent structure of his knowledge.

In the measure in which each individual action is thus viewed in its bearing on some portion of our lasting welfare, our doings become unified or consolidated into what we call Conduct. Impulse, as an isolated prompting for this or that particular enjoyment, is transformed into a comprehensive aim and a rational motive. The agent now compares his particular actions so as to make them harmonise one with another and converge towards one permanent mode of self-realisation. Such conscious unification of particular actions manifestly involves a clearer idea of a consistent self, and has indeed as its secondary motive the rational gratification which comes from a sense of harmony or consistency.[1] The clearest expression of this unification of impulses into a consistent motive-idea is the permeation and regulation of action by a general maxim or principle, such as 'maintain health,' 'seek knowledge,' ' be good,' and so forth.

§ 5a. *Nature of Permanent Ends: Desiring Means as Ends.* The pursuit of these permanent ends illustrates in a specially distinct form a common tendency in all states of desire to the fixing of attention not so much on the end itself as on the con-

---

[1] On the pleasure-element in this sense of consistency, see above, p. 40 f.

ditions of its realisation.   As was pointed out above, the desire
for an object begets a desire for the action which is seen to lead on
to the realisation of it.   In order to carry out any line of action, it
seems necessary that we should fix attention on the immediate
result of the act, as that which guides and controls the process.
Hence the tendency to erect this proximate result into a kind of
secondary " end " of the action.   Thus if a person feels cold and
goes to shut the door, realisation of the idea of the closed door
becomes the immediate object of his action.   That is to say, for
the moment he loses sight of the initial stimulus, feeling of cold
and the idea of the desired warmth, and is occupied in shutting
the door.   If an obstacle occurs, as when the latch does not
answer, he becomes wholly absorbed in this secondary end.   In
the case of pursuing a permanent end, as riches, or health, this
preoccupation of the mind with the means of attaining our object
becomes still more marked.   Money represents many alternative possibilities of avoiding ill and realising good.   The mind
cannot, it is obvious, represent even a small part of these at any
one moment.   Hence a specially noticeable sinking back into
indistinct consciousness of the primary end in this case, and the
engagement of the attention by the secondary or derivative end.

While, however, a certain weight must be ascribed to this
tendency to erect what was originally sought as a condition of,
or means to, an end into an end itself, its effect must not be
overrated.   The development of the motive-ideas corresponding
to wealth, knowledge and reputation is not fully explained as the
result of a transference of feeling and its correlative worth from
end to contributory means.   Even in cases where the explanation seems most plausible, as in that of the miser's greed of
mere money, it is doubtful whether the intensity of the passion
can be explained by the principle of transference.   A miser
seems often, both by temperament and habit of life, not more
capable but less capable than other men of representing the
utility, that is, the purchasing-power, of wealth.   Avarice has
in it the look of a blind instinct, a development possibly of that
secretive impulse which, according to some psychologists, is a
common human instinct.[1]   However this be, it is certain that

---

[1] On the nature of the miserly motive, see W. James, *op. cit.*, ii. p. 423 ff. ; also
Mind, i. 330 ff. and 567.

in the quest of other permanent ends, as knowledge, a circle of friends, there is more than the transference of a feeling of worth. Certain instinctive impulses probably take part here from the first leading straight to the pursuit of these ends; and the processes of development ensure a further and a higher estimate of these objects. Thus the boy comes to pursue knowledge not merely because this acquires a borrowed value as contributing to success in life, but because he has a rudiment of curiosity to start with, and because the growth of his intellect and of the correlative emotions tends to make knowledge more and more desirable on its own account.

These permanent or aggregated ends may be said to illustrate the fact that mental development on each of its three sides tends towards that higher form of comprehensive representativeness which in the sphere of thought we call generality. In the desire for health, property, truth, virtue, or happiness, the impulse seems, as already observed, to be enlarged and in a manner generalised. The object desired here takes on the form of a highly representative, or, as Mr. Spencer would call it, a re-representative idea; and this in a double way. First of all, as was just pointed out, the desire for a particular portion of wealth or knowledge implies a kind of condensed symbolic representation of a large variety of pleasures. In the second place, what is known as the fixed desire for any one of these ends involves a readiness to go on pursuing it as a *collective* idea. Since, as we have seen, this pursuit of a collective idea may be viewed as following a general rule, we may say that the action to this extent takes on a general form. This attitude of mind seems to involve the attachment of a certain calm impulse to the corresponding abstract symbol or general name.[1]

§ 5*b*. *Motives and Ideas : Non-Personal Ends.* In the above account of motive-ideas it has been assumed that what the idea stands for or represents is some object of desire or end, that is, according to the view adopted here, a personal satisfaction or gratification of some kind. An idea only becomes a *motive*-idea when it thus comes into such a relation to our active impulses as to constitute what we call a motive or the representation of a realisable end. Ideas and even interest-

---

[1] It has been assumed here that these highly intellectual or rational ends owe their propulsive force as motives to their relation to our pleasures and pains. Their apparent dissimilarity to the lower motives, in which the element of feeling is much more conspicuous, is on this supposition referrible to their highly representative or intellectualised character. For the opposite view that mere intelligence or reason (apart from feeling) may supply a motive-force, see H. Sidgwick's discussion of the relation of pleasure to desire already referred to. (*Method of Ethics*, bk. i. chap. iv.)

ing and pleasurable ideas may exist, *e.g.*, with respect to travel, study, and so forth, which, if they fail to arouse desire and active exertion, are not true motive-ideas.

A further point is whether when ideas become motive-ideas they all conform to the type here described, *viz.*, representation of personal satisfaction. Does a man in devoting himself, say to science, represent the result of his work as an enlargement of life and happiness *for himself?* This question has been answered in part already in dealing with desire. What we call disinterested pursuit of science does, no doubt, involve a great reduction of the personal factor in the representation of end, *e.g.*, the consideration of wealth, comfort, and even, in certain cases, of health. Yet such extreme concentration of conative impulse appears to be no exception to the general rule that we desire to gain what is pleasurable and to avoid what is painful. To the ardent *savant* the idea of a discovery is deliciously exciting, and that of a failure to discover intolerably painful. As disinterested, and devoted, he does, no doubt, lose sight of a large part of self and its gratifications : yet this does not affect the truth of the contention that the object which is here exclusively represented is envisaged as a pleasurable object, that is, as one, the realisation of which will bring a pleasurable satisfaction.[1]

There are other cases where the exclusion of all personal regard in action seems to be still more manifest. The pursuit of posthumous fame may be taken as an example. Here, it is said, a man is knowingly working for a result which he will not be able to enjoy. In this case the element of self-consciousness is, it is evident, still further repressed. All the same it seems undeniable that the motive in this case is similar in its nature and mode of working to that of ordinary ambition. A man who works for a posthumous reputation cannot help representing the realisation of the object *as if it were* a personal satisfaction. There seems, in this case, to be an illusory

---

[1] Such a case, no doubt, illustrates at once the effect of an original instinctive impulse and of habit. That is to say, the end of truth derives a heightened value from the ardent active impulse to pursue it, an impulse growing out of an original bent, which, if in a manner quieted, is yet confirmed by years of habit. (*Cf.* above, p. 204 f.)

antedating of the realisation, the fact of its being posthumous lapsing from consciousness as the desire kindles.[1]

Finally, reference may be made to those forms of devotion *to others* which are specially marked off as disinterested action or self-abnegation. A mother who knowingly wears herself out to preserve a delicate child, and the soldier who willingly meets death in defence of his country, cannot, it is said, be supposed to be representing a future *personal* satisfaction. This case is marked off by the presence of sympathy. As pointed out above, in sympathising with another we make his happiness our own in the sense of imaginatively realising it, and of desiring its furtherance. Devotion to others as the outcome of sympathy is thus a kind of extension of the range of personal end by the addition of analogous, non-personal ends in the interests and happiness of our fellows. Yet while in such mental absorption in another's good the idea of self and personal happiness may disappear, the action is still, in a measure, describable as the pursuit of the pleasurable. To the sympathetic man or woman the realisation of the happiness of one beloved is pleasure-bringing to himself *as contemplator of the state*. The " self-sacrificing " mother finds her happiness in watching the child's prosperity just as the artist would find his satisfaction in watching the growth of a picture. In other words, the benevolent man who is ready to sacrifice personal comfort for others is a man in whom the sympathetic or " altruistic " feelings are so strong and dominant that he finds his happiness in beneficent action, and would be miserable if cut off from this.

It may be added that the pursuit of these non-personal ends is, to a large extent, the result of our social life and education. As called upon to do things for others, and to work in co-operation with them for *common* ends, we tend to let drop the personal factor, *i.e.*, the idea of a particular gratification for self apart, in our representation of end. Thus a man does not necessarily think of a definite personal gratification when aiding a local charity, or exercising his parliamentary franchise. In

---

[1] Even when the idea of personal extinction is present there is the inexpugnable secondary consciousness of self in the shape of the thought of our knowing of the conferment of the honour.

such cases the social consciousness, that is, the type of consciousness produced in all educated men alike by the assimilating influences of social institutions and education, preponderates over the private or personal consciousness: the agent seeks to realise something which he recognises to be generally or objectively desirable. Here the general proposition that action is determined by a representation of the pleasurable may well appear to be contradicted. Yet the exception is only apparent; for it is precisely because the individual finds his keenest satisfaction in this harmonious co-operation that he adopts such common ends as his own.[1]

According to Dr. Bain, the cases here dealt with would illustrate what he calls exaggerated or impassioned end, an effect analogous to the abnormal phenomenon, fixed idea (idée fixe).[2] Thus, in sympathetic action, there is a kind of illusory substitution of another's happiness for our own so that the action is a disturbance of the process of normal volition. There is, however, an important psychological, as also a biological, difference between the cases. In all examples of normal devotion, including the effect of sympathy, there is, as has been pointed out above, a representation of end, which end, moreover, does approximately answer to the kind of satisfaction of which the agent is most susceptible, that is to say, to his strongest and most vital interest. Hence the devoted, self-sacrificing worker is exemplifying (under special conditions no doubt) the typical form of the volitional process. Where, on the other hand, a man overtaken by the effect of fascination is impelled to throw himself over a cliff, there is no such representation of attendant or resultant satisfaction. It is a mere result of an abnormally vivid and persistent idea of *a particular action*. Hence such ideo-motor actions are properly described as non-volitional.[3] From a biological or teleological point of view the difference is even more pronounced. The action of the devoted mother directly subserves the preservation of the race, and her instinct of self-sacrifice may be viewed as the product of unnumbered ages of evolution. Similarly, the devotion of the patriot, the *savant*, and so forth, is normal because subserving the good of the community, and has presumably become a dominant impulse in a certain number because of its utility. Ideo-motor action, that is, the tendency to carry out an action merely because this is vividly suggested, is obviously not only useless but likely to be positively injurious. Hence it is rightly looked on as a sign of incipient mental disturbance.[4]

[1] The above paragraphs should be read as supplementary to what was said above respecting the object of desire.

[2] See his account of the sympathetic impulse, *The Emotions and the Will*, p. 120 ff.; and of the impassioned or exaggerated ends, p. 390 ff.

[3] *Cf.* above, p. 214 f.

[4] The contrast between personal and non-personal ends here touched upon has much more of an ethical than of a psychological significance. On this latter aspect of the subject, see H. Spencer, *The Data of Ethics*, chap. xii.; H. Sidgwick, *The Methods of Ethics*, bk. iii. chap. iv.; and Leslie Stephen, *Science of Ethics*, chap. vi.
       *

§ 6. *Complex Action.* Our action, as we have seen, gains in representativeness as we take remote consequences into account. And this increase of representativeness implies an increase in the complexity of the action. In a special sense we may call an action complex when it is not the result of a single impulse but involves a plurality of impulses, a representation of a number of objects of desire or aversion, and so an expansion and complication of the initial representative process.

This expansion of the representative stage of action assumes one of two contrasting forms. In the first place, the desires or impulses simultaneously called up may be harmonious and co-operative, converging towards one and the same action. In the second place, the desires may be discordant and opposed, or diverging into different lines of action.

(*a*) *Co-operation of Impulses.* The combination of two or more elements of desire or impulse in one conative process is exceedingly common, and may be said, indeed, to be the general rule. Many actions which seem at first sight to have but one impelling motive will be found on closer inspection to have a number. So simple an action as going out for a walk may be motived by a number of concurrent impulses, as desire for locomotion, fresh air, and change of scene.

The most interesting example of this co-operation of desires is seen in the case when, in addition to the primary impulse related to the pleasure following the action, there presents itself a secondary impulse related to the activity itself. As we saw above, we may be said in all cases of voluntary action to desire the action in a subordinate way, as means to an end. But, in many instances, we go further, and distinctly represent the action as intrinsically pleasurable. In all such cases the action becomes complex by a composition of impulses. To the initial impulse to realise some end there is added another, to follow out an agreeable line of action. We are frequently led to carry out certain actions, the results of which we care for insufficiently, by a representation of the agreeableness of the activity itself. In all sportive or play-like action, from that of the boy up to that of the tourist in his mountain " playground," this secondary impulse attains a special degree of prominence.

Although the desire for the pleasure of the action is here spoken of as secondary, it is not meant that it is in all cases the less potent factor, or even that it is always the second in order of time. The proportion of intensity between the desire for the result or the end and for the means may vary within wide limits. In certain cases the representation of the pleasure of doing a thing may be subordinate and semi-conscious. In other instances it becomes the dominant force. This is true of most games where the interest turns largely on the pleasure of physical exercise or intellectual activity, *e.g.*, riddles. Here the desire to do something is the first and the most powerful impulse in the case, and the desire for this or that particular end, though it seems to dominate the whole process, is, in reality, only of subordinate influence.[1]

(*b*) *Opposition of Impulses.* The second variety of complex action, in which two (or more) impulses come into antagonism, is of yet greater importance. Owing to the opposition in this case the representative or reflective stage of the action becomes much more prolonged and complicated than is the case where a number of impulses are co-operant. In addition to its special psychological importance this type of action has a peculiar interest from an ethical point of view; for moral conduct, or obedience to the moral law, is the outcome of this mode of complex action.

§ 7. *Arrest of Action : Inhibition.* This variety of complex action is characterised by the clearer emergence of an element in the conative process hitherto neglected, *viz.*, the arrest or inhibition of action. This element is of course present from the beginning. As we saw in the preceding chapter, particular movements are learnt by selectively preserving them from among a crowd of concurrent movements, which last have accordingly to be repressed. It is, however, only in the higher region of action which we are now considering that this arrest or inhibition takes on a clearly-marked form. It is when we are simultaneously prompted by a plurality of impulses leading in distinct directions, that is, to different external actions, that the

---

[1] Of course, the promptings of a vigorous muscular system or of an active brain cannot in themselves lead to the pursuit of an end, as end, if *no* value is set on this at the moment. The fox-hunter loves the chase mainly for the sake of the chase itself, as Lessing held the pursuit of truth high above the actual attainment of it. Yet the fox-hunter must get up the excitement of pursuit, that is, a keen desire for the result, before he can have the enjoyment. Similarly in the case of the intellectual pursuit. We do not enjoy the intellectual chase of solving a riddle unless we manage to attach a factitious sort of interest to the result at the moment of search. (*Cf.* II. Sidgwick, *The Methods of Ethics*, bk. i. chap. iv.)

process of inhibition becomes manifest. The opposition of motor forces in this case produces an arrest of action which may be temporary only, leading to a delay or postponement of the action, or may end in its complete suppression. This inhibitory effect of one desire or impulse on another is closely analogous to the reciprocal inhibitory effect of competing presentations or representations.

§ 7a. *Physiological Conditions of Arrest.* The nature of inhibition as a nervous process was touched on above.[1] It only remains to indicate the probable neural correlative of the special phenomenon which we are now studying, *viz.*, the inhibition of one active impulse by another.

As already suggested, inhibition seems to be a general condition of psycho-physical action. That concentration of the mental field which conditions clear consciousness is presumably correlated with a corresponding narrowing or concentration of cortical activity, that is to say, the inhibition of all other nervous processes excited at the time. In like manner it is supposable that a special and intense excitation of a certain region of the motor centres, as in steadily representing a particular action, will be inhibitory to the simultaneous innervation of other motor structures. This seems to be proved by the fact that by concentrating our mind on a movement we are able to counteract any tendency of the moment to other and disconnected movements.

To this it may be added that in the case of active psycho-physical phenomena there is probably a further peripheral factor in inhibition. Common experience tells us that when we try our utmost to check an impulse to carry out a movement, *e.g.*, the group of movements constituting laughter, we bring antagonistic muscles into energetic play.[2] According to this view the arrest of impulse by reflexion, to be spoken of presently, involves the transmission of an opposing or inhibitory nervous current from the higher to the lower motor centres, which partly acts on these directly by an inhibitory central influence, partly counteracts the peripheral results of their innervation (movement of limb) by exciting other motor centres, and so producing antagonistic muscular contraction.

---

[1] See Vol. I. p. 45.

[2] See Münsterberg, *Die Willenshandlung*, p. 157 ff.

§ 8. *Action Arrested by Doubt.* The simplest case of arrested or inhibited action is that in which the belief necessary to the carrying out of an impulse is checked. In the early stages of action we are prone to be confident in our powers. We can easily observe in children's first experiments in movement that they are carried out boldly, that is, with a full assurance of success. To these hopeful tyros in the domain of human action failure comes as a shock. The child looks perplexed, confounded, when he first encounters an object too heavy to be moved. These failures suggest uncertainty, and this sense of uncertainty or doubt will serve to arrest or temporarily paralyse the child's action. Thus, after having had experience of his inability to lift heavy bodies, he will probably have his impulse checked the next time he desires to lift a heavy-looking object.

The occurrence of this form of arrest will, it is evident, depend on the teachings of experience. A child that notes its failures as they take place and afterwards recalls them will more frequently experience this inhibition of impulse than one that is unobservant and unretentive.

The occurrence of this inhibition of active impulse by doubt will further depend on certain organic conditions. It follows from what was said above that in cases where the vital energies are high, and the motor system is vigorous and predisposed to activity, there active impulse will be strong and not easily checked. But this is not all. As has been pointed out also, belief itself is conditioned in part by the same organic factor. Where there is a powerful disposition to act, there confidence tends to be high. Distrust of one's powers, on the other hand, is apt to creep in where the vital energies are low, and as a consequence of this active impulse is weak.

It follows from this that desire for an object and doubt as to its attainability rarely constitute an acute state of antagonism and conflict. Strong impulse will tend to dissolve doubt, and, conversely, the presence of doubt is an evidence of a failing of the springs of active energy. This close connexion of doubt and want of vigorous impulse is clearly illustrated in those morbid conditions of volition in which, even when ideas of desirable objects present themselves, there is no keen impulse to realise the same, and the subject loses mental grasp of them as realities.[1]

§ 9. *Recoil of Desire : Deterrents from Action.* A second and in general more effective form of arrest occurs when desire prompts to a certain action with which is associated some painful accompaniment or consequent. In this case the impulse to realise a pleasure is opposed by an aversion to what is disagreeable. And so far as this shrinking from a painful experience frustrates the positive impulse, we are said to be deterred from the action.

The deterring force in this case may reside either in the representation of the action itself as disagreeable, or in the anticipation of some disagreeable result.

The first case is illustrated in the general inhibition of desire by a disinclination to action whether through indolence or through fatigue. Beginning with the former situation, we have to note that in all sluggishness or indolence there is not merely the absence of the common auxiliary incentive to our actions, *viz.*, a desire for the pleasure of action itself, but the presence of a distinctly antagonistic force, the tendency to recoil from or avoid what is disagreeable. An indolent person holds back from performing actions suggested to him, and even desired for a moment for the sake of their results, under the deterrent force of the idea of the exertion or strain.

The outcome of this process of arrest will depend on the relative intensity or strength of the opposing impulses. If the dislike to exertion is stronger than the desire for the pleasure, this last will be frustrated. Now, as we saw above, the intensity of a particular impulse is determined in part by the general degree of readiness of the moment to be active. It follows that when there is great indolence, the desires are likely to remain in a weak, flabby condition. As a consequence of this there will, in this case as in that where action is arrested by doubt, be no violent collision and opposition of forces. Persons of a sluggish temperament rarely experience this kind of antagonism in its full intensity of excitement.[2]

---

[1] On the peculiarities of this pathological condition (known as "aboulie"), see Ribot, *Les Maladies de la Volonté*, p. 38 ff.

[2] A similar remark probably applies to those morbid disturbances of volition

The situation of fatigue is a different one. Here desire may be strong and urgent, and its realisation obstructed by the deterring influence of weary muscles or brain. The last dreary stages of a piece of heavy work, pedestrian, literary, and so forth, illustrates the type of experience. The desire may be at its keenest, intensified by the knowledge of the nearness of the fruition ; but fatigue is at its maximum also, and the weary toiler feels dragged along as if by some compelling hand. The weariness and disgust at the long monotonous labour reinforce by a secondary impulse of aversion the primary desire. Here, it is evident, the consciousness of antagonism and of conflict is all too distinct and prominent.

We may now pass to the second class of cases in which arrest of action and conflict are introduced through an opposition in the results of the action. It is a familiar fact that we are often at once prompted to an action by the idea of a pleasurable result, and deterred from it by the idea of a painful result. Here the collision, the antagonism, is decided and sharp. Two strong impulses, the craving for what is agreeable, the shrinking from what is disagreeable, assert themselves, each tending to issue in an appropriate action. Only the direction of the two is diametrically opposed. The very action which the one impulse bids us do, the other bids us not do.

Here, again, the effect of the prevision of evil in repressing impulse will vary according to a number of circumstances, such as the relative strength of the attractive and deterrent forces, and the strength of the general disposition towards activity at the time. Here, too, we may note marked differences of effect according as the temperament is wary or cautious, and highly susceptible to the deterrent effects of anticipated evil ; or, on the other hand, heedless of unpleasant consequences, and impatient of delay—a contrast well illustrated in the case of Macbeth and his wife when planning their ambitious crime.

§ 10. *Rivalry of Impulses.* As a third type of arrest, we may take the case where there arises a plurality of positive impulses. When a man is at one and the same moment stimulated

which arise from a failure (rather than from an excess) of impulse. The want of vigorous impulses to act, and of the correlative confidence, may be supposed to carry with it a disinclination which is an exaggerated form of what we call sluggishness. On the nature of these disturbances, see Ribot, *op. cit.*, chap. i.

to different lines of action by two disconnected desires, conflict arises through the prompting of incompatible impulses. A student, we will say, is fond of music, and feels strongly impelled to attend a particular concert. But he has to prepare work for to-morrow's class, and he is interested in the subject, and desires to do his best. Here each desire to realise a pleasurable object is opposed and inhibited by a desire for another pleasurable object. Hence we can describe the state as a rivalry of desires or impulses.

The essential element in this state of rivalry of desires is the simultaneous prompting of impulse towards diverse lines of action. The situation takes on slightly different aspects in different cases. In some instances the opposition seems to arise rather through the limitations of our powers and of the means at our command (*e.g.*, inability to be in two places at one time, to do two things at the same time, to employ the same sum of money in procuring different objects, etc.). In other cases the opposition seems rather to spring out of the incompatibility of the objects desired (*e.g.*, the friendship and good opinion of others, and the gratification of individual tastes).

It is to be added that the two cases here distinguished as recoil of desire and rivalry of impulses are not perfectly distinct from one another. As we saw above, desire and aversion are closely related, and tend to pass one into the other. Thus when a boy recoils from a certain action through fear of giving offence to his mother, this deterrent motive would easily pass into the positive desire to retain her favour, in which case the situation would become distinctly one of rivalry of impulses. On the other hand, when there is rivalry of desires the impulse to realise one of the two desired objects tends to assume the form of an aversion to the alternative action.

This rivalry of impulses or desires may assume different forms. Thus two actual feelings may prompt in different directions, as when, tired and hot after a walk, we are at once impelled to rest, and to procure a draught of water; or the desires for two particular objects may come into conflict, as in the illustration given above, or in the case of the would-be lover of two women, the theme of Mr. Meredith's comedy, *The Egoist*. Other examples are supplied in the rivalry of one of the higher motives with a lower impulse, as in the common experience of the competition of the sensual or appetitive and the more intellectual pleasures, and in the rivalry of motive-ideas one with another, as when the pursuit of study comes into competition with that of society or of athletics.

§ 11. *Passive Resolution of Conflict.* When we are thus at the same time strongly drawn towards and repelled from an action, or drawn towards two incompatible lines of action, the antagonism of forces brings about a temporary state of inaction, each impulse alternately moving us in this and in that direction. This is a condition of acute conflict and profound misery, a state exhibited in its most intense and impressive form in the moral paralysis of Hamlet.

Happily such a process of alternate advance and recoil is rarely a very prolonged experience. It tends to resolve itself. Thus, in many cases, one of the contending impulses proves, after its full development, to be stronger and more persistent than the other, and so acquires the mastery as determining motive. We frequently experience such temporary conflicts which are resolved by the emergence of a mastering impulse. This applies more particularly to cases where a higher motive comes into play against one of the lower impulses, for, as a complex ideational product, the former requires time for its full development.

Again, even in cases where the rival impulses are pretty equally balanced, the strife is apt to resolve itself through the co-operation of another factor, *viz.*, the pain of the state of conflict itself. Where this is keen and intolerable it expedites the process by giving an adventitious or borrowed superiority to a particular impulse that happens to be uppermost at the moment.

This purely passive process of resolution answers in certain respects to the effect of a number of mechanical forces acting on one and the same body. Just as the body tends to follow the direction of the stronger force, so action tends to follow the direction of the keener desire. And just as two opposed forces when equal may counteract one another producing a state of equilibrium and rest, so two opposing desires may counteract one another and produce as their resultant not action, but inaction. The process may, indeed, be closely assimilated on its neural side to a mechanical one if we assume that the innervational excitation and the connected inhibitory action on the rival tendency are in their intensity proportional to the conscious desires (and aversions) involved.

As pointed out above, however, such a state of perfect equilibrium of opposing impulses is rarely prolonged, at least in the case of all who have a strong organic bent to action, and are impatient of the state of inaction. In their case the tendency to act somehow, which is now greatly intensified by the growing aversion to the pain of conflict, operates as a powerful factor on the side of action as against in-action. Which of the actions is finally carried out is in cases of approximately equal stimuli very much a matter of accident, depending on which of the impulses happens to be in the ascendant or most distinctly present at the moment when the desire to act somehow, and the aversion to the pain of conflict, reach a certain strength. In this instance, too, a mechanical interpretation seems to be forthcoming. We may suppose that the state of impatience, involving as its neural concomitant a general increase of motor tension, raises the particular innervational current which happens to be the one started (or to be the fullest) at the moment up to the height required for a complete peripheral discharge.

§ 12. *Regulated Conflict: Deliberation.* Thus far we have supposed the inhibition of action, with the attendant conflict of impulse, as also its resolution, to be comparatively passive processes, in which the several contending impulses are acting like so many isolated forces, each tending to determine the result according to its particular degree of strength. This supposition may be said to represent, roughly at least, many of the earlier and cruder forms of what we have called complex action, as those observable in the more intelligent animals, and in children.

What we mean by the development of the will, however, implies a transformation of this passive type of process into one having more of the active character, that is, more of the element of active consciousness. This added element takes on the specific form of an *effort of will.* That is to say, in this higher type of complex action we consciously endeavour to resist impulse and to guide and control the whole process. By means of this new factor, the play of impulses loses something of its simple mechanical character, and takes on more of the aspect of a unified *organic* process.

This form of conative process is seen in its simplest manifestation in the conscious endeavour to resist a particular impulse of the moment, or the effort *not* to do a thing. Here we have, it is evident, a secondary inhibitory volitional process superinduced on the primary impulsive process. Such an inhibitory interference involves a special and difficult exertion of voluntary attention. In trying to check an impulse to do a hasty and foolish thing I set myself to think of the disagreeable consequence, and to banish the idea of the action from my mind. This process again is accompanied by the carrying out of vigorous antagonistic actions, as in clenching the teeth, the hands, holding in the breath, and the like. Such a special exertion, marked off as effort, is wanting in the pleasurableness of moderate activity, and is apt to be fatiguing and disagreeable. Hence it presupposes a strong motive force behind.

*

This motive, like other motives, is the product of experience. It differs from many of these merely in the high degree of its representativeness. To will not to act when impulse urges us on is motived by a shrinking from the evils of rash or impulsive action. This motive is a slow growth presupposing not merely considerable experience, but careful attention to the less obvious and less immediate results of action, and even processes of comparison and abstraction.

We may now pass to the further stages of this volitional control of impulse. Action being thus consciously and intentionally checked, there supervenes a controlled form of the opposition and competition of impulse. This is known as Deliberation. Here the attention is voluntarily directed to the several representations (motive-ideas) so as to ensure a due persistence and full distinctness of each, and a proper carrying out of what we know as a comparison of their values as ends. Thus, if it be simply a question of doing or not doing a particular thing, we seek to carefully count up the advantages and disadvantages, and to set the one against the other; or, if it be a case of two rival ends, we endeavour to measure the value of one object of desire with another.

In addition to this deliberation respecting the relative magnitude or value of our ends, there is a deliberation respecting the relative advantages of this and that mode of securing our end, or what we call means. Here the co-operation of intellect in the volitional process becomes still more distinct. The estimation of this and that particular end, though involving a careful comparison and discrimination, is ultimately a matter for feeling and desire. The process of deliberation on this side merely ensures the fullest development of the several forces resident in the desires. In the case of deliberating about means, however, the estimation is wholly a matter of (practical) knowledge and judgment. In order to know whether in a given case a particular action will effect a desired result, and which of a number of suggested actions will most certainly contribute to this end, we have to accurately recall a number of facts, to detect their relations one to another, to compare, and even to carry out with some care a process of reasoning as to probabilities. Hence such deliberation, as we all know, presupposes a considerable development of the intellectual powers.

The process of active deliberation here briefly described is a higher form of that work of integration or unification in which, as we saw above, the whole development of consciousness consists. To reflect upon our competing impulses and aims is to make them our own, that is, to take them up as elements in a new mode of self-consciousness. Since, moreover, all deliberation aims at developing our several impulses or motives, it directly leads to a fuller investigation of, and reflexion upon, the complex organisation which we call the self. The process of unification here carried out may be said to take its rise in the recognition of a torn or divided self, and the desire to realise its harmony or unity. This form of self-consciousness always enters more or less distinctly into the work of deliberation. In men of a finer sensibility it constitutes a main characteristic of the process.

§ 13. *Choice or Decision.* Where the process of deliberation has been carried out normally, that is, in strict subordination to practical ends, it leads on to what is popularly known as an act of choice or decision. Thus, after duly weighing the pleasure and the pain, the good and the evil which will result from any action, the one may be seen to preponderate over the other; or, after comparing two competing forms of good, say society and the furtherance of science, we recognise the latter as the greater. In such cases we are said to consciously choose or decide upon the particular course of action with its attendant result.

Here, it is evident, we reach a higher degree of organisation of the conative process. The selection of a particular action as the result of deliberation makes this action *my own* in a new and fuller sense. That is to say, choosing to do a thing, by the very slowness and deliberateness of the process, the marshalling of all the motive forces bearing on the point, and the emergence of a particular motive of preponderant and decisive force, gives rise to the clearest form of active self-consciousness, investing the conative process with the form " *I* will ". Such a process of elective decision, with the concomitant consciousness of full self-assertion and self-manifestation, reduces the chaos of competing impulse to order and unity. In deliberately deciding ' I will do this,' the whole process with its final result is invested with the unity and consistency of self-consciousness itself.

As an illustration of this process we may take the decision not to indulge in an amusement because through a process of deliberation we have grasped the more than countervailing evil resulting in the shape of delayed work, or unfulfilled duty; or the decision to do a service to another involving a comparatively difficult and disagreeable action rather than follow out some pleasurable line of activity, because we have through deliberation come to see the greater importance of the former end.

Here, again, the essence of the psychical part of the process seems to be an act of volitional attention. Just as volitional deliberation is a designed maintenance in consciousness of certain motive-ideas so as to secure a due comparison of magnitudes, so the final decision is the volitional maintenance of a particular idea so as to secure its realisation in action. Thus, when deciding to serve a friend or to set about a difficult piece of work in spite of more attractive suggestions, we are specially directing attention to, and so realising, the idea of the greater or more important end.

It is not, however, to be supposed that the process of deliberative or rational decision ordinarily takes the lengthy and orderly form here described. For one thing, there is in no case such a perfect active control of the process throughout as has been here assumed. The organic conditions spoken of above, viz., the impulse to do something, and the impatience of the painful state of conflict, always contribute to the result. That is to say, where any decision is arrived at it is, in every case, determined in part, "passively" or mechanically, by the forces making for action, and apart from any conscious decision or fiat of the form " I will ".

Not only so, it is manifest that a prolonged process of reflexion and decision such as we have imagined is exceptional. Our everyday decisions have to be carried out rapidly to meet the pressure of the circumstances of the moment. These conditions do not allow of a drawn-out, detailed reflexion such as we have described. A man of decision is one who by practice and training has learnt how to shorten the process, to take in the whole situation by a rapid and intense effort of mental concentration, and, what is equally important, has the judgment needed for disregarding a multitude of relatively unimportant points, and so simplifying the problem of choice.

The process of choice or decision takes on a specially rapid and compressed form in all cases where a comparison of ends as more or less important, and of means as more or less suitable, are alike excluded, and the problem is reduced to a discrimination of a particular presentation, and a selective carrying out of a corresponding or appropriate action. This simple type of elective decision may be illustrated by such familiar acts as recognising our own boots among a number of pairs and putting them on, discriminating a friend in a London street and greeting him, and so forth. Here there is no question of such an explicit process of comparison and conscious decision as has been illustrated above. Owing to the psycho-physical bond which connects a particular kind of presentation and a corresponding action, the whole process takes on the form of a *quasi*-mechanical succession of events, in which the " will " (even in the simple form of initiating a movement by a preliminary idea of the same) can hardly be said to be engaged.

This line of remark evidently applies to those experimental investigations which have been directed to the determination of the *time* occupied by volitional choice. In this class of experiments the subject is called upon either to react by a movement on the occurrence of one particular presentation, say a given colour among a number, that is, to decide to move as against not moving, or to react selectively by carrying out one among a number of movements (say those of the separate fingers) answering by previous arrangement to the particular presentation (colour, letter, etc.) which happens to occur. That this experiment may introduce a new complication in the shape of a rudimentary " act of choice " (Wahlakt) following on the discriminative recognition of a particular presentation seems evident, and is confirmed by the results. Thus it was found that the reaction-time increases rapidly as the number of movements is increased.[1] At the same time, as already pointed

---

[1] Oddly enough, according to the researches of Merkel, this increase in the reaction-time, with an increase in the number of movements, begins to fall off in

out, this situation does not necessarily illustrate the process of choice at all. Owing to the preliminary attachment of the several movements M^1, M^2; etc., to the several presentations P^1, P^2, etc., psycho-physical processes may be partially set up which, on the occurrence of one of the presentations, issue in a mechanical and sub-conscious working out of the connected movement.[1]

§ 13*a*. *Intellectual Factor in Conation: Enlightened Will.* The processes of deliberation and choice just described bring into a new prominence the intellective factor in conation. The simplest type of conation involves, as we have seen, a rudiment of intellection (discrimination, representation, etc.). In these complex processes of deliberation, however, intellection becomes a main feature. The very use of such expressions, "I am *thinking* what I will do," "I must *reflect* on this course of action," illustrates the point.

It seems to follow that the higher forms of conation which we popularly describe as the outcome of "an enlightened will" are a product of two factors, *viz.*, active impulse and reflexion, or what moralists call "practical intelligence" and "practical reason". Volition, *quâ* volition, is activity, and the highest no less than the lowest type of conation is a manifestation of active impulse. As has been seen, inhibition or the checking of an impulse, so far from being *in*activity, is a particularly strenuous and difficult form of activity. The man who for the moment says 'I will *not to do it*' of a thing he is inclined to do is carrying out in the fullest sense a volitional *action*. That is to say, he is in a state of tense active strain, as may be seen by the very tonicity of his muscles, a state which contrasts in the plainest and strongest manner with the flabby, inert condition of the man who through disinclination and sluggishness can only say "I have no *will to do it*".

At the same time, this later restrictive and regulative

---

the case of some persons from the first, whereas in that of other persons it grows at first rapidly, and only later begins to fall off. (*Phil. Studien*, ii. p. 73 ff. ; *cf.* Wundt, *op. cit.*, ii. p. 310.)

[1] *Cf.* above, I. p. 450 f. On the measurement of such processes of choice along with those of discrimination, see Wundt, *op. cit.*, ii. p. 305 ff. ; *cf.* Münsterberg, *Beiträge*, heft i., especially p. 106 ff. W. James distinguishes between different types of decision, *Principles of Psychology*, ii. p. 531 ff.

activity is only possible through the higher development of intellectual processes. Thus, as pointed out above, the initial motive to such deliberation is the product of a memory of, and reflexion upon, the results of hasty or impulsive action. A youth who cannot think never feels the desire to pause and deliberate. The carrying out, moreover, of the processes of active recalling, comparing, and selectively discriminating the greater from among the less, evidently presupposes a developed intelligence and a trained power of steady attention.

Since in this work of deliberation we have a properly intellectual factor, some psychologists, *e.g.*, Münsterberg, would exclude it from the volitional process.[1] But this seems to overlook the important point that deliberation is the work of *practical* intelligence, that is to say, intelligence as distinctly engaged with ends and ways of realising these. In deliberating how to get a favour from a not too generous person, I am, as in a state of intellection, attending to certain ideational contents; but the interest underlying the activity is different. I am seeking not to find out a fact for the sake of the fact and its logical implications, but to devise a practical measure, to compass an end.

It may be added that this volitional function of deliberation disappears as soon as the strict subordination of the process to practical end ceases. Hence in all cases where reflexion remains mere reflexion, and so far from leading to decision precludes this, it loses its true volitional character. This holds good of people who cannot make up their minds, thinkers of the Hamlet type, in whom reflexion is in excess of active impulse, and stifles this. It holds good, too, of those states of mental disturbance and distress in which the subject is plagued with needless self-questionings. This form of mental trouble, called by the Germans questioning mania (Grübelsucht), is a pathological exaggeration of the thinker's irresolution, and a caricature of normal and serviceable deliberation. Thus one patient affected by the trouble could not decide to cross a street without hesitating and asking herself whether somebody might not fall from a window at her feet.[2] It is to be noted, however, that in many of these cases there is a distinct impairment of volitional attention, and so of the intellectual process itself. A true thinker may become incapacitated for action by seeing too much, whereas the irresolute imbecile or maniac does not see enough, that is to say, does not attend to the facts in all their vital relations one with another.

§ 14. *Resolution : Firmness of Will.* One other common accompaniment of this higher and more reflective type of conation remains to be touched on, namely, resolution. By

[1] Thus Münsterberg writes : " The voluntary action only begins as soon as a motive is selected from among those presented ". (*Die Willenshandlung*, p. 89.)

[2] See Ribot, *Les Maladies de la Volonté*, p. 59 ff. W. James looks on this state as illustrating a want of belief, or of a grasp of things as real. (See his *Principles of Psychology*, ii. p. 284 f.)

this is meant the formation of a distinct determination to perform an action which is seen to lead to a desired end. It is something more than selectively deciding on an end as good or desirable. Such decision, where the actual circumstances allow, may instantly pass into action, as when, for example, a gambler decides to stake a particular sum and instantly places this amount. Here, it is evident, there is no time for a resolution, 'I will do this particular act,' to distinctly emerge in consciousness. In its completely developed form, resolution, like the state of desire itself, has reference to something not capable of being realised at the moment. Thus we resolve to pay a call some hours hence, or to meet some contingency as a wet day, or another person's treatment of us, in a particular way.

Resolution, on its psychical side, is equivalent to a complete process of volition. There is not only the presence and unopposed preponderance of a motive to action, but a distinct representation of, and desire to perform, an appropriate action. What differentiates it from a fully executed action is that owing to the circumstances of the moment the motor idea does not instantly issue in the conscious action. There is at once a distinct and steady representation of an action and an inhibition of this through another representation, *viz.*, of the unsuitability of the actual circumstances. Or, to express the attitude otherwise, there is a preliminary volitional activity in the shape of expectant attention and a preparedness to act in a definite way when the proper moment arrives. On the physiological side resolution appears to involve a partial excitation of the motor and sensory centres engaged in carrying out the action, an excitation which is temporarily inhibited by a reflective process, though steadily maintained through the psycho-physical process of expectation, and ready to overflow into peripheral discharge as soon as this ideational process of expectation gives place to the sensational process of a perception of the suitable circumstances.

From this brief account of the process of resolution we may readily see that it is in a manner the crowning phase of the conative process. Action kept, so to say, in suspense prolongs the initiative stage to the utmost. Such prolongation or delay of execution allows of full opportunity for the development of the active form of self-consciousness. The state of mind or psychosis indicated by the expression "I will" here reaches its maximum distinctness. Hence the tendency to look on resolution as the most essential factor in the conative process. According to this common view, we only fully assert our will when we definitely and firmly resolve to do a thing.

It follows, however, from what has been said above that there may be a true conative process into which resolution or 'I will do this' does not distinctly enter. Resolution, as just defined, only begins to appear distinctly as soon as the growth of representative power enables a person to anticipate the future and to pre-arrange his actions.

It may be added that resolution enters into all action so far as this becomes complex in the sense of involving a prolonged activity, or a series of combined movements. Thus, in carrying out a mechanical process as carpentering, in looking up a friend, or in preparing for an examination, we must, it is plain, maintain from the outset a state of determination or resolution with respect to the later stages of the performance. The frequency of incompleted action illustrates this point; for the abandoning of things when only partly done means that the attitude of resolution was not strong enough; that is to say, that the desire for the end, the achieved result, and the readiness to carry out the required actions as the proper moments arrive, were not sufficiently persistent.

What applies to resolution on its positive side (as resolution to do) applies to it also on its negative side (resolution not to do). To resolve not to do a thing involves the anticipation of a certain situation, together with the prompting to a certain action, and, on the other hand, a preparedness of mind to curb the impulse. Resolution, though conceived here to follow a process of choice, frequently appears without any explicit conative selection, as when a desire for an object is unopposed by other desires, though its realisation has, owing to the circumstances, to be postponed. Since resolution implies the maintenance of the idea of an end, and further of that of an opportunity of actively realising this end, it is liable to fail through the lapse of this ideational activity. Hence so many of our resolutions are temporary only and abortive. Again, since resolutions are arrived at in the absence of the appropriate circumstances, they are, even when strong and persistent, no perfect guarantee for actual performance. Their future efficiency will depend on the adequate representation of *all* the circumstances. This accounts for the ignominious collapse of so many brave resolutions when subjected to the touch-stone of actuality.
＊

In all such persistent resolution we have a new display of
"will-power". Strength of will is commonly judged by steadiness and pertinacity of resolve. More particularly, it is tested
by firmness, that is, maintenance of the resolute attitude under
heavy and prolonged discouragements, as in the now historical
crossing of the African forest by Stanley and his party. Such
pertinacity in the face of difficulty has a noble æsthetic aspect,
and has often been illustrated in fiction, nowhere perhaps more
worthily than by Scott in the quiet, heroic perseverance of
Jeannie Deans in her journey south. This steady maintenance
of a particular attitude is a remarkable manifestation of nervous
force ; for it may be safely assumed that the persistence of a
definite volitional idea has for its neural correlative a prolonged
tension of one and the same group of cortical elements.

Where, instead of deterrent difficulty, seductive allurement
of any kind comes in to break the spell of a resolution, we have
this pertinacity under the form of what is commonly known as
firmness or independence of will. Here the attractions of other
objects, the suggestions of friends, and so forth, present themselves as competitors with the particular end pursued, as when
the Sirens seek to woo Ulysses from the arduous toils of the
sea. In this case there is a reinforcement of the dominant
resolution by a special act of concentration, involving a turning
away from and a rejection of the alluring idea. It is to be
noted, indeed, that all pertinacity in resolution is akin to steady
mental concentration, and is indeed merely the practical manifestation of this quality.

Such resoluteness or firmness constitutes a particular volitional quality. It is not by any means the same thing as the
ability to choose and decide from among a number of alternative courses. There are men of a light, versatile habit of mind
who can frame decisions with the greatest readiness and yet
show themselves especially incapable in holding to their decisions ; just as there are others who are slow and awkward
in deciding, but having once decided, are as firm as a rock.
It is needless to dwell on the moral importance of the quality.
It is only as men are known to be resolute that they are to be
counted on.

While, however, firmness is the very backbone of what
we call will, it is apt to take on an exaggerated and hurtful

form. This is known as self-will in its undesirable form, love of opposing others, or obstinacy. As pointed out above, a developed will is rationalised in the sense that the conative process is illumined by calm reflexion. Accordingly, obstinacy, when it amounts to a refusal of others' counsels and suggestions, to a determination under no circumstance to reconsider a resolve once taken, becomes irrational. Such rigidity of resolution is manifestly fatal to growth. The wise man combines firmness in ruling principles with a certain modifiability of particular decisions. Here, then, as in so much else, there is but a step from the sublime to the ridiculous, as we may see in the half-admirable, half-amusing perversity of Lessing's Tellheim and of other characters in comedy.

The quality of firmness or "dourness" is a well-known characteristic of a particular variety of temperament, individual and racial. There are men who, like children, cannot keep to a resolution; whereas there are others who find it just as hard and painful to alter a resolution when once made. The difference cannot be set down wholly to a property of the ideational process, *viz.*, fixity of the motive-idea. It is noticeable that during protracted, fatiguing exertions, as in climbing a mountain, or writing a book, the brave initial impulse often exhausts itself before achievement is reached. The man who, half-way up a mountain, when breath is spent and the legs begin to totter, will not 'give in' may almost have lost sight of the goal, or at least have become comparatively indifferent to the prospect of reaching it. His stubborn pertinacity, so far as it has a clear conscious motive behind it, looks like a secondary volitional process, the prompting of pride and a feeling of self-consistency to a new determination not to be beaten. More frequently, perhaps, it loses the distinctly volitional, end-pursuing character altogether and becomes a kind of blind, obstinate going on with the matter in hand ; or, so far as conscious aim comes in at all, as a striving to escape from the fatigue and possibly the disgust which the self-prescribed task has brought with it.

In all this we see a close analogy between pertinacity or perseverance in action and habit. Both are characterised by loss, more or less complete, of the initial pleasure-impulse, and the substitution of a secondary desire to escape pain for the primary desire to reach a pleasurable end. Both appear to show the effect of organisation or nervous setting in reducing the conscious concomitant of action. Hence it seems reasonable to suppose that the two phenomena have a similar neural basis, *viz.*, what may be called the inertia of the nervous mechanism. The man of habit and the man of inflexible resolution are not two types, but one, and they illustrate a special tendency of the nervous substance to persist in old modes of functional activity rather than to undergo modifications. Or one may say that habit and pertinacity of resolve are a special manifestation in the active region of the conservative property of the nervous centres.[1]

§ 15. *Process of Self-Control.* Through the development of

---

[1] *Cf.* above, I. pp. 55 and 186 ff.

*

these higher processes of deliberation, intelligent choice and resolution, we acquire what is popularly described as the power of self-control. This expression, though a somewhat loose one, brings out the fact that the later and maturer stage of volition implies a systematic and intelligent regulation of the earlier and more instinctive impulses through the stable formation of motive-ideas and motive-principles. Action is now no longer determined by the impulse first excited, *e.g.*, appetite, the desire to possess a thing, or to return an insult, but, owing to the complication introduced by the development of higher, more complex and calmer, because more intellectual or representative, motives, follows other directions and aims at other ends.

The character of the motive-ideas or rational motives which, as a result of development and education, thus become predominant in the normal man has been sufficiently described above. It may be added that the operation of these regulative motives is accompanied with a peculiar *dual* form of self-consciousness, *viz.* (as the expression *self*-control suggests), the opposition of a lower and more passionate to a higher, more thoughtful, and more judicial self.[1] It is only as the higher motives grow so predominant that the force of the lower impulses is habitually kept down that this divided self-consciousness gives place to a new and a larger consciousness of a united self.

The attainment of this condition in which ideational motive preponderates over sensuous or instinctive impulse presupposes a process of physiological development. As pointed out above, the rise and expansion of the ideational life is conditioned by the growth of higher nerve-centres. The motive-ideas, which, as *general* aims or principles, guide the actions of reflective men, are presumably correlated with those higher nerve-centres which, there is reason to conclude, are concerned in the formations and the maintenance of general ideas.[2] This inference is supported by such facts as that when overtaken by fatigue and loss of vigour, as through over-exertion, or ill-health, the power

---

[1] This idea of a duality of self is well brought out in popular descriptions of the process. Thus Shakespeare speaks of being passion's slave, of fighting against, and, again, of bridling passion, and so forth.

[2] *Cf.* above, I. p. 390.

of self-control is sensibly impaired, that it is manifestly interfered with by an excessive use of stimulants and other practices
injurious to the nervous system, and that its decline is one of
the earliest symptoms of the on-coming of *senile* decay and of
mental disease.   This liability to a weakening of self-control is
explained by Hughlings Jackson and others on the principle
that the higher and later developed nerve-centres are less stable
than the lower, the order of impairment or dissolution being
the reverse of the order of evolution.[1]

The processes of self-control have a positive or stimulative,
and a negative or inhibitory aspect.   The first is illustrated
when we do a difficult action ; for example, leap out of bed on a
cold winter's morning under the prompting of a representation
of work to be done.   The development of the higher ideational
motives means an extension of our incentives to action, and the
introduction upon the volitional scene of promptings which are
urgent and effectual in cases where natural inclination points
to inaction.   In this case we may suppose that the higher
nerve-centres supply the initial impulse in the process of
" innervation " which lower centres fail to supply.   The second
or inhibitory aspect is illustrated whenever we restrain an impulse to act, or decide *not* to do a thing, as in declining a tempt-
ing invitation on the ground that its acceptance would throw
us behind with work.   Here it is commonly assumed that the
higher cortical centres exert an inhibitory influence on certain
lower centres which are partially excited in connexion with
the prompting of the impulse to act.

The volitional processes known as self-control assume a
somewhat different form in three cases.   In the first place,
the effect of the psycho-physical process in the operation of
the motive-idea may be to excite or to repress action, as in the
examples given above.   This we will call the control of action.
In the second place, its effect may be to modify the phenomena
of feeling, and more particularly emotion or passion, as when
we repress anger.   This is commonly spoken of as the control
of the feelings.   Lastly, its effect may be to modify the course

---

[1] On the connexion between self-control and the integrity and efficiency of the
higher cortical centres, and on the effects of narcotics, disease, etc., see Carpenter,
Mental Physiology, bk. ii. chaps. xvii., xviii.   *Cf.* Ribot, *Les Maladies de la
Volonté*, chaps. i. and ii.

of intellectual phenomena, percepts and ideas, as when a student under the stimulus of a love of knowledge deliberately concentrates his attention on a particular subject to the disregard of other and attractive matters. This is known as the control of the thoughts. Although, as we shall see, these forms of the process of conative regulation are not radically distinct, it will be convenient to deal with them separately.

§ 16. *Control of Action.* (1) After the above account of the process of self-control in general little need be added as to the particular form of it marked off as control of action. Remembering that the process is at once an excitation and a repression of action, we may proceed to briefly indicate the successive stages of its growth.

The simplest manifestation of the process here considered occurs when the prompting or deterrent effect of an actual feeling is counteracted by the mere anticipation of another particular or concrete feeling, as when a child overcomes his indolence and sets about preparing his lesson in order to avoid punishment, or when he stops his noisy play in obedience to a command. Closely connected with this variety is that in which a gratification directly suggested by the surroundings of the moment and known to be immediately realisable is declined in favour of a remote one, as in all the earlier and simpler forms of abstinence or putting by things.

A higher stage of control is reached when intelligence is developed and motive-ideas representing the enduring ends, or interests, such as health, reputation and knowledge, come into play. Here, as has been sufficiently explained above, action comes under the control of more complex and intellectual conative products, particular actions are brought into a relation of consistency and unity one with another, and so take on a more reflective or rational character. The subordination of particular and temporary ends to these general and permanent interests is an exercise of control. Thus a youth reaches a higher form of control when he sets the maintenance of his health before him as an enduring end, and represses all desires inconsistent with this. In this way he practises the virtue of temperance.

This process of unification is carried a step higher, and action consciously brought into a relation to a single consistent self in what has been called prudence and, better, (practical) wisdom, *viz.*, the intelligent pursuit of the highest personal good. By this is meant that the several aims and interests of life are thought of together, compared as to their relative magnitude or importance, and adjusted in such a way as to yield the greatest sum of happiness to the individual. This implies the subordination of each of the enduring interests, as health, knowledge, to a still higher, more comprehensive, and more abstract principle of action, in which the consciousness of self as an organic unity grows more distinct and prominent.

As a crowning stage in this development of self-control we may take the superinducement on the motive-idea answering to the individual's personal or private good of the yet larger motive-idea representing the common or general good. The psychological character of this motive-idea has been touched on above. While, however, as we have seen, it can to some extent be assimilated in its form and mode of action to other motive-ideas, it stands out from among these partly by its greater psychological comprehensiveness and partly by its peculiar practical or ethical function. The collision between the egoistic impulses and consideration for others is not only frequent and sharp in early life, before education has done its work, but constitutes the capital circumstance in all moral experience, and has furnished the standing theme to all the more elevated forms of fiction. The chief aim of ethics, according to the modern view, is to determine the proper adjustment of the claims of the individual and of the community. All normal development of volition involves the strengthening of the other-regarding as against the self-regarding motives. This is effected to some extent by the growth of those motives answering to 'common' interests, as knowledge and art, since these lift the individual above the thought of merely personal good and attach him to an object of common attachment and pursuit. The motives, however, which most effectively restrain and limit the action of the personal aims and desires are those of duty, benevolence, and generally what we call humanity. It is through the subordination of interest to duty, of egoistic to 'altruistic' prompting, that the highest attainments in the art of self-control are reached.

The operations of self-control here described when carried

out with full explicit consciousness assume the form of acts of obedience to a self-imposed command. Thus a man restraining appetite, or speaking the truth in face of serious risks, may be said to be applying to himself the rule or maxim ' Be temperate,' ' Be truthful '. In this way, as moral development advances, we pass from mere obedience to an external authority to obedience to the inward voice of reason and conscience.

§ 17. *Control of Feelings.* (2) We saw above that all feeling is organically connected with muscular action or movement. In this way it is brought into a certain kinship with action, especially with the earlier and cruder forms of this last. Accordingly we may expect to find the regulation of the feelings allied to that of the actions. Observation verifies this expectation. All control of passion appears to take its rise in an inhibition of the motor factor in emotional display, as the movements of the limbs, the actions of the voice in an outburst of passionate grief. The effect of this inhibition is clearly illustrated in the endurance of pain, *e.g.*, under a surgical operation, which is effected primarily and mainly through an energetic repression of movement by means of an innervation of particular groups of antagonist muscles, *e.g.*, of the jaws, of the hands in clenching the arms of a chair, and so forth. We may suppose that in all such cases, just as in that of restraining an impulsive action, an inhibitory influence is transmitted from the centres engaged in the higher motive process to those concerned in the process of emotive discharge. Owing to the organic continuity of the whole emotional process, it is found that this arrest of external movements tends, to some extent, to allay the whole excitement. We seem, in a measure at least, to crush a passion by a convulsive closing of the hand, or pressing of the foot against the floor. Similarly when we control a feeling of laughter in unsuitable circumstances by carrying out energetic inhibitory actions of the respiratory and other organs.

What the exact effect of this inhibition of external movements will be in any given case depends on a number of varying conditions, as the strength of the feeling, the history of its rise, and the particular emotive temperament of its subject. If an emotion, say of animosity, is intense, and if, further, it has been gathering in force for some time, the suppression of its external signs may do but little to reduce the intensity of the feeling itself. The man or woman may, while outwardly calm, go on nursing the passion internally by brooding on ideas of satisfaction. Temperament will affect the result through the difference between the volatile or readily changing, and the tenacious or stubborn character of the feelings. In the former case, which is generally exemplified by the young, a slight amount of motor inhibition may suffice to stem the affective current, whereas, in the latter case, the effect of a considerable restraint of movement may be but small.

So far we have touched only on the inhibition of the feelings. But here, as in the case of the control of action, there are at once a process of restraint and of excitation. Paradoxical as it may sound, the volitional regulation of feeling includes an intentional rousing of emotive states. The simplest form of this volitional excitation of feeling is seen in all histrionic imitation. Here, it is evident, we have the counterpart of the inhibitory motor factor in the restraint of passion. The mimic takes on the outward motor manifestations of the feelings he represents, such as the facial movements, gestures, bodily pose, and modifications of vocal action. Such assumption of motor concomitant, by introducing a part of the bodily resonance of the feeling, tends in a measure to develop it.

How far a mere assumption of motor concomitants arouses the corresponding emotive experience is doubtful. In all ordinary cases of histrionic imitation there is along with and behind the bodily simulation an intense imaginative realisation of the situation, and of the state of mind of the hero. Hence the fact attested by a number of actors that they are really affected by the grief, terror, or other emotion which they assume, even if it were not opposed by the statements of others, would not be conclusive on the point. A decision of the question might be obtained by observing the effects of suggestion on the hypnotised subject. Here it looks at least as if the inducement of a particular bodily attitude, e.g., that of anger, tends to bring on the feeling. But unfortunately we know as yet too little of the patient's state of mind in such cases to enable us to say whether and how far he actually experiences the feeling he appears to express.[1]

[1] Some account of the effect of excitation of the emotive attitude in the hypnotised subject by inducing a particular pose of limb, e.g., fingers closed and arm stretched out, is given by Carpenter, *Mental Physiology*, p. 605 ff. W. James points out that in the hypnotic state the passive production of particular motor ingredients in the whole expression of an emotion will suffice for bringing on a full sympathetic excitation of the other movements, so that "at last a *tableau vivant* of fear, anger, disdain, prayer, or other emotional condition, is produced with rare perfection". (*Op. cit.*, ii. p. 603.)

When we consider that the action of the voluntary muscles is after all but one among other somatic factors in the embodiment of feeling, we need not be surprised that the control of this factor only carries us some way in the regulation of our feelings. As we shall see presently, the complete and efficient control, in respect both of repression and of excitation, is carried out in a less direct manner, *viz.*, by a control of those presentative and ideational phenomena to which, as has been shown, feeling is so closely related as concomitant or effect. Thus, in the cultivation of moral and religious feeling, it is through fixing the mind on certain ideas that the desired emotional changes are secured. Similarly the mastering of passion is always more than a mere suppression of a muscular concomitant, and includes the withdrawal of the attention from certain excitant ideas.

Enough has been said to show that the control of feeling is a more difficult action than that of movement. It is easier in general to restrain an impulse to do a thing than to subdue the excitement of emotion. This is explained by such considerations as the following : (*a*) Emotion as such is a state of psychical commotion or excitement, and so opposes that steady mental fixation of particular motor representations which is involved in the inhibition of its display. (*b*) The motor discharge in the case of emotive manifestation is rapid and immediate, not depending on a preceding representation of movement as a properly voluntary action does. (*c*) The muscular reaction in emotive states is diffused and comprehensive, and not definitely restricted as in the case of a voluntary movement. Indeed, as already suggested, a true voluntary movement may be regarded as in itself the result of a selective restriction of motor action, so that it lends itself to a further inhibitory process.

This greater difficulty in the control of the display of feeling is seen in the lateness of its acquisition by the race and by the individual. From what we know of the lowest stages of human culture, primitive man must have been but little capable of a repression of turbulent passion. Similarly we find that children are mastered by their feelings, and appear to need a certain amount of exercise in the inhibition of movement in the comparatively calmer processes of gaining ends, before they can carry out any restraint of muscular action in these excited states. The lateness of the mastery of the turbulence of passion further suggests that nature has not endowed man with an instinctive incentive to this form of self-restraint, as is supplied in the case of restraint of impulse by the instinct of self-preservation, or of avoidance of what is hurtful.[1] The desire to keep passion under control presupposes experience, and the development of comparatively late feelings, more especially a regard for appearances, and for others' comfort. Lastly, the backwardness of this form of self-control in children and among uncivilised races illustrates the fact just brought out, that the repression of feeling through the medium of the motor factor is only partial repression, that perfect mastery includes the ability to regulate the thoughts, an ability which is only attained as the result of special intellectual discipline.

§ 18. *Control of the Thoughts.* (3) Just as the higher processes of conation involve a regulative modification of the feelings, so it includes a control of the remaining group of psychical processes, *viz.*, the intellectual. As has been illustrated above, the work of intellection is essentially one that is actively instituted and maintained. The processes of volitional attention constitute a main factor in all that we understand by thinking. This applies alike to the selective concentration of the attention upon particular sense-presentations in observation, to recollection or the active recalling of past experiences, to the constructive rearrangement of imaginative material in what is specially marked off as imagination or invention, and to the varied and complex forms of elaborative activity which enter into the processes of thought. After the consideration already given to this point we need only add a few words on the distinctly conative or volitional aspect of the process.

The regulation of the ideational processes, as in seeking to recall an idea, illustrates the two characteristics of conative phenomena, *viz.*, active consciousness, and initiation by representation of end.

---

[1] Viewing the matter biologically, we may say that control of feeling is less urgent in the interests of life-preservation than that of impulse. The hurtfulness of violent childish passion is as nothing compared with that of many of its impulse-prompted actions.

That volitional attention is an active state has already been sufficiently illustrated and explained. An effort of thought, as in trying to localise a quotation, or to find the explanation of a difficult and crabbed fact, has much in common with an effort to repress a furious impulse, or to remain still under a painful operation. There is a factor of muscular tension in each case, and this seems to some extent to consist of the same constituents, *e.g.*, in my own case at least, strong compression of the jaws, the inhibitory process known as holding one's breath. How far the psychical concomitants of this muscular tension exhaust all that we mean by an effort of thought, or of attention, is, as pointed out above, a question not easily determined. We have found reason to conclude that the known effects of voluntary attention in modifying the course of presentations and ideas, giving special prominence and persistence to certain members of the series, and repressing others, require for their explanation another psycho-physical factor in addition to the muscular. What may be confidently said is that we should not call it effort and so assimilate it to muscular exertion were there no concomitant sense of muscular tension in the case.

Let us now inquire whether these processes of volitional attention come under the general description of the conative process as conscious representation and pursuit of end. A word or two will suffice to make this point clear. We will begin with a simple crude form of the process. An interesting idea "strikes" us and we proceed to fix our attention on it and to trace out its suggestions. Here, it is manifest, there is pleasure immediately realised. The interestingness, that is, the pleasurableness, of the idea acts as a stimulus to the attention, very much as when eating or drinking the pleasure of the experience leading to a desire to continue and increase it prompts and sustains a simple mode of conative process. It may be added that just as the pleasurableness of the muscular activity itself is a common contributing motive in action, so, where the requisite organic conditions are present, the exertion of attention may in itself prove agreeable, so that we go on attending in part for the sake of the activity.[1]

We may now pass to the higher and more difficult action

---

[1] *Cf.* Stumpf, *Tonpsychologie*, ii. p. 281.

commonly understood by 'volitional' attention. The term "effort" is generally restricted to those intenser or more prolonged exertions which, instead of being pleasurable as moderate activity is pleasurable, involve a measure of painful consciousness (sense of over-strain, of fatigue). It is thus, as we shall see more fully by-and-by, a very marked manifestation of what we mean by will. We only make an extreme effort when the requisite conditions, *viz.*, a sufficient amount of organic vigour, and of intensity of initiative stimulus in the shape of a strong desire or motive, are present. A little reflexion shows that what is variously called mental effort, effort of thought, and effort of attention, illustrates these conditions. When we make a special effort to find an answer to a question, that is, to discover an idea, or, on the other hand, to banish a group of ideas from the mind, there is in each case what we mean by a need, or craving for something. The intellectual search for a suitable word, an apt metaphor, and so forth, is determined by a desire analogous to the bodily need which urges a man to go out at the dinner-hour in quest of a restaurant. This state of intellectual desire, moreover, as we saw above, involves a vague representation of that which is needed, and a strong impulse to realise, that is, to bring into clear consciousness, the required idea.

While there is this general agreement between the form of the conative process in volitional attention and volitional (external) action, there is a sufficient amount of difference to account for the fact of their being commonly distinguished. To begin with, the muscular element, though present and easily recognisable in attention, is not definitely localised and patent to view, as in the case of carrying out particular actions. Still more important, there is no definite preliminary representation of the movements involved as there is in carrying out an action. In saying ' I will give this matter my thought ' I am not imagining a particular group of muscular experiences, as when I say " I will walk over and call on you ". The process is further differentiated by the nature of the motive. To seek to recall a fact, or to solve a problem, is an action that grows (immediately at least) out of our intellectual life and interests, the desire to know rather than the desire to do. With this is closely connected another difference in the result of the process. Whereas an action when successful substitutes for the idea the actual sense-presentation, volitional attention substitutes for a vague fluctuating idea a definite stable *idea*.[1]

[1] It must not, however, be supposed to imply that there is less gained in intellectual search than in the pursuit of practical ends. In many cases the attainment of the object of intellectual pursuit is manifestly much more of a realisation, than that of the object of a course of action. That is, the transition from the condition

§ 19. *Connexion between Control of Thought, Feeling and Action.*
While we have thus distinguished between these three forms of
control, we may easily see that they always involve one another
to a greater or to a less extent. This follows, in the first place,
from the mutual opposition of the three psychical functions.
Since feeling, thought and action are in their intenser and more
fully developed form mutually exclusive, it is evident that the
positive furtherance of any one by a process of conation will
involve the inhibition of the others. A word or two will suffice,
after what has already been said, to make this clear.

To begin with the effects of feeling, since all emotional
excitement agitates and disturbs the whole psycho-physical
system, disarranging the mechanism of attention, and substituting a capricious feeling-determined succession for a logical
order in the flow of the thoughts, it follows that the perfect
command of the intellectual processes presupposes the control
of the feelings. In order to think we must first be calm and
self-possessed. Inasmuch, further, as emotion takes possession
of the muscular system, it is plain that the inhibition of feeling
is a necessary pre-condition of a full command of the motor
organs, and consequently of action. We cannot act so long
as we are excited by merriment, shaken by fear, and so forth.
In like manner, so far as external (muscular) action and
internal thought are opposed states of mind, the perfect
command of the intellectual processes will include the inhibition of movement. As was pointed out above, the very
attitude of attention, even when directed externally to objects
of sense, is one of bodily stillness or cessation of movement ;
and the internal direction of the attention in the processes of
reflexion or thought illustrates this inhibition of movement
still more plainly.[1]

of need to that of satisfaction involves a greater contrast. The two processes are
closely assimilated by Münsterberg in his account of volition. (*Die Willenshandlung*, p. 63 ff.)

[1] It is remarked by Dr. Ferrier that the internal diffusion of nerve-energy as
involved in processes of thought, and the external diffusion of it in muscular action,
vary in an inverse ratio. Consequently, " in the deepest attention, every movement
which would diminish internal diffusion is likewise inhibited. Hence, in deep
thought even automatic actions are inhibited, and a man who becomes deep in
thought while he walks may be observed to stand still." (*The Functions of the
Brain* (first edition), chap. xii. § 103.)

We may now look at the relation between the control of the thoughts, of the feelings, and of the actions, as determined, not by the opposition, but by the connexion between these mental states. And here we have to do with two cases, namely, the dependence of feeling on intellection, and of action on intellection and feeling.

(1) We have seen that all emotion is excited in connexion with certain intellectual phenomena, presentations and representations. We are glad and sorry, laugh and sigh, because of the occurrence of certain corresponding ideas, as when we anticipate a success, imagine a painful or a ludicrous situation, and so forth. Hence the importance of a command of the intellectual processes as a condition of a regulation of the feelings. As was pointed out just now, we can only very imperfectly control feeling by mastering the accompanying movements. The only effective volitional action on the feelings, whether for purposes of inhibition or of excitation, is indirect, *viz.*, through a control of the intellectual processes concerned. Thus, if I want to escape from a grief, or to quell a fit of anger, what I have to do is by suitable acts of volitional attention to modify the underlying ideational processes, inhibiting certain exciting ideas, as in confuting a groundless supposition. In like manner, where the object is to arouse a sluggish emotion, as æsthetic admiration, the volitional process consists in a special fixation and intensification of particular presentative or representative elements, *e.g.*, harmonies of form and colour, sublimities of poetic suggestion.

(2) Again, it has been sufficiently shown that both feeling and ideation are involved in action. To desire, to consciously exert oneself in the pursuit of an end, is to be under the influence of an affectively coloured idea, *e.g.*, that of a delightful intercourse with a friend, of a holiday tour, of a rise in reputation. It follows that a full regulation of action includes that of the feelings and of the thoughts so far as provocative of these. Here, again, we may see the connexion illustrated alike in the excitation and in the inhibition of the phenomenon. We only rouse ourselves to arduous and worthy deeds by steadily fixating the connected ideas. It was by the growing imaginative realisation of the coveted power that Macbeth's faint heart was excited to the pitch of effective volition. In

like manner, it is only through the volitional repression of those ideational and affective processes out of which conation takes its rise that the perfect inhibition of action is attained. A vindictive impulse cannot be said to be completely mastered until the feeling of anger out of which it springs is stifled, and consequently until the idea of the injury which excites the feeling is banished from the mind. Hence the importance assigned in the best ethical systems to the control of the desires and thoughts 'of the heart'. The process of deliberation described above implies a considerable ability in controlling the thoughts and along with these the feelings depending on them. It follows that in order to postpone action and to consider calmly and with judicial impartiality the advantages and disadvantages of a particular course, a man requires a highly developed and well-trained power of volitional attention.

§ 19a. *Volitional Control of Belief.* The close organic continuity of the several processes described as self-control is illustrated in the volitional regulation of our beliefs. In the account of this psychical state given above, it has been pointed out that, although primarily an intellectual phenomenon, it is profoundly influenced by the affective factor of mind, and to some extent by the active also. It may be expected, therefore, that a volitional regulation of the process will include that of feeling and of active impulse.

At first it may seem paradoxical to say that we can to some extent believe and cease to believe what we will. The logician, whose concern is with the due regulation of thought, has so accustomed us to look upon belief as a rational process determined in the case of all alike by certain laws of thought that we are apt to overlook the fact that each of us volitionally chooses and determines his own set of beliefs. Nevertheless, popular language suggests that believing is to a considerable extent a matter of volition. Thus, to mention but a single writer, Shakespeare makes his characters bid one another believe and not believe, speaks of making oneself and others believe, and so forth ; that is to say, applies to *believing* the same forms of expression that we all apply to *acting*. A word or two, in addition to what has been said above respecting the nature and conditions of belief, may suffice to illustrate the co-operation of a conative factor in the process.

Belief is apt to accompany all ideational groupings which reach a sufficient measure of vividness and stability. Now, as we have seen, one chief condition of this is feeling or interest. Thus, in our imaginative forecastings of the future, we are apt to fixate and raise to a belief-generating stability ideas of this, that, and the other pleasant experience and situation. Such a process clearly involves a rudimentary form of conation. That is to say, we attend to particular ideational complexes because of their greater interest, and of the pleasure which they yield us. In believing we are foretasting that which we desire. This co-operation of a rudimentary conative impulse is especially apparent in cases where we half-knowingly indulge for the nonce in an illusory belief.

From this there is but a step to a more distinct volitional process, viz., the conscious endeavour to believe a thing. The influence of the emotions, as personal affection, and the religious sentiment, upon belief, is apt to assume this form. That is to say, the person is representing a particular state of conviction as desirable, and is volitionally aiding its realisation by appropriate acts of attention, purposeful or half-purposeful detention of the ideas favourable to the belief and ejection of those unfavourable. Religious systems recognise the duty of a volitional development and maintenance of faith by a suitable direction of the thoughts.

In these cases where particular non-logical feelings come in to influence belief the volitional influence is comparatively simple. Far otherwise in those processes which we call the logical control of belief, that is to say, its adjustment to the requirements of an objective standard. Here the volitional process includes the prompting of a highly representative motive, viz., the love of truth, and is directed, on the one hand, to the subjugation of the forces of passion and of impatience to act which favour hasty and erroneous belief, and, on the other hand, to the securing of that large impartial reflexion on all the relevant circumstances which is the only guarantee for the emergence of a just and true conclusion. Here there is still a desire and a will to believe, only that it is a will to believe what is true, that is to say, a volitional process initiated and sustained by a logical feeling or a regard for truth as such.

It follows from the above that all conative regulation of belief is effected through volitional attention. That which we attend to becomes a contributory factor in the generation of belief, that which remains unattended to fails to acquire this influence. Here, again, we see that the fundamental psychical factor in the conative process is attention. In so far as belief is thus determined by attention it evidently takes on something of the character of a conative process. Hence some writers, as W. James, would, in the last resort, identify belief and volition, the affirming of an idea and the affirming of a resolution, as two forms of one process.[1] This, however, is to carry the process of psychological simplification too far. The selective and exclusive fixation of particular ideas is, no doubt, common to the processes of belief and volitional action. Not only so, as we have just seen, there is frequently a rudiment of volition, *viz.*, the aiming at an end, in belief, just as there is a rudiment of belief (in the attainability of the end and the efficiency of our action) in volitional action. These facts do not, however, serve to obliterate the distinction between the state or attitude of belief, as an intellectual grasp of the fact of reality, and that of conation proper, *viz.*, a striving towards the attainment of a desired reality.[2]

§ 20. *Limits of Control : Measurement of Volitional Force.* Each of the forms of volitional control just illustrated has its limits.   Thus there is a state of lethargy or depression of active energy, out of which even the most powerful motive may fail to rouse the subject.   At the other extreme there is a strength of instinctive or " organic " impulse which no ideational motive can overcome.   The story of the horrors of shipwrecked mariners, and so forth, illustrates the fact that in the case of most men no moral or other consideration will hold back a man from slaking thirst when the appetite reaches a certain intensity, and the means of appeasing it brought within tanta-lising distance of his lips.   In like manner the control of feeling has its limitations.   There are hurricane blasts of passion, as when Lear first takes in the fact of his daughter's perfidy, passion against which the will is, for the moment, powerless; and, on the other hand, there is a depth of emotional languor or insensibility which defeats every effort to feel. Lastly, in the region of thought and belief the same fact of a limitation of volitional effort meets us.   There are certain organised forms of experience variously called "inseparable associations," "necessary beliefs," and so forth, which set bounds to our powers of thinking.   Thus the scientifically

---

[1] See W. James, *The Principles of Psychology*, ii. p. 320 f.

[2] On the relation of volition and belief consult, further, Bain, *The Emotions and the Will*, p. 525 f., and E. Rabier, *Leçons de Philosophie*, chap. xxi.

trained mind rejects the idea of an event which contradicts an ultimate "law of nature" as impossible and absurd. There is a certain amount of patent contradiction that the strongest desire to believe in a friend's honour, or in the truths of religious system, will not surmount.

These limitations are not the same in the case of all individuals. The limit to control of appetite in the case of the drunkard and of the temperate man is obviously a widely different one. What we call strength or force of will is, indeed, manifested in the "height" or maximum degree of intensity of the force counteracted. Thus we say a man has a strong will when he can rouse himself to face a cruelty of destiny that would reduce many to a state of impotent cowardice, as in the Hebrew and Greek stories of those who, in obedience to a divine behest, have sacrificed their own offspring on the altar. Similarly we attribute exceptional strength of will to those who resist a preternatural force of temptation, as the valiant heroes of *The Faery Queene*, martyrs who refused to save themselves by recantation, and the like.

Judged of in this way, strength of will is proportionate to the amount of resistance overcome. Now the counteraction of an opposing force is precisely that which necessitates an "effort of will," that is, a moral as distinguished from a physical or muscular effort. We make an effort when we do that which is difficult and "unnatural," in the sense of not being the thing we are first impelled to do ; and this, whether in acting, notwithstanding the inertia of sluggishness, or the certain prospect of suffering, or, on the other hand, in declining to act, when to do so were to realise a considerable gratification. We may say, then, that strength of will is proportionate to the intensity of the effort producible.

This, however, plainly represents but one factor of the case. Conscious effort is a concomitant and symptom of psychophysical friction, that is, of the want of a perfectly smooth adjustment of the processes involved. Repetition and practice tend to diminish this resistance and sense of effort. Hence a man of high volitional power is one who is less sensible of the effort of great and difficult actions than the ordinary man, and the strength of his will may indeed be measured as inversely proportionate to the intensity of the sense of strain.

It is evident that we have here to do with two aspects or dimensions of the volitional process. Of these the first is the amount of conscious, active exertion (in doing or withholding from doing) which is forthcoming under circumstances of great difficulty, that is, involving much *resistance to be overcome*. This may be called the psychical, or, so far as it is the basis of our judgment of the virtue and merit of an action, the ethical aspect. Here, it is to be noted, we assume that there are acts of self-control, such as in holding down the craving of intense appetite, which involve approximately equal amounts of conscious effort in the case of all persons alike—a proposition which we cannot, of course, be sure of, since the opposing force, even in the case of the expansive tension of an elemental passion, or the impulsive energy of an appetite, may differ greatly in the case of different individuals. This is the aspect of self-control which is considered by the plain man in estimating the virtue of actions, as also by the legal judge, and even by the moralist so far as he is dealing with human conduct in the lump. The second aspect or dimension of self-control may be marked off as the biological or econo-mical. This varies indirectly as the other, in so far as it is now the *smallness* of the conscious effort which is taken into account. Here we look on the process in respect of its ease and effectiveness, estimating these by the ratio : quantity of effect produced (*e.g.*, intensity of passion suppressed), to quantity of conscious expenditure of energy, or sense of effort.[1] In considering this economic aspect we are, it is evident, measuring the effects of practice and training, the economic result, *viz.*, reduction of friction, effected by previous "efforts of will". This is the point of view taken by those who want to test practical efficiency, and the results of education or discipline.

§ 21. *Habit and Conduct : Deliberation as Habitual.* The point just touched on, *viz.*, the effect of repetition and practice on the higher volitional processes, leads on to the influence of habit. What has been called the law or principle of habit applies not only, as we have seen, to the simpler form of the conative process in voluntary movement, but also to those complex forms just considered. By this is meant that these, too, grow more perfect, more easy, and more fixed or un-alterable through long-continued practice.

The influence of habit is seen in the deliberative process itself. Although deliberation is a slowing and complication of action, a substitution of a reflective for an impulsive and *quasi*-mechanical process, it comes under the modifying influence of practice or habit. Thus in the early stages of volitional development, when action is first arrested by an apprehension

---

[1] Strictly speaking, of course, the effect produced diminishes here, seeing that by repeated exercises of self-control we virtually diminish the force of passion or impulse which has to be counteracted. It is assumed, however, in these cases that the *organic* tendency survives and would make itself felt if the regulative effort were relaxed.

of evil consequences, there ensues conflict and confusion of aim. But successive efforts to master the conflict, and to decide according to reason, serve to render subsequent acts of reflexion and decision easier. The vehement forces of impulse have now been reined in to some extent. Every new exercise of the power makes the pause, the consideration, the final calm decision a less arduous or difficult process. As the final stage of this self-educative process, we have what we call a habit of deliberation. When this stage is reached the promptings of impulse, including the effect of indolence, are checked without any appreciable sense of effort. The temporary postponement of action, and the carrying out of the preliminary steps of deliberation and choice, have become, in a manner, natural. "Custom hath made it in him a property." Such a habit of unfailing reflexion brings with it deliverance from what the ancients spoke of as the tyranny of appetite, or of the passions.

§ 22. *Moral Habitudes.* But the principle of habit produces other effects in this region of conduct. The final decision after deliberation, if a rational and good one, does not need to be arrived at again and again in all similar cases. A particular exercise of self-control, say the quelling of a feeling of annoyance, or the determining to do some unpleasant duty, which, in the first instance, was the outcome of a process of reflexion, will, in succeeding cases, be shortened or compressed into control without such preliminary reflexion.

Here we may see that the process of self-control is becoming habitual in a new sense. Certain motives are acquiring a fixed place in the mind as ruling forces, organically connected with appropriate actions, while other and lower forces are losing ground. Every repetition of the situation calling out this particular variety of action (that is, of action having this particular motive or reason) tends to fix conduct in this direction, that is, to establish a habit of doing. The prevailing motive, for example, consideration for others, now passes into the form of a fixed inclination or active disposition. Or, to express the result another way, we may say that conduct is brought more fully under the sway of a general rule or maxim, so as to be immediately determined by the recognition

of this.[1] This result is what is known as moral habitude.[2]
The more frequently this subordination of impulse to a higher
motive has been carried out, the more perfect is the organisation
of this motive, that is to say, the more certainly and instan-
taneously does it assert itself, and produce its effects. Hence,
where the process of organisation is complete, the vanquished
impulse survives in a nascent form only : that is to say, the
man disciplined in the control of his appetites never experiences
the full force of their prompting.

It is obvious from this brief account of moral habitudes that they illustrate the
psycho-physical process which underlies all habit. Thus veracity, with its confirmed
disposition to speak the truth, implies that this particular motive or tendency is
instantly called up by the appropriate circumstances, *viz.*, the situation of being
called on to state something to another. That is to say, there is an organised
connexion between a group of presentation-complexes and an impulse to follow out
a particular line of action. The perfection of the moral habitude depends on this
instant excitation of the higher motive before the lower impulse, which would
impede its realisation, has time to assert itself. To this extent, therefore, a moral
habit has the mechanicality and unalterableness of all habit. At the same time it
is manifest that we have not in this case to do with a process so little conscious
and automatic as the carrying out of a habitual movement. What is the force of
habit does here is to fix the motive-idea as a general or permanent conative
incentive. The application of this ruling motive or principle to particular cases
always involves some amount of reflective consciousness. Thus in following out
the impulse of veracity a person has to consciously determine what is the truth in
the particular case. In many instances, too, as we shall see presently, the right
application may become a matter of doubt, as in deciding what is beneficent or
just in difficult and complicated circumstances.

§ 23. *Volition and Character.* The word character (from the
Greek χαρακτήρ, mark or stamp) is used in everyday language
to mark off any sort of difference in mental or moral qualities.
Thus we are wont to speak of a person's intellectual peculiarities, special tastes, and so forth, as constituents of his character.
In a narrower and stricter sense the term involves a special
reference to qualities belonging to the *active* side of the mind.
Volition, in its rationalised form, conduct, being the final and
most important outcome of mind as a whole, the word character

---

[1] On the way in which the dominant motives become developed into conscious
principles or maxims of conduct, see some good remarks of Waitz, *Lehrbuch der
Psychologie*, § 56, p. 646 and following.

[2] This term seems best to answer to the ἕξις of Aristotle, which exactly
expresses this effect of persistent, repeated action in developing fixed inclinations.

has naturally come to connote in a peculiar manner those qualities, as active energy and deliberation, which go to constitute the higher type of will.

According to the more popular use of the term, every individual has his own stamp or character. This individual character is fixed partly by the peculiarities of the person's psycho-physical 'nature,' or what we call temperament and idiosyncrasy. Thus the contrast of the volatile and fickle, and the pertinacious and obstinate temper of mind, is, as we may see from its early manifestation, a congenital difference based on certain organic peculiarities. At the same time it is evident that even individual character is a growth, and as such illustrates the interaction of organism and environment. Each man's character may thus be said to be a product of particular environmental influences acting upon a particular set of congenital properties or tendencies. Such action, it is to be noted, while presupposing the existence of particular congenital tendencies, in its turn serves to select from among a whole group of such tendencies particular constituents for special developmental expansion or realisation.

In addition to this everyday meaning the word character has acquired an ethical significance. As employed in the science of ethics, it refers not to variable individual peculiarities, but to certain moral qualities which it is supposed to be the special business of social discipline and education to cultivate in all alike. In this ethical sense 'character' has come to stand for 'good character'. This may be defined as a morally disciplined will, including a virtuous condition of the whole mind, that is, the disposition to think and feel (as well as to act) in ways conducive to the ends of morality.

We thus see that every good or moral man possesses a character in a double sense. In the first place, he has a particular group of intellectual, affective and conative peculiarities which constitute his individual character. In the second place, he possesses certain virtuous principles and dispositions which make up the typical moral character and which assimilate him to other moral men. This moral character, though it presupposes the connate organic base of normal human development, may be spoken of as an acquired product, the result of the action of that set of external influences which constitutes the educative action of a civilised and moral community upon a normal human mind.

(*a*) *Character as Organised Habit.* Confining ourselves now to moral character, we see at once that this consists in the possession of certain acquired tendencies or habitudes which we call virtues, both what moralists distinguish as private ones, for example, temperance and prudence, and as public ones, such as veracity, justice, and benevolence. The excellence of the character can be estimated by the fixity and the preponderance of these dominant dispositions. As we have seen, in all comparatively simple and recurring situations where a lower impulse is opposed to a higher motive, the degree of perfection of the moral habitude is indicated in the completeness of the control and the promptness of the right or good action. The less the disturbing force of the instinctive factor (passion, appetite), the more highly developed the character. Thus our idea of a perfectly temperate man ($\acute{o}$ $\sigma\acute{\omega}\phi\rho\omega\nu$) is of one who does not fully come under or realise the temptation to excess. The height of moral character attained in any case is thus determined by the fixity and the commanding influence of the virtuous disposition, which again, as we saw above, is measurable in terms of the facility, or absence of conscious effort, of the controlling process. Such a fixity of certain controlling motives or dispositions and the correlative principles involves the quality of resoluteness as defined above. A man of character is one who follows what is reasonable, just, and virtuous with persistence.[1] This is brought out in the Stoic definition of character ($\mathring{\eta}\theta o\varsigma$), the disposition to be ever willing and unwilling in relation to the same actions ("Semper idem velle atque idem nolle").

(*b*) *Character as Conscious Reflexion.* While, however, moral character is thus woven out of fixed habitual dispositions (Aristotle's $\H{\epsilon}\xi\epsilon\iota\varsigma$) it would be an error to conceive of it as merely a cluster or group of such habitudes. According to the biological or teleological view of mind the habitual, that is, the relatively *un*-conscious, organic process comes in only so

---

[1] On the interesting question how far a good character implies susceptibility to temptation, and sense of effort in doing right, see some valuable remarks by Mr. Leslie Stephen, *Science of Ethics*, chap. vii. § 3, 'Effort'.

far as environmental features and situations recur in like form, and so require similar modes of reaction. Now while it is true that the external conditions of human life, physical and social, are so far recurrent that our actions may be organised into a certain number of persistent norms or types of conduct, as thrift, temperance, fulfilment of promise, and the like, they are not so uniform in their actual, concrete combinations as to allow of our particular actions becoming in the complete sense habitual. Thus, as pointed out just now, even in the particular working out of a general motive-idea or principle, we have in all new cases to carry out some amount of reflexion in discriminating and recognising the nature of the situation and the particular principle which is applicable to it. To act honestly only tends to become automatic in familiar, oft-recurring situations, as in exchanging coin for commodities over the counter of a shop: it may grow into a problem for the most patient reflexion as soon as the situation becomes exceptional, as when we discover a coin in some public place.

Still more important than this, the selection of the wise and the morally good action will always remain a conscious and reflective process, because of that complexity in the conditions of our life which involves the appearance of opposition and collision between what we see to be equally good and valid principles. The finding out of the right thing to do means, in all such cases, the adjusting of the claims of this and that principle; for example, health and intellectual achievement, public service and promotion of family interests. Here we have a permanent demand for those processes of deliberative reflexion which have been described above.

It follows that the ideal of a wise and a good man, or a perfect character, is one that combines promptitude and even an impatience of reflexion in cases allowing of, and calling for, rapid and partially automatic responses, with a certain reserve of wariness, or a readiness to pause and reflect as soon as new features, and especially an unfamiliar complexity, present themselves. In other words, the perfect character is the one that exhibits just that proportion of the reflective to the impulsive in its actions as is required for the fullest, exactest, and most economical adjustment to the circumstances of the environment.

It follows that so long as there is development or life-progress human action cannot become automatic. This applies at once to the history of the individual and to that of the race. Where automatism sets in, as in senile decay, and in unprogressive communities, it is because life is not expansive, that is, cannot take on further modification of adjustive process. As pointed out above, the formation of the automatic and sub-conscious in human action owes its psychological value to the fact that it liberates the energies of the highest cortical centres for new *conscious* processes. From a psychological point of view, we can never regard automatic habit as our end or τέλος. Nor does a sound biology compel us to think of the processes of organic development as issuing in such automatism. On the contrary, just as the historical fact of the genesis of consciousness at a particular point in organic evolution marks a higher stage of this evolution, just as the more highly organised species of animals have more of this consciousness than the less highly organised, so we may say that consciousness plays a larger and a larger *rôle*, alike in the progress of the individual and of the race, as development advances.[1]

§ 24. *Relation of Higher to Lower Volition.* In the above account of the higher and more complex processes of volition it has been assumed that it is continuous with, and developed by known psycho-physical processes out of, the lower and cruder forms. We have not found any abrupt break in the developmental process. Just as the highest and most complex thought-products arise by gradual stages out of the earlier forms of sense-perception and imagination, so what we call rational choice, resolution, self-control, may be seen to be only more complex forms of the fundamental type of the conative process. In the later, as in the earlier, stages of volition, we have alike, as the essential initiative part of the process, desire for end, and representation of a suitable action.

At the same time, there are certain obvious differences between the earlier impulsive and the later reflective volition. Thus it lies on the surface that this last is differentiated by certain features of special dignity and moral value, *viz.*, consciousness of self as agent, of power, and of freedom. These features, though present in a rudimentary form in all conscious volition, only grow distinct and prominent in its later and more complex stages.

These later characters have been supposed by some to

---

[1] This is admirably stated and illustrated by Guyau, *Education and Heredity* (Engl. transl.), chap. ix.

constitute a difference not merely of degree but of kind between impulse and volition, in the complex sense, *i.e.*, conscious
choice and resolution.   According to this view there intervenes
in the higher stage of action a new principle lying outside the
chain of psycho-physical events altogether, though breaking in
upon and modifying this from time to time.   This principle is
known by such expressions as free-will, the self-determining
ego, and so forth.[1]   We have now to review the processes in the
light of this metaphysical theory.   Is there, it may be asked,
anything in the processes of rational selective decision to
justify the psychologist in betaking himself to this hypothesis
of an extraneous active principle ?   Here two main phenomena
will engage our attention, *viz.*, the peculiarities of the psychosis
(*a*) in what is called effort of will, and (*b*) in the sense or consciousness of freedom or free-will.

§ 24*a.   Form of Psychosis in Volitional Effort: Consciousness
of Power.*   The characteristics of the sense of effort have been
touched upon above.   In a wide sense " effort " is used as
synonymous with active consciousness, and so is coextensive
with the whole field of conation.   In a narrower sense it is
confined to those energetic, severe, and more or less painful
exertions which are required in circumstances of special difficulty.   Effort is thus intensification of active consciousness
above the moderate pleasurable limit.   Its main conditions are
sense of difficulty or obstacle, and a sufficient strength and
persistence of motive to prompt and sustain the exceptional and
painful exertion.

The experience of effort occurs in different forms.   The most
familiar one is that of special muscular strain.   As was pointed
out above, this arises when an action to which desire impels us
is excessive relatively to the power of the organ at the moment,
and so irksome and fatiguing.   The disagreeable feeling of
strain in lifting a heavy weight, in walking when fatigued, and
so forth, are examples of such muscular effort.   In all such
severe muscular exertion there is a considerable range of motor
innervation.   Thus, as pointed out by Ferrier and others, a
special exertion of the arm-muscles is accompanied by a con-

---

[1] See Carpenter, *Mental Physiology*, bk. i. chap. ix. § i.   *Cf.* W. James, *op. cit.*,
ii. p. 569 ff.

siderable innervation of those of the chest, of the head, and so forth.[1]

Along with muscular effort there goes the related phenomenon mental effort, that is, effort of attention.   Here, too, the essential circumstance is the energising, under the stimulus of an urgent desire, to an extent which is excessive in relation to the power of the moment, and so involves the disagreeable accompaniment, a feeling of strain.   It follows from the analysis of mental activity and volitional attention given above that this is closely related to, and has a factor in common with, muscular effort.

In the case both of muscular and of mental effort the irksome and disagreeable feature in the action tends as a mode of pain to arrest the impulse.   When, however, the impelling motive is strong enough the action will be sustained in spite of the disagreeable concomitant.   Not only so, in many cases where the motive is an exciting one, as when a boy is performing a muscular feat in order to win admiration, or a pupil is trying to answer his master's question before the other members of the class, there is no distinct representation of the pain before acting, and consequently no shrinking from it.   In such a case there may be the consciousness or feeling of effort during the action, but there is not an effort of will or " moral effort " in the full sense of this expression.

This last occurs when the painful feature or circumstance is distinctly anticipated and resolutely confronted by the mind. Thus the tired labourer who goes on facing his dreary task carries out an effort of will.   Here, it is to be noted, the consciousness of effort does not arise first of all in connexion with the actual doing of a thing, but appears in the preliminary or initiative stage.   It is, in fact, an effort of decision and of resolution.   It is most strikingly illustrated in the higher kinds of moral effort, as when a boy persists in befriending an unpopular boy in spite of ridicule.   Such effort to make good the defi-

---

[1] See Ferrier, *The Functions of the Brain*, chap. ix.   Though all muscular effort so far as it is excessive is painful, the painfulness is often disguised by other pleasurable feelings, such as the gratification of pride.   It is to be added that when indolent we have to make an effort or special exertion in order to do things which are not beyond our muscular powers, and which may even turn out to be accompanied by pleasure.

ciencies of prompting impulse may be called *assistant* or supplementary.

As we have seen above, however, an effort of will is required not only when a difficult ' non-natural ' thing has to be done, but when an alluring and eminently natural thing has *not* to be done. Thus, in turning away from an enticing prospect, say that of a day's excursion in the country, in deference to a behest of duty, we carry out what is called an effort which may be marked off as *resistant* or inhibitory.

Effort of will, then, in all its forms, whether to do or not to do, appears to be specially connected with an initial deficiency of motive force. The feeling arises as a concomitant of the relatively slow development of some new force distinct from the impulses primarily engaged. In making an effort we seem to throw in our strength on the weaker side, either encouraging and aiding a weak ' I will,' or reinforcing a feeble, half-hearted ' I will not '. Thus the effort involved in jumping out of bed on a frosty morning seems to have as its object to neutralise the momentary preponderance of certain agreeable sensations. Similarly in mastering some explosive impulse, as to resent insolence. Effort comes in to alter the natural condition of the scales, to make the more irksome and disagreeable course weigh down the pleasanter and more natural, or to determine action to follow ' the line of greatest resistance '.[1]

The explanation of this apparent exception to the general principle of willing, that action is the result of the desires (and aversions) excited at the moment, has been hinted at above. This effort of will, appearing in cases of insufficiency of stimulus at the moment, is due to a secondary conative process, *viz.*, a preliminary volitional attention to the representations concerned. This act is best described as one of reflexion. It implies a special direction and concentration of attention, either on an idea fitted to rouse the reluctant action (*e.g.*, the value of a prize), or on one fitted to excite aversion to, and so abstention from, the action (*e.g.*, the evils of self-indulgence). In either case it has as its effect the rendering of particular constituents of the ideational group called up at the moment

---

[1] This is well shown in W. James's interesting account of the phenomenon of moral effort, *op. cit.*, ii. p. 535 ff., especially 548.

more distinct, prominent, and persistent, and, as a result of this, of securing the maximum development of their conative force as motives. According to this view, what we mark off as moral effort is ultimately reducible to mental effort, that is, the sense of strain accompanying an act of voluntary attention carried out under peculiarly difficult circumstances.[1]

As implied also in what has been said above, this act of attention, like all other actions, is prompted by its proper motive, which may be called the motive of reflexion, and which is a product of thought about our experiences. It has more of the characteristics of aversion than of positive desire, since it is a recoiling from the evils or pains incident to hasty action on the one side, and neglect or hasty abandonment of it on the other. It is a motive which presupposes a high development of intelligence ; for it implies that the mind has again and again gone back on its actions and found out by a process of comparison that the momentary prompting may lead to ill results, that the actual present or proximate tends to shut out from view the remote, that the presentative has an unfair advantage in competition with the representative. In ' making an effort ' to fix our mind on a distant good or a remote evil we know that we are acting in the direction of our true happiness. Even when the idea of the immediate result is exerting all its force, and that of the distant one is faint and indistinct, we are vaguely aware that the strongest desire lies in this direction.[2] And the resolute direction of attention in this quarter has for its object to secure the greatest good by an *adequate* process of representation.

This motive assumes its highest form in deliberation. Here we may at the outset be far from sure that the good lies

---

[1] It has been assumed here that effort of will is always in the direction of the morally best. But this is not necessarily so in all cases. Even a lofty moral motive (*e.g.*, patriotism) may reach such a pitch of inflammatory excitement, possessing the mind so commandingly and exclusively that an effort of will may be required to call up what are commonly spoken of as comparatively unworthy prudential considerations.

[2] This knowledge of the motive-value of a representation not fully developed at the moment is clearly analogous to other phenomena already touched on, as the awareness of the representative function of the idea, of the universality of the conceptual idea, and of the inferiority of the representation to the presentation, which last, as we saw above, is involved in all desire.

away from the direction of the desire which is at the moment uppermost. But experience has taught us that this is frequently so, and thus the motive underlying the process is the apprehension of the possibility or risk of acting (or not acting) hastily, that is, from an insufficient review of the factors of the situation.

Finally, it may be said, with some degree of confidence, that the phenomena of volitional effort lend themselves to a psycho-physical description. The intervention of effort as an altering, reversing factor, analagous to the addition of a scale-turning weight, probably means in physiological language the co-operation of certain highest or latest developed nerve-structures, the action or discharge of which makes good the biological deficiency, or effectually inhibits the · biologically excessive action of certain lower nervous planes. That the special intervention of these in the case should have as its concomitant a sense of effort, of obstruction to be overcome, is just what we should expect from our general conception of the relation of consciousness to the hierarchy of the nerve-centres. That the action of these centres, whether supplementary and reinforcing or inhibitory, should be capable of producing such great results as appear in the more astonishing and heroic forms of self-control seems consistent, to say the least, with what we know of their physiological function. Speaking biologically, it may be said that the utility of these supreme structures depends on their being able to counteract the inertia of the subordinate centres. That the maintenance of this function costs much, and is readily lost, is illustrated in the lapse of controlling power, including that of making an effort, which immediately ensues on a lowering of the cerebral energies through fatigue, ill-health, etc., and on that impairment of function which marks the on-coming of old age, and still more manifestly of mental disease.[1]

It remains to add a word on the consciousness of power which is specially developed in connexion with this experience of volitional effort. As we saw above, the consciousness of self as agent capable of producing certain effects or changes

---

[1] On the whole subject of volitional effort, see James, *loc. cit.* On the pathological aspects of the phenomenon, see Ribot, *Maladies de la Volition*, p. 64 f.

in the environment begins to arise, from the first, in connexion with the experiences of voluntary movement. As the process of volition expands by the enlargement of the ideational initiative stage, this consciousness is enriched by a fuller representation of the self, its feelings, and active tendencies. It is, however, only when the process of development is carried far enough to allow of the experience of consciously reinforcing a weak, or mastering a too strong impulse, that the consciousness of power grows distinct. Here the effect is seen to depend not only on a desire or purpose recognised as 'mine,' but on a special exertion of 'my' consciousness. This concentration of mind on certain motive-ideas presents itself, by reason of the intensified consciousness it involves, as well as of the extended processes of ideation or reflexion which it calls in, as emphatically an act of self, or as a self-assertion. Hence the experience is precisely fitted to develop a clear consciousness of the self as efficient agent, or as endowed with a power of causally determining the carrying out or not carrying out of a particular outward action, together with the results which are immediately dependent on this.

§ 25. *Consciousness of Freedom : Free-Will.* The culminating phase of volitional development is action accompanied by the consciousness of freedom. Since this is among the most subtle and complex of psychical phenomena, and has given rise to prolonged and heated discussion, we must take special pains in the definition and analysis of it.

The idea of freedom as an aspect of the volitional process appears to be borrowed or derivative. Freedom, in its primary meaning, and as popularly understood, is the opposite of external compulsion. We are free so far as we are not physically constrained to do things. In this sense it is obvious that all voluntary acts are 'free'. The child becomes free in the measure in which it grows independent of external restraint and is allowed to act for itself.

We reach a secondary and narrower idea of freedom when we contrast our calm and deliberate actions with those carried out under a stimulus of unusual and all-mastering strength, as in the case of actions issuing from explosive passion, *e.g.*, anger or jealousy, or those carried out by an all-possessing instinctive impulse, as recoiling from imminent danger of life.

In these latter instances there is, as pointed out above, an analogy to the overpowering effect of compulsion. A man yielding his purse to a highwayman who presents a loaded pistol is, if not physically, at least morally 'coerced'.[1] The all-possessing, all-mastering form of vehement passion has, in ancient and modern times, been likened to the compulsion of tyranny.

In contradistinction to this *quasi*-coercive impulsive action deliberation presupposes a certain calmness, and consequently immunity from the agitating, disturbing effect of a preternaturally strong and monopolising impulse, as intense fear, or rage. As pointed out above, the process of deliberation allows of a full, impartial representation of our several desires, and so involves a special amplitude of self-manifestation. For a like reason deliberation, with its normal result, a calm choice or preference of one among a number of alternatives, gives rise to a particularly distinct sense of freedom and self-determination of action. Such reflective choice clearly substitutes for the *mechanical* type of impulsive action, in which a single impulse is monopolising consciousness and determining action with the mastering force of compulsion, an *organic* type, in which impulses react one on the other, and show themselves within certain wide limits to be modifiable or alterable, a type of action which is coloured by a full and vivid realisation of the self. Action initiated by rational choice thus exhibits the characteristics of conscious action in their highest intensity and distinctness. The sense of freedom is the realisation of the function of consciousness in its most complex and impressive manifestation.

This realisation of the ego as freely intervening and determining the result grows specially distinct in those cases where the consciousness of effort, that is, exceptionally severe

---

[1] The analogy is plainly seen in the fact that we commonly speak of the actions of children, soldiers, etc., in obedience to commands strongly backed by penal sanctions as compelled, though, strictly speaking, since disobedience is not only possible but can occur at any moment when the temptation is strong enough, the process is volitional and selective. The closest analogy, perhaps, to physical compulsion by another and stronger person is seen in the commanding effect of a strong on a weak will, an effect which becomes abnormally exaggerated in the phenomena of suggestion as observed in the hypnotic state.

exertion, takes part.   To diminish the effect of a hasty impulse by a strenuous effort of attention to ideas of which it took no account, is to assert oneself in a specially emphatic and striking manner, is to strike at the tyrannous monopoly of impulse by a special intensification of consciousness.   Hence it begets a particularly clear apprehension, not only of self as agent or cause, but of self as selecting or as freely determining the issue.   That is to say, it generates in its most distinct form the consciousness of freedom or free-will.

. While the processes of deliberation and rational choice, especially under conditions where a severe volitional effort is involved, are the main generative conditions of the consciousness of freedom, they are aided by other experiences.   Oddly enough, we regard ourselves as free not merely where action is particularly difficult, requiring the reinforcement of effort, but where it is particularly easy.   Thus in the ability to move a limb when we wish to do so we have an experience that helps to develop this mode of consciousness.   Here, it is evident, the distinguishing characters are absence of strong impulse, and the maximum effect of a vivid idea of movement in bringing on the actual movement.   The very absence in this case of a powerful, easily recognisable motive gives to the action the appearance of a mere freakish interposition of the self.   Hence it serves to sustain that belief in the modifiability of action by conscious processes which is the essential factor in the sense of freedom.

In this brief account of the psychological genesis of the consciousness of freedom no reference has been made to the philosophical aspect of the question. The philosophical or metaphysical doctrine of free-will contends that our actions are not wholly determined in the sense in which events in the physical world are determined, by particular sets of phenomenal conditions (here psycho-physical processes), but are in part the outcome of a transcendental principle, *viz.*, spiritual activity, or the assertion of self as hyper-phenomenal or "noumenal".   Since this doctrine is commonly based on the supposed "deliverance of consciousness," that every man knows himself to be free, it becomes important before considering the question to examine into the nature and origin of this consciousness of freedom. The above is a slight attempt to show wherein this peculiar mode of psychosis consists. According to our view, a consciousness of freedom, that is, of an analogue of non-compulsion, develops in connexion with the whole process of the higher and more complex type of conation, into which enter deliberation, choice, and effort. This view, however, manifestly finds a meaning for the sense of freedom without abandoning the fundamental assumption of a scientific psychology, *viz.*, that all actions, just as all other psychical processes, are ultimately determined by groups of psycho-physical conditions. The sense of freedom, according to this way of envisaging the subject, is a *tertiary* psychical product, begotten on those secondary volitional processes of deliberation which have been dealt with above. A man learns to regard his actions as free, that is, as having the maximum of modifiability, just in proportion as he acquires and exercises the power of reflexion and rational choice, and subsequently reflects on these reflective processes.

Whether this psychological view of freedom, or modifiability within the circle of psycho-physical events, exhausts all that is meant by the idea, is one of the points which are still a matter of dispute. To this it may be added that the metaphysical doctrine of freedom has sought to base itself not merely on the introspective observation of our actions, but on certain objective observations, such as the absence of perfect uniformity in men's conduct. This part of the argument again has a psychological reference, inasmuch as it is for the physiological psychologist to enumerate all the co-operating conditions of human action, physical as well as psychical. And here it becomes important to point out that our everyday actions are not merely the result of clearly conscious motive-forces. The stirrings of blind instinct, the working out of vaguely realised organic impulses, as seen, for example, in the on-coming of sexual love, the mechanical action of habit : these must, it is evident, prevent the actions even of the wisest and most highly evolved of mankind from being wholly explicable in terms of the conscious processes of desiring and aiming at ends. Again, the very complexity of the psycho-physical organism, and the changes it is continually undergoing, necessarily tend to give our actions a certain appearance of capricious fluctuation. Thus, those complex and subtle changes which we call our passing moods, and which involve alterations of feeling and of active impulse, are, as we all know, disturbing to that perfect consistency of action which is the wise man's ideal. All this suggests that a truly scientific view of action, so far from requiring the appearance of perfect uniformity and predictability of conduct, is, in fact, distinctly fatal to this result. Here, again, then, the psychologist finds a meaning for the alleged freedom or " indeterminateness " without abandoning the scientific position that all action is psycho-physically determined.[1]

REFERENCES FOR READING.

On the more complex processes of volition and their relation to primitive action, see Bain, *The Emotions and the Will*, pt. ii. chap. vii. ; James, *The Principles of Psychology*, ii. chap. xxvi. p. 528 ff. ; Höffding, *The Outlines of Psychology*, vii. B. The German reader may also consult Waitz, *Lehrbuch der Psychol.* § 43 ; Wundt, *Physiol. Psychologie*, ii. cap. xx. ; and (as more fully presented) *Ethik*, lect. iii. chap. i. ; and Schneider, *Der Menschliche Wille*, theil iii. Moral habit and character are specially dealt with by Carpenter, *Mental Physiology*, chap. viii. ; Bain, *op. cit.*, pt. ii. chap. ix. ; Volkmann, *Lehrbuch der Psychologie*, ii. § 154. The educational aspect of character is handled by A. Martin, *L'Éducation du Caractère ;* and Guyau, *Education and Heredity*, pt. i.

[1] The psychological analysis of the sense of freedom has been carried out by H. Spencer, *Principles of Psychology*, pt. iv. chap. ix. § 219; Bain, *The Emotions and the Will*, pt. ii. chap. xi., especially par. 9, " Consciousness of Free-Will ". Further references to the psychological aspect of the question are to be found in my essay, " The Genesis of Free-Will," *Sensation and Intuition*, chap. v. The reader may also consult Wundt, *op. cit.*, ii. cap. xx. 2 ; Münsterberg, *Die Willenshandlung*, p. 147 f. ; Höffding, *Outlines of Psychology*, vii. B. 4 and 5. On the connexion between the psychological and the metaphysical question, see below, Appendix M.

CHAPTER XIX.

CONCRETE MENTAL DEVELOPMENT : INDIVIDUALITY, NORMAL AND ABNORMAL PSYCHOSES.

§ 1. *Unity of Mental Development.* The three movements of the mental life just traced, *viz.*, that of intellectual, affective, and conative growth, though capable by an artifice of abstraction of being traced out separately, are, as already hinted, intimately conjoined. That is to say, they are not three currents of change which run on side by side, but rather different aspects of that one current which constitutes the "flow of consciousness". A word or two in addition to what has been said above on this organic unity of our conscious experience may fittingly close our detailed examination of psychical processes.

We have found each of the three psychical factors, the cognitive or presentative, the affective, and the conative, as a primitive element, asserting itself to some extent as a separate functional tendency. The baby in the first week of life manifests the rudiment of intellectual activity, of feeling, and of motor impulse. The actual concrete form of the mental life is developed by the conjunction and ever-varied interaction of these psychical forces or tendencies. Without attempting to follow out the endless variety of product thus resulting, we may just refer to the two more important modes of combination or interaction: (*a*) that of intellectual activity with feeling, and (*b*) that of intellectual activity and feeling with conation. Since each of these has been already touched on, a few supplementary observations will suffice.

(*a*) *Interactions of Intellect and Feeling : Interests.* That the development of feeling and of ideation are parts of one total process of mental growth has been already implied. The idea of an intellectual life without any tinge of feeling, and of an emotional life unsupported by presentations, are both fictions. Our 'horde' of ideas, *e.g.*, that corresponding to our social experience, our professional pursuits, our art-studies, etc., is vitalised and to a large extent penetrated and cemented by feeling. On the other hand, as we have seen, the most violent passion, even that of a maniac, invests itself with the semblance of an ideational process.

It is evident that we have here to do with a growth, *pari passu*, of two concurrent factors, each of which is affected by, and in its turn affects, the other. This is seen in the formation of those fixed presentative-affective groups with their attractive and repellent tendencies which we call interests. A boy acquires a strong keen interest, say in cricket, by acquiring knowledge of the game, its practice and its rules, its history, its prominent heroes, and the like. At the same time this knowledge is the outcome of certain pleasurable sensibilities, as love of physical exercise and social instincts. In other words, certain affective tendencies determine a particular grouping of experiences, with the grouping of ideas resulting from this; and the formation of this ideational structure constitutes the base of a new and larger emotive experience. Our interests, together with the particular attitude of attention which they induce, are thus at once a product of an affective and of a representative factor. It may be added that as interests they clearly betray their double parentage; for to have an interest in a game, or in some branch of fine art or science, is at once to be disposed to enjoy or find pleasure in particular lines of mental activity, and to be intellectually alert about, that is, specially ready to throw our intellectual energies into, particular kinds of presentative material.

(*b*) *Interactions of Intellect and Feeling with Conation.* Let us now glance at the other main aspect of this organic interdependence or interconnexion among the psychical factors. In dealing with the development of conation it was pointed out that this is throughout conditioned by the growth of ideation and of emotion. This dependence is clearly recognised in the common order of psychological exposition, and still more plainly in the tendency in certain systems of psychology to treat volition as not co-ordinate in respect of fundamental character with presentation and feeling, but as a secondary and derivative psychical phenomenon. This last view we have repudiated on the ground that the primitive element in volition, active consciousness, is quite as distinct and independent as the other two factors. At the same time, it is manifest that the expansion of this active germ into what we mean by a strong and capable will implies at once the deepening and widening of our feelings and the establishment of the higher sphere of intellection. That is to say, will is a product of active impulse enriched by additions of the stimulative element of feeling and the illuminative element of intellection.

Here, too, however, we have to do with a reciprocal action. Not only does the growth of feeling and of ideation thus minister to the normal expansion and consolidation of activity, but the production of a firm enlightened type of volition reacts on feeling and thought. As pointed out in our account of the processes of self-control, the highest kind of cultivation of emotion and of intellect alike includes a regulative volitional factor. The intellectual man is *ipso facto* the volitional man so far at least as the voluntary direction of the attention is concerned. Similarly, though less manifestly, does the highest realisation of the emotional nature depend on a volitional factor. The modern epicurean who thoughtfully plans out his life so as to get the greatest variety of refined pleasure with the least possible amount of disagreeable drawback must have, not only some of the range and precision of intellectual view of the man of science and of the philosopher, but some of the indomitable firmness of will and completeness of self-control of the ascetic himself.

It is thus evident that in spite of the fact that intellection, feeling, and active impulse are distinct psychical forces or tendencies, and that in their most energetic forms they assume the aspect of hostile or incompatible tendencies, they are organically implicated, so that there can be no normal and complete development of one without a concurrent and correspondent development of the others. In other words, the highest development of feeling, of thought, and of volition is a phase of a complete typical development of mind.

§ 2. *Typical and Individual Development.* The complete harmonious development of mind just described, which involves a proportionate fulness of each of the constituent phases, and of each constituent in its several distinguishable varieties, is an ideal never perfectly realised. The actual concrete minds which we know all exhibit deviations from this typical scheme in one or another direction. Thus, as already hinted, we find now a special intensity of feeling, an emotional excitability or passionateness which is in excess of the powers of thought and of volition. In like manner we see men who are intellectually great but are lacking in a commensurate power of general volitional control; and others who to much energy in action and firmness of purpose join intellectual narrowness, and emotional dulness.

It follows from what was said in the preceding section that such one-sided development always involves a limitation in the power that seems to be predominant. Thus the passionate, excitable man will be found to be explosive on certain sides only, as anger, amatory passion; similarly the merely intellectual man may be seen to be intellectually quick and vigorous in relation to certain domains of idea only, and the practical man, of vigorous energy and indomitable purpose in certain directions, to be volitionally feeble in others, as in fixing his attention on objects lying outside the narrow circle of his practical interests. This incompleteness or one-sidedness of intellectual emotive and volitional development is a fact of the greatest practical importance. In studying and dealing with men we require to know not merely on which side (intellect, volition or feeling) they are strong and weak respectively, but also in what particular directions of intellectual activity, and so forth, they are so.

The problem of determining and formulating the several modes of variation of mental development is a subject which has hardly begun to be seriously grappled with. The innumerable varieties of individual character among civilised men appear indeed to defy any attempt at scientific analysis and classification. In a work on general psychology like the present this problem can only be touched upon.

§ 3. *Varieties of Mind.* If we regard the human beings we know, we are struck at once by the fact of numerous diversities. Each individual has a mind which, while it is an example of the common type of mental structure, is at the same time something unique, a peculiar group of psychical or, if we include the correlative nervous factor, psycho-physical tendencies.

A little inspection shows us that these variations are measurable in different ways. Thus, to begin with an obvious distinction, we may arrange men according to their place in the scale of mind or of cerebro-psychical development. In any community there is a scale of mental power and of correlative brain-power, from arrested development and imbecility up to the highest manifestations in the few great minds. This comparative mode of estimation may be said to give the evolutional height of a particular mind. It is the mode commonly employed by the anthropologist in determining the place of particular races in the evolutional scale.

In the second place, we may distinguish and classify minds according to the relative distribution of the several psycho-physical forces or tendencies which constitute mind. In this way, for example, we get the contrasts of temperament and character which appear in general between the two sexes, and among different races at the same level of culture, and among individual members of a community having similar amounts of cerebro-psychical force. Here we are concerned with the varying ratio of psychical tendency.

These variations in mode of distribution, again, are more general or more special and individual. As an example of the former, we have the preponderance of the emotional over the intellectual, and so forth (in the sense already defined). The more special individualising differences are illustrated by the particular bent of intellectual activity, the relative strength of the several emotional susceptibilities and aims. It is a familiar fact that men of good intelligence are not, as a rule, equally capable in all directions of observation and thought. There is the special intellectual aptitude for this or that group of ideas, and a corresponding ineptitude in other directions. Thus there are men observant and retentive in respect of certain classes of sense-impressions, e.g., visual presentations, and, more narrowly, of one group of these, as colours. Not only so, we see a further limitation of the special intellectual bent connected with a peculiar interest in a group of concrete objects, as is illustrated in the special intelligence which some persons manifest with reference to human faces, and generally what may be called personality, a specialised mode of intelligence found in many women. This mode of variation may be marked off as mental grouping or configuration.

These differences in combination of elements or configuration doubtless have their nervous correlatives. Just as evolutional height is correlated with the degree of cerebral development as a whole, so psychical configuration presumably corresponds with the particular structural configuration, and the *relative* development of the several parts of the brain. The later researches in cerebral anatomy and physiology enable us to perceive, to some extent, wherein such differences of cerebral configuration consist. At the same time, our knowledge does not as yet enable us to determine these correlations with any degree of certainty.[1]

The two modes of comparing minds here distinguished answer, it is evident, to a common, everyday distinction. We judge of a man's mind by its size or magnitude, that is, by its place in the scale of typical development. Here we are conceiving of minds as arranged in a linear series or ascending scale, and do not think of deviations or oscillations to the right or left of this line. We mark off regions of this scale by the expressions ' high,' ' low,' and ' medium ' or ' average '. When, on the other hand, we take in configuration we are thinking of such deviations. The expressions ' typical ' and ' individual '—especially, in the extreme sense, ' peculiar ' and ' eccentric '—obviously refer to the second mode of comparing minds. In the case of each mode of estimation there is a standard of reference. In that of evolutional height we commonly set out with an average or customary magnitude. Thus a man of ' great ' or of ' small ' mind is one who rises considerably above, or falls considerably below, the average standard of mental height. A " peculiar " mind, on the other hand, is appreciated by a reference to a complete typical form of mental structure. It may be added that both kinds of dissimilarity are ultimately reducible to quantitative differences. This is obviously the case with inequalities of mental height. A great or lofty mind is one which excels the average magnitude of mental powers. Differences of psychical configuration, moreover, although, at first sight, they appear to contrast with the first as qualitative dissimilarities, will be found to be constituted by the ratio of intensity or strength of particular psychical constituents.[2]

[1] *Cf.* what was said above on the localisation of brain-function, I. p. 50 ff., and, again, on the physiological basis of temperament, p. 57 f.

[2] It is to be further noted that, since we always look on intelligence as the main constituent or aspect of mind, we tend, in comparing mental heights, to fix attention more particularly on intellectual power. A " great " mind is more than anything else a great intellect. This way of estimating height, however, though

§ 4. *Scientific View of Individuality : Measurement of Psychical Capacity.* It follows from the above definition of individual variation, as a peculiar combination or mode of grouping of elements, that a scientific treatment of the problem must set out with the elementary or fundamental psycho-physical constituents. These are the tissue out of which the mental organs which we call faculties are formed. Every observable difference of individual minds ought to be susceptible of being exhibited as a result of particular groupings of these elements in certain ratios of strength. In order to solve this problem two conditions have to be satisfied. First of all we must know what is elementary or fundamental. In the second place we must be able to measure these elementary forces with something approaching to scientific exactness.

With respect to the first of these conditions, it may be confidently said that our psychological analysis, aided by the physiology of the nervous system, enables us to some extent to solve this problem. Thus we have seen that all intellectual ability is in general determined by the perfection of the senses together with the closely conjoined intellectual function known as discrimination. Fineness, or delicacy of discriminative sensibility, is thus one constituent element of intellectuality. Further, we have learnt that power of attention, as illustrated in prolonged fixation, rapid transition from object to object and grasp or comprehension of a plurality of elements, assimilation or readiness in detecting similarities, as also retentiveness and cohesive (integrative) power, are fundamental constituents of intelligence. In the case of feeling, again, we have traced down all emotional susceptibility to certain organically determined sensibilities to pleasure and pain ; and similarly we have found in active impulse under its two forms, attention and (psychical) movement, the fundamental element in conation.

If now we turn to the second condition of the problem and ask how far these elementary psychical capacities are measurable, we are, as already hinted, confronted with certain obvious difficulties. Our psychical states are not quantitatively com-

---

of obvious practical value, is not scientific. Even the mind of a Shakespeare is, as we shall see, only properly appreciated when it is regarded as an example of exceptional height of total typical development.
*

parable in the way in which material magnitudes are so.   Yet, as we have seen also, the new science of psycho-physics, or, to use a more exact expression, psychometry, has made a promising beginning in carrying out certain simple quantitative determinations.

Psychical phenomena may be said to exhibit three aspects under which they admit of quantitative comparison or measurement.   (a) Of these the first and most obvious is Intensity. The various modes of determining this, in the case of the simpler psychical phenomena, sensations, have been dealt with above.[1]   It is evident that these methods of measurement are fitted to determine individual variations.   Thus individuals can be compared with some exactness in respect both of absolute and discriminative sensibility to intensity of light, sound, and so forth.   There is no reason why, provided a suitable psychophysical apparatus could be devised, the higher psychical phenomena, as the emotional states, and the energy of the psycho-physical process in active impulse, should not be similarly measured in respect of intensity.[2]

(b) Next to intensity we have Duration, or time-magnitude. All psychical phenomena have this dimension, and, as we have seen, important experimental contributions have been made to the measurement of this quantitative aspect of mind.   More particularly, the inquiries into reaction-time have led to some account of exact knowledge of the time occupied, not only by transmission of nervous excitation from periphery to centre, and conversely, but by the central psycho-physical process itself.   Here, too, important individual differences have been noted, and this line of inquiry promises to be a fruitful instrument of a comparative measurement of psychical processes.   Thus the power of adjustment to sensations (attention) of associative suggestion, and even of judgment and choice, may be measured in the case of any individual by means of the

---

[1] The several methods which Fechner has formulated for ascertaining the finest discernible difference of sensation are briefly described by James, op. cit., i. p. 540 f.

[2] The interesting experiments of Ch. Féré and others in dynamometry, that is, the exact measurement of the amount of motor reaction called forth by sensation, suggest that the external phenomena of movement furnish a means of measuring feeling and conative impulse on the objective side, corresponding to the psycho-physical measurement of sensation. (See Ch. Féré, Sensation et Mouvement.)

experiments already devised, that is to say, by comparing the duration of the process and estimating the power as inversely proportional to this.

(c) Lastly, a bare reference may be made to Extensity, and what is only another aspect of this, Complexity. The discrimination of extensive magnitudes may be estimated in the same way as that of intensive. As pointed out above, individuals differ greatly in respect of the delicacy of the tactual discrimination of points, and these differences of local aspect can be numerically determined. Another direction of psychical measurement in which range or complexity comes in is what has been called "span of prehension" or the number of objects clearly distinguishable by a momentary glance. It is highly probable that here, too, important inequalities would be discoverable among different persons. Comparative measurement of this power might be supplemented by that of the power of comprehending a succession of sense-impressions, as those of sound. Some interesting experiments go to show that this capability varies pretty uniformly with age and degree of intelligence, and might with advantage be taken as at least a rough criterion in the estimation of children's mental power as a whole.[1]

Enough has been said to show that a scientific measurement of what appear to be fundamental modes of mental capacity is possible in certain directions and up to a certain point. And although such measurements as are possible would still leave us a long way from our goal, the exact quantitative apprehension of all those secondary or derivative differences which constitute concrete individuality, they would be an important step in this direction. The rapid pace at which psychological, including psycho-physical, experiment is now advancing inspires the hope that the area of this possible comparative measurement will soon be greatly widened. Such an expansion will, it may be conjectured, put us in possession of something like a complete apparatus for a systematic anthropometric observation of psychical peculiarities.[2]

[1] See above, I. p. 280, and for a fuller account of the experiments, under the head "Experiments on Prehension," in *Mind*, xii. p. 75 ff.

[2] A step in the direction of framing a systematic scheme of psychical measurement in its simpler phases has been taken by Mr. Francis Galton in his anthropo-

§ 5. *Causes of Individual Variation.* If mental development in its common typical form is a product of two factors, congenital power, and exercise of function or what we commonly call experience, we may infer that all variations depend on differences in these two factors. That is to say, every degree of general superiority or inferiority of mind, and every special modification of mental configuration, arise from certain differences in the original psycho-physical constitution or in the life-experience of the individual.

That the primitive psycho-physical constitution is variable from individual to individual is a fact of common observation. Just as a glance tells us that no two human faces, even at the age of infancy, are perfectly similar, so we have reason to suppose that no two human brains, and consequently no two sums of mental capacity, are alike. (*Cf.* Vol. I. p. 57 f.) This conclusion is partly reached by observation ; for the most careful observers of infants are able to point to important psycho-physical differences, *e.g.*, in the effect of sense-impressions in calling forth the reaction of attention and motor phenomena generally, which appear in the first week of life. It is further deducible from general biological principles. Thus it seems certain that there is a tendency to individual variation (within certain limits) in the case of the members of every species, and that the result becomes more and more marked as we ascend from the lower to the higher forms of life.[1]

These congenital variations are by some ascribed to ever-varying resultants of the forces of heredity. Thus a child is a new and unique product because it represents a new com-

---

metric scheme. (See his volume, *Inquiries into Human Faculty*, p. 19 ff. ; *cf. A Descriptive List of Anthropometric Apparatus* (1887).) This scheme, though professedly aiming merely at bodily measurements, really includes important aspects of sensibility. On the general problem of dealing psychologically with individual differences of mental character the reader may consult Bain, *On the Study of Character*, especially chaps. x. and xiii. and following ; and L. George, *Lehrbuch der Psychologie*, 1er theil, §§ 7, 9. The student may further consult with advantage a paper by Mr. H. Spencer on " The Comparative Psychology of Man," *Mind*, i. p. 5 ff.

[1] That is to say, the amount of observable dissimilarity is greater as we pass from relatively simple to more complex. organisms having more numerous characters. It is to be added, however, that, as Darwin and others have pointed out, higher organisation as *specialisation* seems to render the species less variable.

bination of ancestral influences. According to this view, individuation is the result of a continually changing mixture of hereditary tendencies. It cannot, however, be said that the theory of heredity has as yet succeeded in making this mode of explaining native individuality or idiosyncrasy perfectly clear. Hence there is room for the supplementary hypothesis that original differences of mental character are to some extent due to what has been called " accidental variation," that is to say, to the variable action of unknown causes which in connexion with the process of reproduction help to determine the character of the germ that goes to make a new individual. It may be added that it appears to follow from the general law of the action of environment on organisms that the earliest observable peculiarities of the infant bear the impress of the action of varying conditions of nutrition, etc., during embryonic life.

The question here touched upon is one of the most perplexing in biology. Darwin was content to view congenital variations as in part at least " accidental," that is, due to the action of unknown causes. More recently, attempts have been made to refer all original differences of individual nature, innate capability, and idiosyncrasy to the forces of heredity. That this is a factor in the case has been shown by some statistical investigations of Mr. Galton, according to which an individual's height, as also the colour of his eyes, is determined by those of each of his parents, and to a less extent by that of each of their respective parents, and so on. To what extent, however, this mixing holds, and how far it can account for the endless diversity of physical and psychical character which we see among individuals, is still unknown. The several physiological hypotheses put forward by Darwin, Galton, Weissmann and others, for the purpose of explaining the phenomena of heredity, do not appear to show as yet how the 'mixing' of ancestral tendencies is brought about.[1]

Whatever the nature and the extent of these congenital organic foundations of individuality they have to be supplemented by our second factor, viz., functional exercise. The biologist's conception of development is that of a process of interaction between organism and environment. In order to

---

[1] On the biological problems of individual variation and heredity, the student should consult the classical work of Darwin, *The Origin of Species*, especially chap. v.; also H. Spencer, *Principles of Biology*, vol. i. pt. ii. chap. viii. and following; and the *Essays* of Weissmann. On the application of the laws of heredity to the explanation of individuality, he may consult Galton's works, *Hereditary Genius* and *Natural Heredity ;* and Ribot's volume, *L'Hérédité Psychologique.*

the formation of any organic product there must be first the requisite germ of organ, and also the appropriate stimulus to excite this to its proper functional activity. In like manner, as we have seen, mental growth is determined by environmental agencies, by the presence of certain stimuli or excitants, fitted to call forth the several psychical reactions. These external conditions vary considerably from individual to individual. In addition to the common physical environment, as determined by such circumstances as climate, locality, and so forth, there is the individual environment constituted by the peculiar group of forces acting on his organism. Thus, no two children, not even members of the same family, come under precisely similar conditions of temperature, nutrition, excitation of movement, etc. The succession of sense-stimuli, with their correspondent motor reactions, making up the external life-experience of an infant, is a different one in every case. Still more evidently is the human environment a variable one. Even twin members of a family have an unlike social *milieu* in so far as the parents and others feel and behave differently towards them.

We may say, then, that individual development is the selective action of what Mr. Galton has happily called "nurture" upon "nature". The possibilities are no doubt organically determined from the first. A child never becomes that for which he has not a native aptitude. Yet, while the broad limits are thus fixed by nature or congenital organisation, the determination of what particular original tendencies shall be developed falls to the environment.

To determine how much is due to what is here marked off as nature or the organic factor, how much to nurture or the environmental factor, seems an insoluble problem. The tendency of the earlier psychology, more especially that of Locke, was to attribute all individual differences of mental character to experience and education, and even so recent a writer as J. S. Mill went a long way in this direction. The development of the biological sciences, more particularly of the modern theory of the origin of species by natural selection, has served to bring out the fact of a universal tendency to individual variation. This conclusion from biological considerations has been confirmed by more careful observation of the infant-mind in its various manifestations.

The scientific conception of growth as the result of organic and environmental interaction divests the question, how much nature and nurture contribute to the result, of a part of its significance. We have not here to do with two mechanical forces combining and producing a resultant in which the effect of each is separately

(b) *Extreme of Normal Configuration : Eccentricity of Character.* We may now pass to the other kind of psychological
extreme, *viz.*, that mode of structural variation of mental
character that varies most widely from the typical form.
All individuality is, as we have seen, a modification of the
common ideal type in particular directions. These modifications are more numerous, if not of wider extent, in the case
of human beings than in that of the lower animals ; and it is
probable that as evolution of the race advances they become
still more numerous. There are, however, certain limits to
what we regard as the normal extreme of these variations.
Where the individual shows such a preponderance of particular impulses and tendencies as to give rise to the appearance
of defectiveness in what we consider the common features
of a human mind, we are apt to characterise it as eccentricity.
Thus extreme concentration even on a worthy aim, as art
or science, when it is accompanied by apparent neglect of
the bodily health, or of the requirements of social life, is
wont to be viewed with suspicion. As is well known, a
number of great men, for example, Newton, Beethoven, have
been marked by such peculiarities, whence the amount of
literature that has occupied itself with the " eccentricities of
genius ". Where what all recognise as greatness of aim is
absent, and the individual shows a complete want of sense of
proportionate value, then popular thought and science alike are
disposed to recognise a still plainer approximation to an
abnormal variation.

§ 7. *The Normal and the Abnormal Mind.* In these extremes
of eccentricity we have, it is evident, to do with phenomena
that lie on the boundary line between the normal and the
abnormal. Our exposition of mental development has occupied
itself with the normal type : yet, in order to give an adequate
idea of mind in its concrete manifestations, we must make a
brief reference to abnormal oscillations from this type.

The distinction of normal and abnormal, and the closely
related distinction of healthy or sane and pathological or

insanity is given by Radestock, *Genie und Wahnsinn.* (*Cf.* Joly, *La Psychologie
des grands Hommes.*) I have summarised the results of inquiry into the subject
in my article " Genius and Insanity " in *The Nineteenth Century,* 1885.

insane, are psychological in so far as they point to actual differences of psychical or, to speak more completely, psycho-physical phenomena. An abnormal mind is one, the particular organic conditions and manifestations of which deviate by a considerable and easily recognisable interval from the typical plan of psycho-physical configuration. Thus an extreme affection for certain animals, amounting to self-denying devotion, and a correlative indifference to human objects of benevolence, is an index to a psycho-physical organisation, a configuration of intellectual and emotive tendencies, which manifestly departs from what we regard as the normal and typical pattern of mind.

At the same time, it is evident that we have to do here with more than a purely psychological or matter-of-fact distinction. The records of our lunacy courts and the admissions of pathologists tell us that there is no wide gap between the normal and abnormal such as the popular contrast suggests, but that the sound type shades off into the unsound by a continuous succession of fine gradations. The significance of this contrast is only understood when we regard it as teleological and practical. The normal is adjustment of organism to environment, the abnormal is mal-adjustment. Our practical instincts lead us to mark off sharply those amounts of individual oscillation from the standard type which render the subject incapable, irresponsible for his actions, a burden if not also a menace to society.

A glance at some of the familiar examples of mental obliquity or unsoundness will at once show this to be so. Thus a disturbance of that fundamental reaction of mind to environment which we call perception leading to individual " illusions of sense " is one of the indications of abnormality most plainly recognised. Similarly where the realities of the surroundings are misapprehended through the rise of bizarre delusions of the imagination. In like manner, perversions of the feelings, such as transformations of what we call ' natural ' affection into its opposite, also of the active impulses, as in the direction of the volitional energies to what is valueless or unattainable, are popularly viewed as manifestations of mental unsoundness.

§ 8. *Abnormal Tendencies in Normal Life.* As already observed, the normal typical mind is a scientific fiction never fully realised. The truth of this proposition is strikingly illustrated in the fact that in the case of all men there are discoverable, more or less distinctly, tendencies which point in the direction of the abnormal. Thus, if we look at the phenomena of sense-perception, we see that there is in the case of no man an exact correspondence throughout between mental percepts and external realities. There are common tendencies to sense-illusion due to the conditions of our sensibility, as in the effect of colour-contrast, or to the overpowering effects of habit, as in the partial deceptions of the mirror, and so forth. Since, however, these errors of sense-perception are of minimal extent and importance, and are, moreover, connected with what on the whole is a normal psycho-physical condition, they are with reason disregarded.[1]

A clearer tendency to the abnormal shows itself in certain individual peculiarities of sense-presentations. Thus in the curious phenomena known as " secondary sensation," as where sensations of bright light (photisms) immediately accompany sensations of tone and certain sensations of touch, we have, it is evident, a disturbance of that complex of conditions which keeps our sense-perceptions in correspondence with external things. Since, however, these oscillations from the normal type of nervous organisation are slight, and lead to no practical mischief, we are disposed to bring them under the border-line group of eccentricities.[2]

The phenomenon of occasional abnormality manifests itself more plainly in those disturbances of sense-perception which arise from temporary organic conditions. Thus the tendency to illusory transformation of sense-impressions under mental excitement and a too powerful imaginative anticipation, as when a person is misled into supposing he sees a ghost or other supernatural apparition instead of the plain homely

---

[1] See my volume, *Illusions*, chaps. iii.-v. I have there tried to show that what we may call normal, that is, common as distinguished from individual sense-illusion, is analogous to the occasional error arising from the deductive application of a general principle which is only approximately true. It is thus the parallel of the occasional want of adjustment discernible in those fixed forms of reflex movement which *on the whole* are adjustive reactions. (*Cf.* Ziehen, *Leitfaden der Physiol. Psychol.* p. 7.)

[2] *Cf.* above, I. p. 136. Bleuler and Lehmann (*op. cit.*) found that the phenomena of coloured hearing are examples of a more general tendency discoverable among a number of persons to have a double sensation. (*Cf.* Ziehen *op. cit.*, p. 144 f.)

object which is actually before his eyes, is clearly a case of
mal-adjustment, and of a kind of mal-adjustment which leads
to wasteful and hurtful action. Still more clearly is there the
phenomenon of occasional abnormality where through a disturbance of either peripheral or central nervous structure the
subject 'reacts' in the form of a sensation where *no* external
stimulus is at work. Of these subjective counterfeits of
objectively excited sensations, again, the most distinctly
abnormal, because the most significant of hurtful organic
disturbance, are what are known as hallucinations, as when
a person imagines he hears another's voice, though no external
sound is present, or is deceived into supposing that he sees
a thing in the dark, and so on. It is well known that temporary changes in the tone of the nervous system, more particularly
in childhood, are apt to occasion such hallucinations.[1] With
these occasional aberrations in the perception of external
objects may be grouped the falsifications of self-observation
which sometimes show themselves in normal life, as in over-
weening conceit, and the opposite state of gloomy self-distrust.

Lastly, a bare reference may be made to these temporary
perversions of the feelings and will which occasionally show
themselves in sane persons, and which are doubtless due to
slight disturbances of functional activity in the nervous system.
Here, again, illustrations may be found in the misdirection of
interest and active energy to worthless, or possibly unreal
objects, which occasionally ruffle the current of youthful mental
progress.

§ 9. *Dreams as Abnormal Phenomena.* Among the abnormal
or *quasi*-abnormal incidents of what on the whole is normal
life are the phenomena of dreams. The phantasms of sleep
are abnormal inasmuch as they assume the character of sense-
phenomena which do not correspond with the real external
world of the. moment, *viz.*, illusions and hallucinations.[2] As
such they are, moreover, deviations of the action of the indi-

---

[1] The frequency of hallucinations in normal life has recently been made the
subject of statistical investigation by Ed. Gurney and others. (See *Phantasms of
the Living,* especially chap. xiii., and the *Proceedings of the Society for Psychical
Research,* pt. xix. p. 259.)

[2] On the question how far our dreams are peripherally excited, how far pure
products of central nervous action, see my volume, *Illusions,* chap. vii.

vidual from the common type of action. A man's dream-experiences are an isolated life. Dreaming, moreover, is evidently connected with temporary disturbances of the ordinary cerebral conditions, *viz.*, those underlying the state of sleep, and probably resolvable into an altered condition of the intra-cranial circulation (slowness of circulation and clogging of veins), with a connected increase of pressure from without inwards.[1]

At the same time we do not regard the dreams of ordinary sleep as abnormal in the sense in which hallucinations in the waking state are so. This is due to the fact that sleep is a regular and periodic incident of normal life, and one, moreover, which is known to benefit the organism by promoting nutrition and recuperation of the several structures. So far as dreaming is connected with and a result of such healthy and beneficial sleep it may be viewed as normal.

How far this is really so is a matter of dispute. It is a question much debated by physiologists and metaphysicians whether all natural sleep is accompanied by a faint remnant of consciousness, or whether the healthiest sleep is not a condition of perfect dreamlessness. Supposing the latter to be the case, dreaming would, of course, assume the aspect of a slightly abnormal phenomenon. However this be, for practical purposes ordinary dreaming may be regarded as harmless, since it involves merely a temporary confusion of the presentative order, together with the connected ideas and feelings, and does not lead to *wrong motor reactions*. Where, however, as in the case of the sleep-walker, such reactions are forthcoming the abnormality of the psychical state forces itself on our attention.[2]

The phenomenon of sleep has always attracted the notice of philosophers and psychologists by reason of its mystery and the curious questions it raises. Sleep is obviously a periodic gap in the continuity of our waking conscious life. Even the best of our dreaming is but a confused chaotic reminiscence of waking experience. Thus, when dreaming, we frequently forget facts with which we have been

[1] This is the view taken by Dr. Cappie in his work, *The Intra-Cranial Circulation and its Relation to the Physiology of the Brain.*

[2] The abnormal aspect of dreaming is well brought out by Radestock in his work, *Schlaf und Traum*, cap. ix. ("Vergleichung des Traumes mit dem Wahnsinn ").

familiar for years and which have the deepest interest for us; our actual surroundings are completely ignored, and the particular conditions of time and place set at
nought.   In some cases only a confused form of self-consciousness is retained, our
personality undergoing the strangest metamorphoses.   All this shows that the
psychological assumption of a continuous, uninterrupted flow of consciousness is a
fiction.   The organisation of our experiences into a consciousness of personal history
is rudely stopped every time we fall asleep.   What we call our representation of
the continuous past is thus an illusory idea.   We *construct* an idea of personal
continuity by mentally leaping over the gaps of unconsciousness and attaching the
tracts of experience on either side of the gap.[1]

Again, the apparent reality of the pseudo-perceptions of sleep naturally suggests the question wherein consists the difference between dreams and waking
perception.   If, when asleep, I am just as certainly and immediately cognisant of
the presence of objects as I ever am when awake, how can I be sure *in either
case* that there is anything more real than the "stuff" of which dreams are
made?   It will be found that this significance of dreaming has played a considerable part in the philosophical discussions respecting the independent existence of
things, the problem of Realism and Idealism.[2]

Lastly, it may be added that in the older systems of spiritualistic philosophy
great stress has been laid on the continuity of consciousness during sleep.
According to a fundamental conception of Descartes and his school, mind always
thinks, and consequently cannot be wholly unconscious in sleep.   Our modern
familiarity with other modes of interrupting the flow of consciousness, as by a
concussion of the brain, the use of anæsthetics, and so forth, leads the physiologist to regard the idea of a dreamless sleep not merely as a possibility but as a
probability.   It may be added that, since confessedly our consciousness during
sleep is detached from, and out of *rapport* with, our waking experience and
frequently exhibits a surprising amount of disorganisation and degradation, the
doctrine of incessant spiritual activity cannot, one would say, receive much assis-
tance from any theory of dreaming.[3]

§ 10. *Artificial Sleep : The Hypnotic State.*   Closely related
to the phenomena of natural sleep is that artificially induced
state which has been variously described as the Magnetic,
the Mesmeric, and, more recently, the Hypnotic trance.   The
hypnotic state, which has of late been made the subject of a
vast deal of experimental research, is a particular cerebro-
psychical condition inducible in a certain number of persons
having the requisite susceptibility by various means, as by
getting the subject to fix his eyes steadily on a bright button
held near the forehead, by stroking the skin of the head, face,
etc.   The result of these operations is a state of sleep in which

[1] See above, I. p. 480, also my volume, *Illusions*, p. 283 ff.

[2] On the nature of this philosophical problem, see below, Appendix C.

[3] On the question whether we always dream during sleep, see my article
" Dream " in the *Encyclopædia Britannica; cf.* W. James, *op. cit.*, i. p. 199 ff.

the subject is, as in natural sleep, though much more profoundly, insensible to sense-stimuli generally, but continues to react to stimuli coming from the operator, as in following out a command, however absurd and even hurtful.

The mental phenomena observable in the hypnotic state are highly curious and of great value to the student of normal psychoses. They are, however, too complicated, and as yet too imperfectly understood, for one to attempt a full description of them in a work on general psychology. A word or two on some of the more important aspects of the state must suffice.

(a) *Nature and Conditions of Hypnosis.* In the first place, then, reference may be made to the question as to the nature of the hypnotic trance. Different views have been held on this point. According to one theory, it is a distinctly pathological condition or neurosis into which persons fall who have the required nervous predisposition. Opposed to this is the view known as suggestion, *viz.*, that the phenomena of the hypnotic trance involve no particular pathological state or neurosis, but are explainable as the outcome of mental tendencies which we all possess. Without discussing the point here, it may be enough to say that the true view of the matter will probably embrace and reconcile these opposite theories. The manifestations in hypnosis are presumably conscious phenomena, and due to the psycho-physical process of suggestion. At the same time, it is certain that so profound a modification of normal consciousness, involving a reduction of psychical activity to a limited group of special reactions, presupposes an exceptional and abnormal set of cerebral conditions. What these are we can as yet only surmise. There seems, however, some reason to suppose that the physiological basis of the hypnotic state is an altered state of the intra-cranial circulation, in consequence of which only certain regions of the psychical centres remain active.[1]

(b) *Psychical Manifestations in Hypnotic State.* As the name hypnotism suggests, the state has a close analogy to that of normal sleep. Thus there is a rupture with, and forgetfulness

---

[1] This is the view of A. Lehmann, as set forth in his volume, *Die Hypnose*, where, as we have seen, vaso-motor changes are viewed as constituting the physiological base of all restricted and intensified consciousness, that is to say, attention.

of, waking experience, and so an interruption of the current of normal self-consciousness. As in natural sleep, so here there is a tendency to a narrowed intensified imaginative consciousness or mono-ideism. Since the mass of external sense-stimuli remains inoperative the subject is wholly possessed by the particular images suggested. Again, as in much of our dreaming, the emotions are roused to a preternatural degree of intensity.

At the same time, the hypnotic state contrasts with that of normal sleep. This contrast already shows itself, as pointed out above, in the fact that the hypnotic patient remains sensibly awake and particularly alert to one region of impression, viz., that answering to the actions of the operator.[1] Among other points of difference one of the most important is the circumstance that hypnosis includes as one of its main manifestations the action of the voluntary muscles. This activity of the motor apparatus gives greater volume, as also greater clearness of manifestation, to the emotions aroused in the state, and, what is more important, renders the actions called forth organic reactions to environment in the complete sense. It is under this aspect that they show themselves most distinctly abnormal. The hypnotic patient who is ready to drink a nauseous liquid, or even to commit a crime, in obedience to the suggestion of the operator, is manifestly wanting in adjustment—is, in fact, from a biological and practical point of view, as much a mis-shapen thing, and one that ought not to be, as a madman.

Under another of its aspects the hypnotic state contrasts with normal sleep and has a closer analogy to that of alternating consciousness or double personality which is known to have been induced in a number of cases by certain forms of injury to the brain. Our successive lapses into the dream-world do not constitute a connected experience. To-night's dream contains no reminiscence of its predecessors.[2] In the

---

[1] There is, however, as pointed out by A. Lehmann, a point of analogy to this in the phenomena of normal sleep. The sleeping mother who hears no sounds but that of her child's cry preserves a restricted cerebral responsiveness to stimuli similar to that retained by the hypnotised subject.

[2] This statement is not exactly true. Many persons, like myself, have a recurring form of dream, which presumably involves the effect of a cumulation of traces. Not only so, I sometimes catch myself on waking from sleep with the

case of the hypnotised subject, however, there is the curious phenomenon of the recollection during a trance of what happened in preceding trances. That is to say, the successive trance-states form a second connected experience or life, isolated from the normal, since in neither is there a recollection of the events of the other.

This aspect of hypnotism is probably the most important to the psychologist. The several stages of the hypnotic sleep, and the concomitant splitting up of the consciousness into a number of detached mental streams, form a valuable supplement to the phenomena of interrupted consciousness by normal sleep, anæsthetics, etc., and those pathological disturbances of self-consciousness or sense of personality which are a common incident in states of insanity. More particularly the experiments by which the particular personality suppressed or submerged at the moment has been 'tapped' by being made to carry out certain sub-conscious actions ("automatic writing") are highly significant. They serve to suggest that even our normal consciousness is not so simple a structure as we are apt to think, that there are different strata (*couches*) of personality, some of which remain for the most part sub-conscious, though exceptional circumstances may suffice to raise them to the level of clear and predominant consciousness.[1]

One word must be added on what are perhaps the most remarkable phenomena of the hypnotic state, *viz.*, the alterations of sensibility. Loss of sensation (anæsthesia) and abnormal intensification of sensibility (hyper-æsthesia) are both producible in the hypnotised subject. Here, again, we have phenomena analogous to certain familiar incidents of pathological disturbance. The psychological significance of these modifications in the hypnotic trance is that they are brought about by the central psycho-physical agencies of suggestion.[2] The same line

images of persons and places which still wear the aspect of old acquaintances, so that at the moment I think they are waking realities; and this accompaniment, the sense of old acquaintance, not improbably points to recurring forms of dream-consciousness. Lastly, I, at least, not infrequently on growing sleepy find myself vaguely recalling some dream of the previous night. How far this fragmentary reproduction is due to a resumption by the brain of its previous condition under the influence of the changes brought about by sleep, how far to the suggestive action of the sum of bodily sensations connected with the lying posture, etc., is doubtful.

[1] The cases of alternating personality due to mental disease are described by W. James, *op. cit.*, i. p. 379 ff.; *cf.* Ribot, *Les Maladies de la Personnalité*, chap. i. and following. On the curious multiplications of personality in the hypnotic trance, see James, *loc. cit.*, p. 384 ff. and p. 202 ff.

[2] The explanation of these modifications of sensibility is not clear. The effect of suggestion seems to be to induce cerebral inhibitions analogous to those

of remark applies to the striking phenomena of rigidity of body and paralysis of muscle which are well-known accompaniments of the state.[1]

§ 11. *Transition to Pathological Psychoses.* The disturbances of normal psycho-physical equilibrium hitherto considered constitute an intermediate region between healthy or sane and morbid or insane psychosis. In this last we have to do not merely with a temporary dislocation of the physical base of mind, the result of changes of circulation, but with a more or less permanent functional and structural disorder, due either to organic predisposition (inheritance) or to injury or " stress ". Here we find the normal type of psychosis disintegrated and biologically degraded by the loss of certain normal psycho-physical factors, or by what we may call disequilibration.[2]

(*a*) *Peripheral and Central Disorders.* Such loss of equilibrium may be introduced by peripheral injuries, as in the case of the destruction of a sense, the loss of a limb. These peripheral injuries are, however, as we know, compensated by these new psycho-physical co-ordinations which a healthy brain is capable of carrying out. Thus, as we have seen in the case of the blind, and even of a child with the sense-apparatus so seriously impaired as Laura Bridgman, the higher development of the remaining senses, more particularly active touch, has

---

brought on by anæsthetics. It is to be added that the disturbances of sense-perception point to, and illustrate in a peculiarly striking manner, the co-operation of an ideational (imaginative) factor in sense-perception. Thus when the subject is made blind to a particular object, but not to the other objects present, it is evident that we have to do primarily with an inhibition of the imaginative factor. (See James, *op. cit.*, ii. p. 606 f.)

[1] In this brief description of some of the more salient and psychologically important features of the hypnotic state I have followed W. James's interesting account. (*Vide, op. cit.*, ii. chap. xxvii., also i. chaps. viii. and x.) The student who desires a fuller setting forth of the facts may consult Binet and Féré, *Animal Magnetism* (International Scient. Series); A. Moll, *Der Hypnotismus* (American translation, 1890); and A. Lehmann, *Die Hypnose und die damit verwandten normalen Zustände.* Paul Janet's work, *De l'Automatisme Psychologique*, is a storehouse of interesting observations.

[2] As already hinted, sanity shades off into insanity through the intermediate stages of eccentricity. The sharp demarcation of insanity from sanity is even now a problem beset with difficulties. See, for a single example, the definition of melancholia furnished by Griesinger, and quoted by Bevan Lewis, *A Text-book of Mental Diseases*, p. 116.

served to secure the required sensuous basis for a fairly complete development of mind.[1]

It is far otherwise where the central structures are impaired. In this case we see a real process of mental disorganisation. Psychical factors which are essential to the normal pattern of consciousness are lost, other constituents acquire abnormal strength and preponderance, and the whole arrangement is distinctly a mal-adjustment to the circumstances of the environment.

Without attempting in a work on the normal workings of mind a systematic account, even the most concise, of the intricate phenomena of mental disease, we may indicate one or two of the more important characteristics of these morbid changes. In this way we shall be able to include them in our general view of the variations of mental organisation.

(b) *Restricted and Extended Brain-Disturbances.* The disturbances of normal psychical configuration brought on by disease of the central organs may be comparatively restricted, involving local deterioration only, or, on the other hand, extended and deep-reaching, involving a more general and profound disorder of the central organs. Modern pathological research supplies us with many curious and interesting examples of the former. These form a valuable supplement to experimental physiological investigation into the special functional activities of the several regions of the cortex. Thus pathologists have brought to light the fact that localised injury to the cortical substance may issue in a certain specialised form of amnesia, such as loss of a particular group of words, *e.g.*, proper names, or in word-deafness, *i.e.*, loss of the ideational connexions by which we apprehend or understand words. Similarly in the case of a restricted group of sense-disturbances, as in recurring auditory hallucinations which appear to point to abnormal excitability of particular sensory regions of the cortex.

(c) *Varieties of Central Nervous Disorder.* The extended disturbances of psycho-physical action are of even greater

---

[1] For a full and interesting account of the blind deaf-mute, Laura Bridgman, and the marvellous results of her careful education, see Professor Stanley Hall's article in *Mind*, iv. p. 149 f.

interest to the psychologist in their relations to the normal pattern of consciousness. Here we see two apparently contrasting tendencies, *viz.*, to depression (melancholia) and to exaltation (mania). The former, as recently pointed out by Dr. Bevan Lewis, has points of analogy with the hypnotic state.[1] The tendency of the most recent scientific opinion seems to be to regard these apparently opposed states as modifications of one and the same kind of profound psychophysical disturbance. In each alike we have to do with the suppression, more or less complete, of certain factors of the normal mind, and more particularly the impairment of the higher intellectuo-volitional processes. This impairment of the regulative factor in normal states leads in certain directions to loss of energy with a correlative impairment of mental action, lack of interest, and imperfect grasp of the realities of life ; in other directions, to an excessive and dangerous explosiveness of action in certain lower centres which have now lost the salutary inhibitory control of the higher central arrangements.[2]

(*d*) *The Course of Cerebro-Psychical Dissolution.* According to the classical researches of Dr. Hughlings Jackson the general course of this psycho-physical degeneration is the reverse process to that of nervous evolution. The highest and latest-evolved nervous arrangements, being the most unstable, are the first to be thrown *hors de combat* by the inroads of general cerebral disease : the successive changes of dissolution retrace the path followed by those of evolution.[3]

It seems to follow from this that mental disease tends to reduce the type of consciousness from the complex mature pattern of the sane and educated adult to that of the child or of the imbecile. It is allowed that there is a certain general

---

[1] *A Text-book of Mental Diseases*, p. 152.

[2] See Bevan Lewis, *op. cit.*, p. 162 ff. The writer thinks that mania is the result of a greater depth of dissolution than melancholia or stupor.

[3] A curious illustration of this is given by Bevan Lewis, *op. cit.*, p. 164. In the reductions of mania the psychical integrations answering to assimilation are dissolved before those of contiguity : that is to say, the classifications and abstractions reached latest and with most difficulty are lost before the relatively early associations of time and place ; or, to use Mr. Spencer's language, the re-representative is lost before the representative.

analogy in the case. Insanity has its affinities to mental
infancy and the closely related phenomena of senile decay, as
also to the ill-regulated and confused psychoses of the dreamer.
Yet the analogy must not be pressed too far. Mental disease,
attacking the mature or developed brain, does not bring about
an exact counterpart to the cerebral condition of infancy.
Dissolution does not work in so complete and orderly a way
as this supposition would require. Thus in the delusions of
the monomaniac, in the homicidal and other impulses of the
maniac, we have to do with phenomena which have little in
common with infantile psychoses, with shreds and tatters of
adult experience, that is to say, with a psychical state of
things which suggests that certain central regions have retained
something of their later-acquired modes of functional activity.[1]

(*c*) *Forms of Mental Disturbance.* Without trying to follow
the pathologist into the difficult and almost impossible task
of classifying the various types of mental disease, we may just
glance at one or two of their main features. Here we may conveniently follow our triple scheme of mind as a guide. In
doing this, however, we must not be misled into thinking that
disturbances of intellectual function and so forth appear as
isolated phenomena. All mental disorder implicates mind as
a whole, though it may show itself as *more particularly* a derangement of intellect, of the feelings, or of the volitional factor.[2]

(1) *Intellectual Disorders.* Among the more important intellectual disturbances are those hallucinations or subjectively
occasioned pseudo-perceptions which show an encroachment of
the life of imagination on that of sense, as when the patient
' sees' imaginary beings, ' hears' imaginary voices, and so
forth. Here we have to do with a morbid excess of action of
particular sensory regions of the cerebrum in consequence of
which the normal distinctions between the percept and the
image become effaced. Much the same line of remark applies
to those fixed delusions (" compulsory ideas," " *idées fixes* ")
which are so common a feature in certain states of insanity.
From this, again, we may infer a morbid super-excitation of par-

---

[1] Bevan Lewis expresses the fact by saying that the denudations of the
nervous system in insanity are not uniform. (*Op. cit.*, p. 176.)

[2] *Cf.* above, I. p. 67 ff.

ticular cortical regions resulting in a preternatural vividness and persistence of certain ideas. Among these delusions special mention must be made of disordered ideas of self, as when the patient imagines himself a king, a criminal, and so forth. Here we recognise ruptures of the normal apprehension of personality, dire confusions of self-consciousness, which are fitted to lead the subject into the most extravagant forms of maladjustive action. Such disturbances in the consciousness of personality are highly instructive to the psychologist by showing how closely this psychical product is related to the bodily life and its concomitant organic sensations. Dérangement of the digestive and other vital processes leading to profound alterations of the cœnæsthesis is a common cause of this confusion of self-consciousness.[1] This group of pathological phenomena further illustrates the organic connexion between the normal consciousness of personal identity and memory; for one effect of the disturbances here referred to is to abolish the memory of the past, and so to isolate the present self, as in natural sleep and hypnotic trance, from the past self.

One other common intellectual feature of these morbid conditions must be referred to, viz., the loss of " object-consciousness" and the correlative excess of subject-consciousness. A liability to hallucinations and delusions involves to some extent the upsetting of the normal equilibrium here, that is, the due subordination of ideation to presentation. In certain cases there is a further disturbance, viz., the inability to grasp external objects as realities. The patient ' sees' but at the same time does not fully realise what is before him.[2] This pathological condition serves, as was pointed out above, to strike at the roots of a healthy strong belief in things such as is necessary to vigorous action.[3]

---

[1] On the disturbances of organic sensation in sanity, see Mercier, *Sanity and Insanity*, p. 323 ff. ; *cf.* Bevan Lewis, *op. cit.*, p. 229.

[2] It is pointed out by Bevan Lewis (*op. cit.*, p. 118 f.) that this impairment of the normal object-consciousness involves the motor apparatus and those muscular sensations which are so important an element in the perception of objects.

[3] *Cf.* above, I. p. 492. Closely connected with this pathological enfeeblement of belief is the rise of a morbid inquisitiveness, leading to a questioning of what is obvious—as " Why do I stand here ? " " Why is a glass a glass ? " This excess of inquisitiveness, called by the Germans Grübelsucht, or questioning mania, comes

(2) *Affective Disturbances.* The disturbances of the feelings are quite as marked a feature of mental disease as those of the intellectual processes, and commonly accompany these last. Thus in the decline of the object-consciousness, and the intellectual torpor that accompanies this, we see the co-operation of affective changes, a morbid hyper-trophy of self-feeling, lack of objective interests. So, in the disturbances of cœnæsthesis which commonly underlie ruptured personality, we have to do with alterations of feeling. It is to be noted that the morbid aberration is not always in the direction of the gloomy and the miserable. There are morbid exaggerations of the agreeable feelings of gaiety also, leading the subject to a grotesquely disproportionate exaltation. This suggests that both extremes of melancholy despondency and exuberant high spirits are cases of mal-adjustment, the realities with the correlative claims of life requiring that moderate tone of cheerful serenity which has been regarded as characteristic of the wise man.[1]

(3) *Volitional Disorders.* In the region of action we see the double tendency to depression and exaltation observable in the other departments. A common form of pathological disturbance is loss of impulse and motor vigour and intensification of the sense of effort, the correlative and in part the result of the loss of belief and of objective feelings or interests. On the other hand, we see in insanity striking illustrations of a morbid intensification of particular active phenomena connected with the loss of the higher regulative apparatus. This is seen in those "fulminating psychoses" illustrated in the outbursts of maniacal passion. It is further illustrated in the unchecked exuberance of movement (pantomime) observable in certain states of "exaltation," as also in that loss of the power of volitional attention which is discernible in the invasion and overpowering of the patient's mind by a torrent of ideas.[2]

from the loss of a firm mental grip of realities. (See above, p. 259; *cf.* James, *op. cit.,* ii. p. 284; and an article on "The Insanity of Doubt," by Dr. P. C. Knapp, in *The American Journal of Psychology,* vol. iii. p. 1 ff.)

[1] *Cf.* my volume, *Illusions,* p. 323 ff.

[2] The connexions between normal and pathological mental conditions are dealt with by Maudsley, *Pathology of Mind;* Mercier, *Sanity and Insanity;* and Bevan Lewis, *A Text-book of Mental Diseases; cf.* also Ribot's monographs, *Les Maladies de la Mémoire, Les Maladies de la Volonté,* and *Les Maladies de la Personnalité.*

APPENDICES.

APPENDICES.

APPENDIX A.

THE DIVISION (CLASSIFICATION) OF MENTAL PHENOMENA.

The tripartite or "trichotomous" division of mental function expounded above is a comparatively recent addition to psychological doctrine. The first attempts at a classification of psychical manifestations were hampered by metaphysical presuppositions. Thus, in the elaborate scheme of Plato, the mental powers are arranged in a hierarchy, a certain lower province of mind being handed over to partnership with the body while a purely spiritual part is marked off as detached, independent, and eternal. The same view of mind as a graded series of powers reappears in Aristotle, only that here it becomes incorporated into a *genetic* view of mind, and so prepares the way for the modern developmental theory.[1]

The first essay in distinguishing between co-ordinate mental functions led to a bipartite or dichotomous division, *viz.*, into an intellective and an active or conative factor. The germ of this bisectional view may be found in Aristotle, who, while he gave independent functional value to intellect or thought (νοῦς) and to desire or appetite (ὄρεξις), subordinated feeling to these.[2] This twofold scheme remained the prevailing one up to comparatively recent times. It survives in the classification of Reid, *viz.*, (1) Intellectual and (2) Active Powers, and in the popular psychology of everyday life. The separate recognition of feeling as a co-ordinate phase or function of mind is due to the German psychologists of the Wolffian school who wrote about the middle of the last century, more especially Moses Mendelssohn and Tetens. The tripartite division was adopted from these by Kant, and by his authority fixed as the permanent one.

The long overlooking of feeling as a separate phase of mind is accounted for by its special inaccessibility to observation. Feeling is in a peculiar sense subjective, internal, and detached from external objects. Knowing and willing are concerned with these directly or indirectly, but feeling is cut off (save through its

[1] See Siebeck, *Geschichte der Psychol.* vol. i. p. 201; ii. p. 21 ff.
[2] See Siebeck, *op. cit.*, ii. p. 89.
*

(327)

external manifestations) from the external region, and can only be observed therefore by careful *subjective* observation.　To this it must be added that feeling, as we
have seen, is closely bound up, on the one hand, with sense-impressions and other
matter for intellection, whence the common view of Leibniz and others that feeling
is a vague or imperfect cognition; and, on the other hand, with desire and action,
for which reason it was, for a long time, confused with these.[1]

Even after the recognition of the fundamental distinctness of feeling, the three
modes of mental function were not regarded as equally primordial and independent.
Thus, not only Leibniz, Wolff, and their followers, but the more recent psychologists of the Herbartian school, regard intellect or the power of presentation
(Wolff's *vis repræsentiva*) as the primary and fundamental one, on which feeling
and desire are dependent, and from which they are, in a manner, derived.
Hamilton, who strongly insists on the generic distinctness of the three classes,
feeling, knowing, and willing, may be said to go a certain way in the same direction
when he writes : " The faculty of knowledge is certainly the first in order, inasmuch
as it is the *conditio sine qua non* of the others ".　By this he means that we only
have feelings or desires in so far as we are conscious of them, and that consciousness
is knowledge.　He adds that, though he can conceive a being all cognition, he
cannot conceive one all feeling and volition.[2]

The precise relation of the three psychical functions is well brought out by
Lotze.　He shows, on the one hand, that they do not resemble three branches
shooting up side by side from the first.　Feeling is mostly called forth by intellectual
states (presentations and representations), and desire and will have feeling as their
antecedent condition.　Yet, on the other hand, this does not entitle us to say that
representations are the *cause* of feelings, or feelings the cause of volitions.　By
merely considering the mind as capable of having presentations we could never
discover any reason why it should pass into the new mode of manifestation, feeling
of pleasure or pain.　Similarly we cannot derive the active element of striving from
feeling.　The later mode of manifestation, though presupposing the earlier as its
antecedent condition, implies an independent and pre-existing capability.[3]

While feeling has thus been denied by many the status of an independent
mental function, an attempt has recently been made by Horwicz and others to
regard it as the primordial type of mental manifestation.[4]　This assertion is based

---

[1] This is well shown by Höffding, *Psychology*, chap. iv.　For an account of
the history of the subject, see Hamilton, *Lectures on Metaphysics*, vol. i. lect. xli. ;
Brentano, *Psychologie*, buch i. cap. v. ; and Wundt, *Physiologische Psychologie*, i.
p. 13 and following, and p. 538 and following.

[2] See his *Lectures on Metaphysics*, vol. i. lect. xi.　Since Hamilton allows the
existence of unconscious mental states, " latent mental modifications," it might be
more correct to say that he makes the cognitive element fundamental in all the
overt or *conscious* processes of mind.

[3] *Microcosmus* (Eng. transl.), i. p. 177 *seq.*

[4] See Horwicz, *Psych. Anal.* theil i. abschnitt vi. and theil ii. hälfte i. ; also
Carneri, *Gefühl, Bewusstsein, Wille*, and E. Kröner, *Das Körperliche Gefühl*.

on the fact that in the early stages of mental development both of the individual
and of the zoological series the element of feeling (sense-feeling) is conspicuous
and predominant. This fact has been recognised above. On the other hand, it
has been pointed out by more than one writer that in the simplest sensational
consciousness there is involved a rudiment of intellection.[1]

While feeling has thus been viewed now as a secondary now as a primary
psychical phenomenon, a somewhat similar fate seems to have overtaken the third
psychical factor, conation or will. Thus we have on the one side the view of
Schopenhauer and his followers that will is the innermost and deepest element in
our nature. This idea assumes a new form in the doctrine of Wundt, that active
impulse (Trieb) constitutes the primordial and fundamental psychical phenomenon.[2]
In opposition to this view we see a tendency to deny volition any distinctive rank
among psychical functions. This tendency appears most plainly among those
psychologists who resolve all the active phenomena of conscious movement, effort,
etc., into a particular group of peripherally initiated sensations together with representations of these.[3]

[1] G. H. Schneider (Der Menschliche Wille, kap. ix. p. 190 seq.) argues against
Horwicz that a feeling always involves the discrimination of a favourable from an
unfavourable state. This statement, however, if by 'favourable' is meant anything
more than agreeable or pleasant, is doubtful. Ward urges against Horwicz the
fact that every sensation contains a presentative element. (See his article, Mind,
viii. p. 472; cf. article "Psychology," Encyclop. Britann. p. 40.)

[2] Physiol. Psychologie, vol. ii. cap. xxiv. § 2. Höffding adopts much the same
view, Psychology, iv. e (p. 996).

[3] The elimination of a fundamentally distinct volitional function appears very
plainly in H. Spencer's analysis of mind into 'feelings' and "relations between
feelings," where 'feeling' seems to include both the presentative and the affective
element in sensation and its derivatives, and "relation" marks off the intellective
function, Principles of Psychology, pt. ii. chap. ii. The resolution of volitional
processes into complexes of sensation is explicitly attempted by Münsterberg, Die
Willenshandlung, p. 62.

APPENDIX B.

THE QUESTION OF VISUAL SPACE.

As pointed out in the text, the nature of our visual perception of space has been the subject of much discussion. We may here first give a historical *résumé* of some of the main theories on the subject and define the present condition of the question. After this we will briefly epitomise some of the more important observations on animals, children, and those born blind, which directly bear on the issue.

1. So far as we know, antiquity adopted the naïve view that in sense-perception we immediately apprehend the qualities of things, their " primary qualities," as form, as well as their secondary qualities, as colour. Democritus, following a crude materialistic conception of psychical processes, supposes that in perception emanations of things penetrate through the organs of sense into the body, as a consequence of which presentations arise. In the case of seeing the emanations (images) retain the form of bodies. The effect of distance in rendering objects indistinct is accounted for on the supposition that it is not the emanation or image itself, but the air which takes its impression, that reaches our eyes.[1] The theories of sense-perception elaborated by Plato and Aristotle hardly touch the point here dealt with.[2]

Coming to modern times, we find that Descartes regarded extension as external, forming indeed the real nature of bodies, and so distinguished from time, number, and the like, which are relations, and therefore *modi cogitandi*. Yet though this view, taken with his dualistic conception of the absolute apartness of mind and body, should, one supposes, have led him to give special attention to the way in which the mind perceives extension, he seems to have made no contribution to the subject.[3] Hobbes came near psychologising on the point in his idea that the projection of sensations into external space is due to the fact that the inward motion, received by way of the sense organ, calls forth an outward reaction from the brain or heart.[4] Hobbes seems further to have caught a glimpse of the Kantian idea that

---

[1] Zeller, *Pre-Socratic Philosophy*, ii. p. 265 ff.; Siebeck, *Geschichte der Psychol.* i. p. 102 ff.

[2] For an account of their doctrines, see Siebeck, *op. cit.*, i. p. 209, and ii. p. 21 ff. Aristotle recognised that figure and magnitude (like motion, rest and number) were perceptibles not peculiar to one sense as are colour and sound. But how far he was away from the modern psychological problem of space is seen in the fact that he co-ordinates the perception of extension with that of colour, each being regarded as a real quality in things. (See Siebeck, *loc. cit.*, and Grote, *Aristotle*, ii. p. 186 ff.)

[3] See Erdmann, *Hist. of Phil.* ii. p. 19. According to Descartes, space and matter were identical.

[4] See Croom Robertson, *Hobbes*, p. 125.

space and time are subjective.[1] According to Locke, our idea of space is derived from sight and touch. He held that we directly perceive by sight a distance between bodies of different colours and between parts of the same body. But that he had an inkling of the dependence of sight on touch is seen in the fact that when his friend Molyneux put to him the question whether a person born blind, and recovering sight when adult, could by sight alone distinguish a cube from a globe, he answered, in agreement with the questioner, 'No'.[2]

The psychological question, so far as this country is concerned, was precisely defined by Berkeley in his epoch-making *Essay towards a new Theory of Vision* (1709), in which is clearly laid down the proposition that sight is incompetent to reach the perception of distance, and that our apparently immediate visual intuition of space is a product of experience and association, the interpretation of a visual language. Berkeley did not go beyond the visual problem, and, so far as we can tell, regarded the tactual intuition of space as immediate and requiring no explanation.[3] Hume's treatment of the subject of visual perception, coming after Berkeley's discussion, is disappointingly meagre. According to him, the idea of space is an abstract idea derived from "the impression of colour'd points, dispos'd in a certain manner" after omitting reference to the particular colours so disposed. He adds that we find a resemblance between impressions of sight and of touch in the disposition of their parts.[4]

The next important contribution to the psychology of space was made by Brown, who followed up Berkeley's reduction of visual space by maintaining that not only is distance, together with real magnitude and figure (*i.e.*, solid figure), 'tangible,' but even length and breadth also. It was Brown who first brought into prominence the part played by the muscular "feelings," *i.e.*, sensations. He recognises them as a factor in the ocular signs of distance, and, what is more important, in dealing with tactual perception ("sensations commonly ascribed to touch") shows that the idea of extension, as made up of co-existent parts, is derived from the successions of muscular experience accompanying our tactual impressions.[5] Brown's method of dealing with the perception of space as an outgrowth of the successive experiences of movement and touch combined has been further developed by Bain and Herbert Spencer.[6]

---

[1] He calls space an "imaginarium, quia merum phantasma," *Elem. Phil.* vii. 2 and 3, quoted by Volkmann, *op. cit.*, ii. p. 5.

[2] See Essay on *Human Understanding*, bk. ii. chap. ix. § 8.

[3] *Cf.* with the *Essay*, the Dialogue, *Divine Visual Language*. Berkeley's psychological work, which was destined to give so powerful and lasting an impulse to subsequent research, was carried out in the interests of his philosophic idealism, the doctrine that matter apart from mind is non-existent, and that objects are merely regularly recurring conjunctions of 'ideas,' *i.e.*, sense-impressions, which the creator causes thus to arise together concurrently. (See Fraser's *Selections from Berkeley*.)

[4] *Treatise of Human Nature*, pt. ii. bk. i. § 3.

[5] *Philosophy of the Human Mind*, lects. xxii. ff. and xxviii. ff.

[6] See *The Senses and the Intellect*, p. 234 ff.; and *The Principles of Psychology*, vol. ii. pt. vi. chaps. xiii. and xiv.

The tendency observable in the Berkeleyan theory of vision, *viz.*, to make the idea of space fundamentally tactual, has been opposed by a number of writers. In this country Reid contended in a lukewarm manner that we get the idea of space from *each* of the two senses, sight and touch, separately, though he allows that the former is only a 'partial conception'.[1] Hamilton, while he accepted Berkeley's derivation of distance from touch, argued vigorously against the idea that the perception of visual extension is derived, maintaining that the perceptions of colour and extension necessarily involve one another, and that "vision is not only a sense competent to the perception of extension, but the sense κατ' ἐξοχήν, if not exclusively, so competent". With respect to the latter suggestion (which reverses the Berkeleyan mode of affiliation or derivation), he seems disposed to accept the idea of Platner that the blind have no proper idea of extension.[2] A more direct attack on the Berkeleyan theory has been made by recent writers, among whom may be named S. Bailey, Mahaffy, Abbot, and W. James. The two latter—James with the fuller light of recent optical research to direct him—make a noteworthy attempt to show how vision furnishes us with the perception of space in three dimensions.[3]

In Germany the psychological question has also received much attention. Kant's treatment of space as a subjective form necessary to the very constitution of a sense-intuition naturally directed attention to the psychological aspect of the subject. J. Müller, the eminent physiologist, sought to carry Kant's idea over into the psycho-physical domain by teaching that the retina has an immediate knowledge of its own local arrangements. Müller's speculations started what is called the "nativist" view of physiologists and psychologists, a view represented among recent writers by E. Hering and Stumpf, *viz.*, that in seeing we have, by help of certain organic arrangements, a means of intuiting space (in three dimensions) immediately, and independently of experience. To this way of treating the visual perception of space is opposed that of the "empirist," of whom Lotze, with his doctrine of local signs, Helmholtz and Wundt are the best-known representatives. The intricacy of the optical phenomena dealt with in these discussions precludes the possibility of even roughly sketching the several theories put forward, and the reader who requires further information must have recourse to more technical works.[4]

[1] *Essays on the Intellectual Powers*, ii. chap. xix.

[2] *Lectures on Metaphysics*, vol. ii. xxviii. He held, too, that sight is competent to give us the idea of externality. For a criticism of Hamilton's views, see J. S. Mill's *Examination*, ch. xiii. p. 266 ff.

[3] Bailey's views are given in his *Review of Berkeley's Theory of Vision* ; Prof. Mahaffy's in his *Critical Philosophy for English Readers*, vol. i.; Mr. Abbot's in his interesting and masterly volume, *Sight and Touch ;* and Prof. W. James's in his *Principles of Psychology*, ii. chap. xx.

[4] I have summarised the discussion among German physiologists in the article "The Question of the Visual Perception in Germany," *Mind*, iii. p. 167 ff. *Cf.* W. James, *op. cit.*, chap. xx., where copious references to recent researches in physiological optics are to be found; also Wundt, bd. ii. cap. xiii. 8.

2. Leaving the difficult field of optical research, we may glance at those more readily intelligible observations which have been made on children and the blind.

Common observation tells us that the young animal excels the child in those adjustments, such as the pecking of a chick, which appear to involve a rudimentary consciousness of space.[1]

Recent scientific investigations have shown how perfect are these adjustments in certain cases. Thus Mr. Spalding, who carefully kept chickens hooded for two or three days after they left the shell, and then observed what they could do, found that they were able to peck with perfect precision from the outset at so small an object as a worm.[2]

The child is less richly endowed than the young animal with instinctive ability, agreeably to the theory that the higher the place of a species in the zoological scale the more helpless the progeny at birth. With respect to vision, the infant has a far more complicated apparatus to master than the chick, which does not possess our binocular vision, and is not required to select particular movements. Hence we need not be surprised to learn that recent careful observation has established the proposition that the child only begins to attain to complete vision and to co-ordinate sight and touch towards the end of the first half-year. Without going too much into detail, it may be enough to give some of the main results of this observation as summarised by a recent writer.[3]

There are two main epochs in the development of the visual perception of space : (*a*) that ending in about the fifth week, (*b*) that ending in about the fifth month.

(*a*) By the fifth week the child learns to fixate an object when it is in the line of vision. Associated movements of the eyes, horizontally and vertically, are learnt. There is no fixation when the retinal image falls away from the yellow spot (macula lutea), *i.e.*, on the side region of the retina. The child has now acquired the reaction of the blinking movement when an object approaches along the line of vision, but does not yet blink when an object approaches from the side. Certain co-ordinations of movements, those of the eye-ball and the lid, are acquired about the same time.

(*b*) The child acquires by the later date (fifth month) the power of orientating in the field of vision, so that it can direct its glance to objects in the side regions. It now acquires, further, the lid-closing reflex when objects approach from the side of the field. More important than all, it is about the same date that we observe

---

[1] The facts were noted by Adam Smith, who urged them as favouring the supposition that children, too, have some instinctive perception of distance. (Quoted by Hamilton, *Lectures on Met.* ii. p. 182 ff.)

[2] *Macmillan's Magazine*, Feb., 1873. The results of Spalding's experiments are summarised by Romanes, *Mental Evolution in Animals*, p. 161 ff.

[3] E. Raehlmann, *Zeitschrift für Psychol. und Physiol. der Sinne*, bd. ii. p. 53 f. This *résumé* is largely based on the careful work of Preyer, recorded in *Die Seele des Kindes* (p. 25 *seq.* and p. 122 *seq.*). The results of different observers do not, however, fully agree, and suggest that there are important individual differences. (See Stumpf, *Ueber den psychol. Ursprung der Raumvorstellung*, p. 295.)

the earliest attainments in tactual exploration. Thus the hands are first noticed in the thirteenth week to move from side to side, the eyes being *fixed* on them ; and the first movement of grasping takes place about the end of the fifth month, though it is not till about the end of the sixth, or even the seventh, month that the child can put out his hand straight (without circuitous movement) towards an object.

These facts seem to show conclusively that particular space-intuitions have, in the case of the human being, to be acquired. At the same time, the rapidity and the degree of regularity with which they appear suggest that inherited arrangements are a co-operant factor in the development.

We may now pass to the more difficult phenomena which present themselves on the removal (by the operation of couching a cataract) of a congenital blindness. Here we have a number of carefully observed cases. At the same time, the fact that the patients were able to some extent to see the difference of light and dark (though no definite outline) in the case of near objects before the operation diminishes the value of these as perfect examples of the inductive operation known as method of difference.[1]

The two best-known cases are those of Cheselden's and Dr. Franz's patients. Cheselden's patient was a boy of about twelve. After the operation, when able to see objects, he showed at first no discriminative perception of distance. " He thought all objects touched his eyes, as what he felt did his skin." He could not at first distinguish the shapes of objects by sight alone, thus verifying the conjecture of Molyneux and Locke, and only succeeded in doing so after again and again going from the visual impressions to the corresponding tactual experiences. Two months after he was couched he discovered that pictures (which he had previously viewed as ordinary surfaces) represented solid bodies, though now he fell into the error of taking them for the actual objects themselves, as children take the shadow of an object on the wall for a body in relief.

The account of Dr. Franz's patient is much fuller and more exact. The patient was a youth of eighteen. His sense of touch had attained a remarkable degree of perfection, the lips being specially employed in the minute inspection of objects. After the operation he was subjected to careful observation. The purpose was at first to discover how far he could discriminate the elements of form, as lines, which are presumably cognisable by sight alone. He was able after a little inspection to distinguish between a vertical and a horizontal line as such, that is, to tell which was the horizontal, and which the vertical. He also distinguished a circle, square, and triangle, as such. After this the inquiry was directed to ascertaining how much he could discern with respect to the distance and solidity of objects. He took solid objects such as a cube and a sphere to be flat surfaces. He could not distinguish between the position of an object floating on the surface of some water, and another object sunk one foot below. All objects appeared so near to him that he was afraid of coming in contact with them. He had no idea of per-

[1] This is well shown by Preyer to apply to the case of one of Ware's patients, *Die Seele des Kindes*, p. 381 ff.

spective, and could not understand pictures. He saw even a familiar object of touch, such as the human face, as a flat plane. Other and later observations tend to confirm these results. Thus, one of Home's patients, just like Cheselden's, thought objects touched his eyes, and was unable to distinguish and name forms by sight alone, calling, for example, a triangle 'round'. The like has been found to hold in the case of more recent observations, as that of the boy Rubens, carefully recorded by Raehlmann,[1] who could not for some time distinguish a cube from a sphere, or a solid cube from a flat board, who thought objects at a greater distance smaller (thus showing that he had no perception of distance or of real magnitude), and that when an object was placed in different positions, *e.g.*, a bottle upright and then end-on, it was taken for two objects. The power of estimating distance was carefully measured by Raehlmann in the case of another patient, a girl of fourteen years, three weeks after the operation, by making her look through a tube at two objects at unequal distances. She judged rightly which was the further only eight times out of thirty. Other points noted by Raehlmann in these cases are the inability to move the eyes at first, movements of the head being substituted, and the losing of an object, as happens with infants, when the retinal image passes away from the yellow spot.

These observations, though not perfectly harmonious or conclusive throughout, seem, as Raehlmann points out, to show pretty conclusively that those born blind develop their visual perception in the same order as normal children, and that seeing objects as normal adults see them at definite distances, and of a certain real figure and magnitude, is acquired.[2]

It may be added that some of these conclusions are borne out by what we can ascertain respecting the space-consciousness of the blind. Platner's idea that the blind have no proper idea of space does not agree with our present knowledge of their mental state. No doubt the difference in the case of tactual and visual perception, more especially the fact that we construct tactual space by help of a succession of partial co-apprehensions, serves to differentiate their space-consciousness from that of normal men. But there is abundant evidence to show that they acquire all the fundamental constituents of our space-ideas.[3]

[1] *loc. cit.*, p. 71 ff.

[2] For the several accounts and interpretations of the experiments, see Hamilton, *Lectures on Met.* ii. p. 176 *seq.;* Mahaffy, *Crit. Phil. for English Readers*, vol. i. pt. i. p. 122 ff. ; Stumpf, *op. cit.*, p. 288 ff. (where the ambiguities of the results are emphasised) ; Preyer, *op. cit.*, p. 381 ff. ; and Raehlmann, *loc. cit.*

[3] On Platner's view of the space-ideas of the blind, see Hamilton, *Lectures on Met.* ii. p. 173 ff. (*Cf.* J. S. Mill, *Examination*, p. 278.) On our present knowledge of their ideas see some interesting quotations from a work written by W. H. Levy, himself blind, given by W. James, *op. cit.*, ii. p. 204 f.

APPENDIX C.

PSYCHOLOGY AND PHILOSOPHY OF PERCEPTION.

In the foregoing account of the perceptual process we have been concerned with the genesis and development of our perceptions as subjective or psychical processes having certain physiological concomitants. In other words, we have been asking by what steps a child comes to acquire sense-cognitions of external objects. Closely connected with the question, and in a manner growing out of it, is another, having to do with the objective validity of perceptual cognitions. Having traced the growth of a percept we may go on to ask : " What does this percept in my mind stand for and guarantee ? What is this thing localised in space which I seem to perceive as something apart from me and existing independently of me ? What is the real material object in itself, and how is it related to those sensations of active touch, etc., by which, as the psychologist tells me, I have come to know it ? "

These questions reach over into the domain of philosophy or theory of know-ledge, that branch of science which determines the objective value or validity of our cognitions.[1] The problem raised has filled a large place in modern philosophical discussion. It is known as the problem of realism and idealism. The realist [2] is one who considers that in sense-perception we have the assurance of the presence of a reality which is absolutely distinct from the percipient mind, and exists indepen-dently of our perceptions. The idealist, on the other hand, asserts that the idea of an absolute existent out of relation to a conscious mind is a contradiction, that the meaning of the terms ' thing,' ' object,' ' existence ' has a reference to percipient mind. The reality of the external world is perceptibility, or, as Berkeley, the founder of modern idealism has it, the *esse* of things is *percipi*.

To trace out the origin and growth of idealism lies beyond our present province. We must confine ourselves to a bare indication of the relation of the psychology of perception to this extra-psychological question. It is evident, then, to begin with, that our psychological account of the perceptual process cannot help us straight away to determine the precise objective import and validity of its result. Whether to the percept when formed there does correspond any absolute reality in the philo-sophical sense, *i.e.*, independent of, and out of relation to, mind, and how we are to interpret the ' objectivity ' which is certainly one aspect of sense-perception, these are points which the psychologist who concerns himself merely with the mental process itself is incompetent to decide. At the same time, we have con-

---

[1] *Cf.* above, I. p. 4 and p. 12.

[2] That is, the perceptual realist. He must be distinguished from the concep-tual realist, who holds that our *general ideas* have their self-existent objective counterparts.

tended that the psychology of perception is closely connected with the philosophy, and it remains to make this connexion a little clearer.

The drift of the above psychological narrative of perception has been to show that we come by percepts as the result of having what we call sensations, the power of distinguishing these, and of connecting them in certain ways. Since all our minds are formed on a common type, we come to build up such percepts in a similar manner. We find, moreover, that the percepts thus produced in our several minds tend, and do so more and more as we observe carefully, to agree or harmonise, so that we acquire a common presentative view of things. It seems natural to suppose that the facts emphasised above, such as the fundamental distinction between the sensation or percept and the image, the regular recurrence of sensations in fixed arrangements or groupings, the special significance of the experience of resistant body, and the (approximate) agreement of the sense-experiences of different persons, would of itself, whether there is any independently existent ' thing ' or not, give rise to a distinction *within the sphere of consciousness* of a mind perceiving and a reality perceived, a subject or self, and an object or not-self.

This being so, it would seem to follow that the philosophical discussion of a material world will do best to set out at the point to which psychological analysis brings us, and to take up the question under the form : " Is material reality *wholly* resolvable into a mode of our conscious experience ? When we say that a thing is real, and exists when we do not perceive it, do we mean anything more than that we and others would have certain modes of sense-experience, *viz.*, contact together with obstacle to movement or resistance, by satisfying certain preliminary conditions of locomotion, etc. ? "

This is, in fact, the way in which Berkeley and his successors have approached the question.[1] They have laboured to show that what ordinary men understand by matter, by a stable external world, is translatable into terms of sense-experience. The service of psychology in this connexion has been most signal in the case of the English empirical or associational idealists, that is, the writers, from Berkeley to Mill and Bain, who seek to resolve the idea of material object and of externality into a product of association on sensation-elements. From this extreme psychological treatment of the problem, which seems to leave our uninvestigated belief in physical reality half illusory, must be distinguished the new way of Kant and his followers. Kant held that our perception of objects is a joint product of sensation, and the mind's receptive and constructive activity. The manifold of sense (*i.e.*, sensations) must be first received into certain general *a priori* or transcendental ' forms,' *viz.*, space, and time, then constituted into objects by the synthetic activity of thought (judgment). According to this view, objective reality (in a phenomenal or relative, as distinguished from a noumenal or absolute sense) is secured by the co-operation of this constitutive intellective activity. This theory of externality and objective reality, though having certain points of contact with the " psychological " theories

---

\ Mill boldly calls his idealism a psychological theory. (See his *Examination of Sir W. Hamilton's Philosophy*, chap. xi.)

of the external world put forward by English associationists, differs from them in the material circumstance that it regards the external order, the system of permanent objects, not as an apparently illusory result of a fortuitous concomitance of particular sense-elements and the ideational reflexion of this concomitance through the mechanism of association, but as grounded on certain fundamental and necessary conditions of intelligence or knowledge as such.[1]

[1] For a good account of the meaning of the philosophical problem and its psychological connexions, see Prof. Fraser, *Selections from Berkeley*, Introduction ; *cf.* Hamilton, *Lectures on Met.* xxi. ff. ; J. S. Mill, *op. cit.*, chap. x. and following ; and Bain, *The Senses and the Intellect* (" Of External Perception "), p. 364 ff. I have touched on the bearing of the psychology on the philosophy of the subject in my volume, *Illusions*, p. 344 ff. The distinction between the Kantian and the English associational view of perceptual knowledge is emphasised by Mahaffy, *The Critical Philosophy for English Readers*, vol. i. pt. i. ; and by Adamson, *On the Philosophy of Kant*, especially lect. iv. *Cf.* Green's criticism of Hume's theory of perception, Introduction to the *Treatise*, p. 189 ff. On the other hand, the compatibility of Kant's view of space with a psychological account of its genesis out of particular psychical elements is well brought out by Adamson, *op. cit.*, lect. i.

APPENDIX D.

MEMORY AND LAWS OF ASSOCIATION.

The phenomena of memory have been the subject of special attention in ancient and modern times. Different views have been held respecting the nature of the conservative and reproductive process, as also respecting those laws of association by which images are called up or suggested. We may begin by a reference to some of the main theories of memory and then enumerate the principal versions of the process of associative revival.

1. Memory has at once a sensuous or bodily and a mental or intellectual side. As a representative copy of a sensation (presentation) the mental image stands in a manifest relation to the senses ; and as a factor in all internal processes of thought it ranks as a phenomenon of intellect. The several theories may be distinguished according as they emphasise one or the other of these aspects.

In the theory of Aristotle we have a recognition of the double-sidedness of memory. Both memory and imagination or phantasia are phenomena, not of the cogitant or intelligent, but of the lower sentient soul. In memory pure and simple ⟨μνήμη⟩ we have to do with an organic process, viz., a movement from the central organ (the heart) and the organs of sense to the soul : in reminiscence (ἀνάμνησις), i.e., mental reinstatement actively induced by a methodical following of the lead of association, we must suppose a counter-movement from the soul to the organs of sense.[1]

The clear reference to an organic base in memory characterises most of the modern theories of the subject. Thus Descartes in his dualistic conception of psycho-physical action supposes recollections to be ideas called up on the occasion of certain bodily processes, viz., the movements of the vital spirits through the traces of images on the brain. Hobbes, after Aristotle, conceived of imagination ⟨phantasm⟩ and memory as due to a surviving motion in the organs of sense. In like manner, Hartley, the founder of English associationism, referred the phenomena to the "miniature vibrations," i.e., those of the small medullary particles of the brain, which appear to be left over by the vibrations set up in sensation.

The tendency to emphasise the organic side of memory has led to an extension of the idea to the organism as a whole. This view of a memory, that is, a retentiveness, of the effects of previous action as a common property of organic structures, was broached among others by Malebranche, but has only recently been developed into a scientific doctrine, more especially by E. Hering. According to him, conscious memory is only a particular case of a general physiological fact, viz., that all organs are modified by repeated action so as to acquire a special disposition to one particular mode of action, a fact illustrated in the phenomena of habit, and

---

[1] See Grote's *Aristotle*, p. 474 ff.

those of heredity and of instinct.[1]   Hering's view is followed out by Horwicz, Ribot, and others.[2]

On the other hand, we see a tendency to emphasise the intellectual side of memory, to deal with it as a conscious affection.   Here two points of view offer themselves : (1) We may regard memory as a unique faculty, a power which is wholly mysterious and inexplicable.   This way of conceiving of it is set forth with special clearness by Reid.   To him the knowledge of things past certified by memory is unaccountable, not to be explained by vividness of idea or association.[3]   (2) We may seek to explain the phenomena of memory or reproduction by psychological principles.   Here past presentations are supposed to persist as obscure or sub-conscious when no longer present to consciousness, and reproduction is merely the raising of these above the threshold of consciousness.   This is the direction taken by Leibniz in his system of clear and obscure representations. The tendency is worked out by Herbart and his school into a mechanical theory of reproduction by the interaction of presentations.   A like disposition to conceive of reproduction as emergence of what was pre-existent in a sub-conscious form is seen in Hamilton's theory of the conservative faculty, and in Ward's development of memory from the presentation-continuum through movements of attention.[4]

In connexion with these attempts to account for the retention and reproduction of presentations a bare reference may be made to the work of Fechner and others in bringing into prominence the facts of primary memory as forming an intermediate stage between the percept and the (revived) image.[5]

2. The recognition of certain laws of reproduction dates from the time of Aristotle.   We find, however, as might be expected, considerable divergences in the mode of conceiving and of formulating these principles.   It is evident that the contrast in the views of memory just referred to involves a further contrast in the view of the associative or suggestive process.   In this brief summary we shall content ourselves with a reference to the more important of these differences.

Aristotle, dealing with association in connexion with the higher and partly intellectual act of reminiscence, distinguishes three orders of reminiscent advance, *viz.*, from the similar, the contrary, or the coadjacent of the required idea.   Here we have the germ of the laws of association, since named similarity, contrast, and contiguity.   Hobbes, who gives interesting examples of the workings of association, adds little to the theory of the subject.   He seems to confine himself to contiguity,

---

[1] See his lecture *On Memory as a Universal Function of Organic Matter.*

[2] See Horwicz, *Psych. Analysen*, bd. i. p. 265 ff.; Ribot, *Les Maladies de la Mémoire*, chap. i.

[3] *Essays on the Intellectual Powers*, iii.

[4] See Hamilton's *Lectures on Metaphysics*, vol. ii. lect. xxx.; and Ward's article, p. 60.

[5] See Fechner, *Elemente der Psycho-physik*, ii. chap. xliv.   A reference to some of the historical conceptions of memory may be found in Hamilton, *Lectures on Metaphysics*, ii. lect. xxx.; a more complete *résumé* is given by Volkmann, *Lehrbuch der Psychologie*, i. p. 395 ff.

and attempts a vague kind of physiological hypothesis to account for the same. Locke, who fixed the classical expression " Association of Ideas," gives no statement of the law, and, like Hobbes, refers only to contiguity, connecting this with '' trains of motion in the animal spirits ". It is noteworthy that Locke only speaks of association as explaining *chance* sequences of ideas as distinguished from the controlled processes of thought.[1] A more important contribution to the doctrine of association was made by Hume, who distinguishes three laws, *viz.*, resemblance, contiguity (in time or place), and cause and effect.[2]

An important step was taken in the development of the doctrine of association when Hartley, working out hints of predecessors, Hobbes and Locke, attempted to derive association, *i.e.*, contiguity, from physiological laws. Hartley's theory (though not quite as clear as it might be) explains the reproduction of an idea as the result of the mechanism of vibrations. Thus, he tells us, the vibrations A, B, and so forth, when associated together get power over the small vibrations *a*, *b*, *c*, so that A may excite *b*, *c*, etc. The English doctrine was further developed by Jas. Mill, who, with little reference to nervous concomitants, followed Hartley in recognising only contiguous association, in its two branches, synchronous and successive. Mill's exposition is remarkable on account of its attempt to reduce similarity to contiguity on the very weak ground that we are accustomed to see like things together.[3]

A further step was taken by Brown, who distinguished between primary and secondary laws of suggestion. By the former he meant the three commonly recognised principles : resemblance, contrast, and contiguity ; by the latter—Brown's real and important addition to the doctrine of association—he indicated the effects of vivacity, recency (recentness), and frequent repetition.[4]

Recent writers have occupied themselves in reducing the number of the laws of association. Hamilton (who regards the process of suggestion as re-emergence into consciousness of cognitions existing as unconscious), after reducing contiguity to simultaneity, and similarity and contrast to affinity, seeks to subsume these under one law, that of redintegration or totality. " Those thoughts suggest each other which had previously constituted parts of the same entire or total act of cognition "—a simplification which seems to involve Jas. Mill's strained supposition that a thing reminds us of its like only when they have been previously thought of together. Bain more modestly contents himself with eliminating contrast as a fundamental law by reducing it to the other two.[5] Herbert Spencer, reversing the procedure of Jas. Mill, attempts to resolve contiguity into similarity. According to him, the only mode of association is the cohering of feelings (*i.e.*, sensations) with previously experienced feelings of the same class. Contiguity is always re-

---

[1] *Essay*, bk. ii. chap. xxiii.

[2] *Treatise of Human Nature*, bk. i. pt. i. § 4.

[3] *Analysis of the Human Mind* (J. Mill and Bain's edition), vol. i. chap. iii. p. 111.

[4] *Philosophy of the Human Mind*, lects. xxxiv.-xxxvii.

[5] *The Senses and the Intellect*, p. 564 ff.

solvable into likeness of relation, *viz.*, in time, in space, or in both.[1] Lastly, reference may be made to the endeavour of Ward, James, and others to resolve similarity into contiguity or "continuity". According to this view, all similarity is partial identity. The like feature in the new presentation is identical with its prototype in the previous presentation-group, so that the whole of the suggestive part of the process has to do with concomitants.

In Germany the labours of Beneke and others have brought association into the foreground of recent psychology. Wundt has developed a scheme of association in which he distinguishes external (including contiguity and passive similarity) from internal or active association. His elaborate scheme has led to an attempt to simplify the theory. Thus Münsterberg, starting with the assumption that all association is a psycho-physical process susceptible of being expressed in terms of nervous action, seeks to eliminate similarity as physiologically inconceivable, and to reduce the whole process of suggestion to contiguity, and, more precisely, simultaneity, that is to say, the organic connexion of different psycho-physical factors in consequence of their occurrence at one and the same time. Efforts have been made by Münsterberg, James, and others to explain the physiological mechanism here involved, though, as observed above, the results are as yet not perfectly satisfactory.[2]

---

[1] *Principles of Psychology*, vol. i. pt. ii. chaps. vii. and viii., especially § 120.

[2] See Münsterberg, *Beiträge*, heft i. ; W. James, *op. cit.*, i. p. 566 ff. For a fuller account of the different views of association, see Hamilton, *Lectures on Met.* xxxi., and Edition of Reid's Works, Appendix D** ; also Croom Robertson's article "Association," *Encycl. Britann.* : Volkmann, *Lehrbuch der Psych.* i. p. 429. Anmerkung 4; Wundt, *op. cit.*, ii. p. 389 ff.; and W. James, *op. cit.*, i. p. 594 ff. A full historical account of the theories of association from the time of Hobbes is given by L. Ferri, *La Psychologie de l'Association*, pts. i. and ii.

APPENDIX E.

THEORIES OF TIME-PERCEPTION.

The psychological theory of time is a later development than that of space. To the naïve consciousness of the plain man time seems to be immediately apprehended as a feature or aspect of his experience, and this unreflective view seems to have been shared by the earlier thinkers. The thought of antiquity gives us nothing better than the idea of Aristotle that time is the " number of motion ". To Descartes time, like space, is immediately intuited.[1] Locke took the first step towards a psychological theory of the subject when he denied that (as Aristotle taught) we get the idea of time from that of motion, referred it to sensation *and reflexion*, and first defined the idea of succession or time-order.[2] The problem of time naturally perplexed Hume in his attempt to refer all ideas to previous impressions. He had to allow that time is not a particular impression but is an " abstract idea " which " is derived from the succession of our perceptions of every kind, ideas as well as impressions, and impressions of reflexion as well as of sensation," or, as he expresses it later on, " is always discover'd by some *perceivable* succession of changeable objects ".[3]

A more serious attempt to analyse the idea of time was made by Reid, who charges Locke with confounding succession and duration, and urges that the notion of duration must be antecedent to its being measured—to succession. He also brings out the fundamental point in the psychology of time, *viz.*, that, strictly speaking, succession can be the object neither of sense nor of consciousness, inasmuch as the operations of both are confined to the present moment. He seeks to explain the development of the consciousness of time as a conjoint product of sense and memory. Sense gives us the present (*i.e.*, the philosophical as distinguished from the vulgar idea—" specious " present as it has since been called), memory the past, and out of these we construct the future.[4] Brown makes a slight contribution to the subject by pointing out that time is in actual reality but a relation. The present moment is the starting-point, and the conceptions (ideas) which arise when complicated with the " feelings " which we call relations, *viz.*, those of priority and succession, give us the idea of the past and of the future.[5] James Mill

[1] Volkmann points out, *op. cit.*, ii. p. 19, that Descartes, agreeably to his conception of mind and bodily object as absolutely apart, regarded time as subjective, a *modus cogitandi*.

[2] *Essay*, bk. ii. chaps. xiv. and xxvi.

[3] *Treatise*, bk. i. pt. ii. § 3.

[4] *Essays on the Intellectual Powers*, iii. chap. iii. (Of Duration) and chap. v.

[5] *Lectures on the Philosophy of the Human Mind*, xli.

deals with the idea of time in connexion with memory. He argues against the supposition of Reid that memory gives us time. "Pastness" is included under memory. Memory "*notes*" antecedents and consequents of the several parts of the chain of remembrance, and "*connotes*" the feelings themselves : when .the connotation is left out we have the idea of pastness.[1] A similar kind of psychological analysis is followed by H. Spencer, who distinguishes between feelings, and relations between feelings, and regards time as the abstract of all relations of position among successive states of consciousness. The relation of sequence with its other aspect, change, is regarded as the fundamental relation to which all others are reducible. In this way time-consciousness or perception of succession appears to be logically and psychologically alike the first relation.[2]

In Germany we see the problem of time receiving special attention. This is due, in no small measure, to the work of Kant, who, elaborating and modifying a doctrine of Leibniz, regarded both space and time as innate, or, as he phrased it, *a priori* mental forms. This view of time, as at once subjective, and an integral factor in the constitution of knowledge, gave a new significance to the partly psychological, partly epistemological problems, touched on by Hume and others, as to the infinite extension and divisibility of time. This result is illustrated in Hamilton's peculiar view of the subject, according to which our thought of time, as of space, is 'conditioned,' *i.e.*, lies between two inconceivable extremes, *viz.*, the infinitely great, or boundless extent, and the infinitely small, or boundless division.[3]

Another result of Kant's view was to direct the attention of psychologists to time as a special mode of consciousness, and to stimulate effort in the direction of explaining the mechanism by which the mind connects its ideas in particular time-relations. The foundation of this branch of psychological research was laid by Herbart. This writer seeks to develop the time-perception out of his theory of "evolution" and "involution" of mental series. Evolution takes place when the series of presentations is revived in the order of sense-perception ; "involution," when a member of a series calls up anterior members, but not successively (as in the case of posterior members) but simultaneously. The time-perception (which implies a presentation of sequence, as distinguished from a sequence of presentations) is differentiated from the space-perception thus : If the presented sequence be one-sided, passing only from *a* through *b* to *c*, it is a time-order, if it starts coincidently from *d* and *a*, passing from the latter through *b* and *c* to *d*, and from the former through *c* and *b* to *a*, we have a space-order.[4] Herbart's way of developing

---

[1] *Analysis*, chap. xiv. § v. (Time). J. S. Mill in a note criticises his father's theory, remarking that time is not an abstract idea of "pastness," "futureness" of our successive feelings, but rather a collective name for our feeling of their succession.

[2] See his *Principles of Psychology*, vol. ii. pt. vi. chap. xv.

[3] *Lectures on Metaphysics*, vol. ii. xxxviii. *Cf.* J. S. Mill's *Examination*, chap. vi.

[4] A summary of this view of time is given by G. F. Stout in his exposition of Herbart's psychology, *Mind*, xiii. p. 334 f. and p. 474 f.

the time-perception is followed with certain modifications and additions by his chief disciples, as Drobisch (*Empir. Psychol.* § 22), Waitz (*Lehrbuch der Psychol.* § 52), Volkmann (*Lehrbuch der Psychol.* § 87 ff.), and Th. Lipps (*Grundtatsachen des Seelenlebens*, p. 588 ff.).

A similar line of study has been followed outside Germany. Thus M. Taine, in his account of the image, seeks, in an ingenious way, to show by what mechanism the mind comes to project its images now in the past as antecedents to the present sensation, now in the future as its consequents.[1] Ward works out a theory of temporal signs somewhat in the manner of Volkmann and Lipps. Lastly, reference may be made to the pleasantly written accounts of our consciousness of time given by Guyau, *La Genèse de l'idée de temps*, and by W. James, *Principles of Psychology*, i. chap. xv.[2]

[1] *On Intelligence*, pt. ii. bk. iii. §§ 7 and 9.

[2] For a fuller account of the history of views of time, see Volkmann, *op. cit.*, ii. p. 19 (Anmerkung 4), and the articles on the " Psychology of Time," by Herbert Nichols, in *The American Journal of Psychology*, iii. p. 453 ff.

APPENDIX F.

NOMINALISM AND CONCEPTUALISM.

The nature of general notions, concepts, or 'abstract ideas,' and their precise relation to names, have given rise to much discussion. This discussion had its origin in a properly *philosophical* question, namely, that respecting the nature of general knowledge. It was asked whether there is any external reality corresponding to our general notion, *e.g.*, 'man,' over and above that of individual men whom we have seen, or we or others might see. Certain thinkers have held that there is such a universal reality, that in the region of external existence there is something corresponding to 'man' as distinct from 'James Smith,' 'John Brown,' etc. These were called realists.[1] In opposition to these the nominalists asserted that the universal or general has no existence in the realm of nature or objective reality, but only appertains to the name as a common sign which is applicable indifferently to this, that or the other object which are found to resemble one another in certain respects.

In modern times the controversy has tended to assume the character of a *psychological* discussion. Instead of the ancient realists we have the conceptualists, who assert that our *ideas* may be general, or that the mind has, over and above the power of picturing individual objects, that of forming general notions, or ideas of classes of things. These general ideas are not 'sensible representations' of individual objects, but abstract ideas, that is, representations of the common features (or the relations of similarity) of many individuals. In opposition to these the nominalists assert that when we use general names we are still picturing or imaging the individual, but in a very imperfect way, that is, by attending exclusively to certain features marked off by the general name. The nature of the concept is only understood by considering the function of general signs. Inasmuch as a name is such a sign, applicable alike to an indefinite number of individual objects, we are able by means of it to view a mental image as presenting features common to it and other presentations. The name man, though calling up a more or less distinct pictorial image of a concrete man, at the same time emphasises the features in this image which answer to common human traits. In this way the image, though in itself as image particular and concrete, becomes representative of an indefinite group of like things, that is to say, of a class.

That it is only the name and not the idea which is truly general is, according to the nominalist, shown in the fact that general names when dwelt upon tend to develop distinct pictorial representations of particular objects. If, for example, I dwell on the name town, I become aware of a gradual reinstatement of a particular image answering to some first impression, or some recent percept.

[1] That is, conceptual not perceptual realists. (*Cf.* above, p. 336.)

The image-content called up by the general name appears to differ in the case of different individuals. According to some, it is a particular image that varies on different occasions ; according to others, it is rather a cluster or series of particular images. As already hinted, differences of individual experience appear to modify the character of the image. Thus Stricker tells us that even in representing an abstract quality he does so by means of a pictorial image. Thus in thinking of eternity he images blue sky, of liberty, a statue of a goddess.[1]

It is to be added that the whole discussion of the nature of the general or abstract idea has been confused by the unreal supposition that we employ general names alone, and apart from those organised verbal structures which we call sentences, or, to use the logical expression, propositions. A true theory of the thought-process sets out with the judgment as the unit. A name is always employed in a relation, and this relation necessarily modifies the character of the accompanying idea. Thus, if I say, "The primrose is paler than the dandelion," the very relation into which I am bringing the primrose (that of a certain difference in colour) serves to give prominence to the colour-element in the generic image. In this way it may be said that there is no such thing as a constant or fixed general idea, but only one that fluctuates within certain limits according to the thought-relation (indicated by the context) into which it is brought.[2]

The recent ingenious theory of Mr. F. Galton as to the structure and mode of formation of the generic image appears to offer a way of reconciling the opposed views here set forth. As generic the psychical product differs in an important respect from the detailed particular image. On the other hand, as image it meets the contention of the nominalist that all ideation is at bottom imagination. As pointed out in the text, the process of accumulative assimilation helps us to understand how it is that the common features stand prominently out from among the whole complex of features.[3] And this circumstance, again, is of material value in the subsequent reflective development of the general idea or concept.[4]

---

[1] Ueber die Bewegungsvorstellungen, p. 2 ; cf. Galton, Inquiries into Human Faculty, p. 182 ff.

[2] Cf. Ward, loc. cit., p. 76, col. 2.

[3] This applies, as Herbart and others show, to the assimilated presentation. In recognising an object, as a flint, a primrose, and so forth, I specially attend to those features in the concrete presentation which are common to the class to which I refer the object.

[4] For an account of the controversy between conceptualism and nominalism, see Hamilton, Lectures on Metaphysics, ii. xxxv. ; Mansel, Prolegomena Logica, chap. i. p. 13, etc. ; J. S. Mill, Examination of Sir W. Hamilton's Philosophy, chap. xvii. ; Dr. Bain, Compendium of Mental Science, bk. ii. chap. v., also Appendix A ; and M. Taine, On Intelligence, pt. i. bk. i. chaps. i. and ii. ; pt. ii. bk. iv. chap. i.

APPENDIX G.

GROWTH OF LANGUAGE IN THE RACE AND IN THE INDIVIDUAL.

The question as to the psychological relation of language to thought is closely connected with the problem of the origin of language in the history of the race. In spite of the series of elaborate researches commenced by Herder, there is still a good deal of uncertainty on this point.[1] Thus it appears to be still a matter of dispute among philologists whether the first verbal signs were conceptual, or merely marks of particular objects ; further, how far the application of these signs to objects grew out of a prior emotional use of vocal sounds, or was distinctly imitative of the sounds of natural objects.[2] We may, however, pretty safely say that both the view which regards the origin of language as due to a conscious process of invention pre-supposing a considerable development of the power of thought, and the opposite view which makes the growth of thought wholly a result of a physiological *differentia*, *viz.*, the possession of an organ of speech, are one-sided and inexact. The mere possession of a suitable organ, though, as the absence of language in the case of the intelligent apes suggests, a co-operant condition, would not alone guarantee the development of language without some correlative development of brain-power and thought, and the social state which this presupposes. On the other hand, thought could never have reached more than a rudimentary or nascent stage without the aid of language. According to the view of the evolutionist, man owes his superiority as a thinker over the lower animals to the double circumstance, the possession of a finer and more highly developed brain, and also a special physiological apparatus for carrying out a system of linguistic, that is, interchangeable, signs. The higher brain would favour the formation of those groupings of image-content which we call general ideas, and the gift of speech would serve to give precision and perma-nence to this rudimentary thought and so aid in its further development.[3]

The progress of language, like its supposed origin, illustrates the organic inter-connexion of the thought-process and its verbal embodiment. In the case of the most backward races known to us we find a low power of abstraction coupled with a paucity of general names. Thus the Tasmanians, though possessing names

[1] For a succinct view of the several theories of the origin of language, see Sayce, *Introduction to the Science of Language ; cf.* article " Philology " in the *Encyclo-pædia Britannica.*

[2] The first has been contemptuously called by Max Müller the pooh-pooh, the second the bow-wow theory. A modification of the first is that of Noiré, *viz.*, that word-sounds came to be associated with particular objects or actions through being first used as concomitants of conjoint or common actions.

[3] For a careful working out of this theory, see Romanes, *Mental Evolution in Man*, chap. xvi.

for particular varieties of trees, appear to have no name for tree in general. Among North-American Indians, again, we read that it is exceptional to find a term sufficiently general to denote oak-tree. A like deficiency in the use of abstract language is shown in the fact that the Tasmanians have no words to express qualities, such as our ' hard,' ' tall,' but have to resort to a combination of words, as ' like a stone,' ' long-legs '. As the mental powers develop and thought-relations grow more distinct, the language becomes less pictorial, more abstract, and takes on more of grammatical structure; and advance in the development of the verbal organism reacts upon and quickens the process of thought-evolution.[1]

It is a somewhat different problem when we consider the relation of the growth of thought and of speech-power in the case of the individual. Here, again, the powers of speech (articulation) and of thought develop *pari passu*. To some extent, as has been suggested above, the child may be supposed to be reproducing the early phase of the development of language in the race by spontaneously uttering word-sounds of his own invention in order to indicate the resemblances which he discovers in things.[2] This spontaneous tendency is strikingly exhibited in the case of born deaf-mutes, who articulate sounds of their own which they are unable to hear.[3] Such unprompted articulation must be regarded as instinctive and the result of inherited nervous connexions between the centres of perception and ideation and those of articulation. Under normal circumstances, however, this spontaneous speech is soon abandoned in favour of that adopted by others and impressed by way of the child's social needs and impulses. And it is plain that this process of learning and reproducing a highly-developed speech-structure, embodying the thought-distinctions and thought-relations of many generations, is widely different from that process of groping the way after new sounds as new ideas arise, through which the race has had to pass. Through this action of the speech-medium the progress of intellectual growth is, as has been pointed out in the text, furthered and expedited to an incalculable extent. The child becomes familiar with advanced or highly abstract concepts such as ' thing,' ' cause,' and so forth, long before his unaided intelligence could have even dimly descried them.

[1] On the language of backward races, see Lubbock, *Origin of Civilisation*, chap. ix., and *Prehistoric Times*, p. 573; Tylor, *Primitive Culture*, vol. i. chap. vii. ; Romanes, *op. cit.*, chap. xii. and following. The reader who desires further information on the early phases and mode of development of language should consult Steinthal's *Abriss der Sprachwissenschaft;* Fr. Müller's *Grundriss der Sprachwissenschaft;* and Paul's work on Language (Engl. transl.).

[2] M. Taine would regard such utterances as analogous to emotional expressions. They express the state of feeling of the observer when he is struck by a resemblance. (*On Intelligence*, pt. i. bk. i. chap. ii.) This view evidently connects the early speech of the individual with the speech of primitive man in so far as this is regarded as having its origin in vocal expression.

[3] See Tylor, *Early History of Mankind*, p. 72. The spontaneous articulations of Laura Bridgman, who was blind as well as deaf, illustrate this tendency still more clearly. (See Tylor, *loc. cit.*, pp. 74, 75.) Some curious observations as to the range of unprompted invention of articulate sounds by children is given by Prof. Hale in an essay, *The Development of Language.*

APPENDIX H.

PSYCHOLOGICAL AND PHILOSOPHICAL TREATMENT OF KNOWLEDGE.

The difference between the psychological and the philosophical, or, as it is now the fashion to say, the epistemological view of knowledge, has already been dealt with in part in connexion with sense-perception. (See Appendix C.) It remains to bring out the relation between the two somewhat more fully in the light of the above psychological account of the higher processes of thought.

The philosophical or epistemological problem considers knowledge objectively in respect of its reality, while psychology considers it subjectively as a mental process. The truth or falsity of the intellectual phenomenon which we call a cognition is a matter of indifference to the psychologist : illusory perceptions are as much percepts, *i.e.*, particular psychical phenomena, as true ones : the most absurd delusion of a maniac is of equal value with a perfectly rational belief for the psychologist's purpose.

Now there is another science which considers cognition on its objective side in respect of its truth or falsity, *viz.*, logic. As already pointed out, however, logic deals with knowledge only so far as it involves processes of reasoning or inference from previous knowledge, which processes it is its special business to regulate.[1] Philosophy, theory of knowledge, or epistemology, goes beyond this, and considers the objective aspect of *all* cognition. Not only so, it considers this objective aspect in another and more searching manner than logic does even in its own sphere. Logic, like the special sciences, stops short at our common notions and our standard of truth and reality. Thus to logic as such objects have a real existence, and universal truths a universal validity, the nature of which it is not its concern to inquire into.[2] Here it is that philosophy steps in. It wants to know what this reality is, how it is possible that the mind can grasp the existence of anything distinct from itself, in what sense we can attain to universal and necessary knowledge, for example, of the relation of events to their generating conditions, and of mathematical relations, which, as we see, necessarily include *every* supposable case.

This properly philosophical inquiry into the ultimate nature and conditions of knowledge set out, not unnaturally, with a consideration of its mental source or origin. A mere glance at our ordinary processes of cognition suggests that there

---

[1] *Cf.* above, I. p. 12.

[2] These distinctions between logic and philosophy have not been always observed. Hegel and J. S. Mill (to select an oddly-mated pair) agree in confusing the boundaries of logic and philosophy.

are apparently two such sources, *viz.*, the senses and thought or reason. The knowledge drawn from these seems to be different, the former telling us only of the particular which comes and goes, the latter of the universal and the immutable and necessary, *e.g.*, the law of causation. The first impulse of philosophy, following this common-sense distinction, was to erect reason into a special and superior source of knowledge, and to regard it as an essential factor in all true or valid cognition. This tendency is known as rationalism or intuitionalism, the word 'intuition' being used to mark off the alleged immediacy and certainty of this rational cognition. This view of knowledge is specially represented in ancient philosophy by Plato, in modern thought by Descartes, Leibniz, and Kant. Opposed to this is the tendency to refer all cognition back to sensation, to regard it as essentially particular or 'contingent' (*i.e.*, not-necessary), and as assuring us only of the facts of our common sense-experience, *viz.*, sensations, and, as seems to be allowed also, the observable temporal juxtapositions of these. This tendency is known as sensationalism, experientialism (or empiricism), and more recently (owing to the part assigned to the laws of association) associationalism. Its doctrine, already put forward in ancient philosophy, was fully developed by Locke and his followers, more particularly Hume.

The dispute, as first carried on in modern times between Descartes and Leibniz on the one side, and on the other by Locke and his followers, directed itself to the question of the existence of certain 'innate ideas'. The significance of this dispute turns on the supposition of the intuitionists that an original idea, not traceable to experience, is implanted in the mind by the Creator and so carries its own validity on its very face. In its earlier form the controversy appeared to concern itself with the date of the appearance of these ideas. Here, it is evident, the philosophical discussion encroached on the psychological domain ; and, in truth, Locke, when seeking to show that the infant mind is not furnished with ready-made ideas of God, and so forth, is really psychologising. Such a historical way of considering the matter could not determine the philosophical question of validity, for rational intuitions in their clear articulate form might come after sense-experience and reflexion on this and still be valid intuitions.

In its later and more guarded form, however, particularly as shaped by Kant, the intuitionalist has differentiated his problem much more clearly from those of psychology. It is now made plain that, according to intuitionalism, the mind does not at birth possess ready-made intuitions ; that, on the contrary, the materials of experience as supplied by the senses are necessary to the proper development of these intuitions. What the modern intuitionalist does assert is that these materials, even when conjoined by the processes of association, do not constitute what we mean by *cognition*, that in order to its constitution the mind must contribute its own proper *a priori* forms, and its own synthetic activity.

As between recent experientialists and intuitionalists, then, the question may be put as follows : Is knowledge a mere outcome of sensations conjoined according to known psychological laws, or does it involve as a further factor the co-operant activity of a rational principle ? To take the case made prominent by Hume, Is my belief in the universality of causation a mere effect or residuum of experience

＊

and habit, or is it a product of the mind itself working according to its pre-established and unalterable forms of activity on sense-experience?

Here, again, we may point out that the psychological treatment, though distinct from the philosophical, naturally leads on to this. Thus the psychological theory of sense-perception, and of general ideas, by deriving the intellectual products from sensations as their psychical elements, naturally *suggests* that the knowledge reached represents and refers to these sensations. Hence the fact that English experientialists, as J. S. Mill and Bain, quite frankly apply their psychology to the resolution of the philosophical problem. Taking this line, they have been met by the philosophical contention that our cognition of things as persistent unities, and, still more, of universal truths, is not in its objective import the same as any conceivable modification of mere sense-phenomena; and, as supplementary to this, by the psychological contention that this cognition properly understood cannot *as such* be accounted for as the effect of the elements and processes assumed by the psychologist.

Here the evolutionist has come in to help the older experientialist, who sought to trace back cognition to individual experience. In the transmitted products of ancestral experience with which Herbert Spencer endows the child, we have, it is evident, a psychological counterpart of the subjective factor of the intuitionalist. That is to say, the individual acquires his knowledge by the co-operation of certain instinctive intellectual tendencies, answering to recurrent modes of ancestral experience, with his own sense-supplied experience. Hence it has been urged by Spencer and others that his doctrine of evolution reconciles the two opposed philosophical views.

Enough has been said to show how closely the psychological and the philosophical problem are connected, and how impossible it is to deal with the question ' What is true cognition?' without some amount of psychological consideration of the intellectual state itself, and the way in which it has come to be. Yet, in the end, the philosophical problem, however much help it may thus receive from psychological doctrine, remains over after the psychologist has completed his doctrine of mind. Thus it has been urged, and with force, that no extension of the range of experience, such as Spencer's theory of transmitted products of ancestral experience attempts, can affect the question: Is *any* experience, in the sense of a connected whole the parts of which are seen to be bound together by real objective relations, possible through the mere play of sense and association? Does not all cognition of reality, even that which we gain in connexion with the use of our senses, involve the work of an active principle of mind or reason? Psychology can, to some extent, trace the lines of development which cognition follows, but it cannot find a ready answer to the question what the cognitive product amounts to, in what sense, if any, it guarantees the reality which it appears to represent.[1]

[1] On the nature of the philosophical problem of knowledge the reader may consult Prof. Seth's article "Philosophy" in the *Encyclop. Britann.;* also Prof. Fraser's Introduction to the *Selections from Berkeley;* and Dr. Bain's historical sketch of the question of the origin of knowledge, *Compendium of Mental and Moral Science,*

APPENDIX I.

THEORIES OF PLEASURE AND PAIN.

Although the theory of feeling is the latest development of psychological doctrine, we find that attempts were made by Greek thinkers to define the most general conditions of pleasure and pain, and subsequent philosophers have followed their example. In this way we have a historical series of doctrines on the subject. These different views of pleasure and pain have been developed in connexion with the writers' respective systems of philosophy, and an adequate account of them in their historical connexion belongs to the history of philosophy. Without attempting to deal with this aspect of the subject, we may point out some of the more important psychological differences in the theories propounded.

We have found that pleasure and pain arise from what appears to be a variety of psychical or psycho-physical conditions, and that it seems impossible to include all the phenomena under one simple principle. Now the tendency of speculation has been to formulate such a principle, and this, as we might expect, has led to a partial and one-sided examination of the phenomena.

1. The first known attempt to generalise on the subject is the theory of Plato. According to this, pleasure is nothing positive or real, but only a filling up of a want of existence, or a restoration to the normal condition. The radical form of feeling is thus the pain of craving, and pleasure becomes purely relative and negative.[1] This view, as Aristotle in his criticism points out, is based upon the phenomena of desire, and more particularly bodily appetite, where, as everybody allows, pleasure is an incident in the satisfaction of a pre-existent want. The pleasures of moderate activity, and the pains of excessive action, appear here to be equally lost sight of.

Plato's theory is pretty closely represented in modern thought by the doctrine of Kant,[2] which views the ordinary state of our vital powers as a neutral condition of comfort (Wohlbefinden), out of which we are constantly urged by pain. Here pleasure is nothing but the liberation from this prior state of pain. A like doctrine has been expounded by Schopenhauer, and made a fundamental feature in his system of pessimism.

Appendix B. The confusion of psychological with properly philosophical questions in the doctrine of Locke and his followers is brought out in T. H. Green's severe examination of the same (in the " Introduction " to Hume's *Treatise*). The relation of philosophy to psychology is further dealt with in the articles already referred to, vol. i. p. 13, footnote 3 ; in Shadworth Hodgson's *Philosophy of Reflexion*, bk. i. chap. i. 2 ; and in my volume, *Illusions*, chap. xii.

[1] Plato's view is unfolded in the *Philebus*, and the Ninth Book of the *Republic*.

[2] *Anthropologic*, § 60.

2. Directly opposed to this way of envisaging the subject is the doctrine which views pleasure as positive and independent, a concomitant of our actions or energisings. This doctrine springs out of a special consideration of the pleasures of activity, bodily, and still more intellectual, a class of pleasures which Plato's theory ignores. Aristotle first gave shape to this doctrine. According to him, pleasure is not (as Plato asserted) a change or becoming (γένεσις), but something perfect from moment to moment, consequently real and positive. Pleasure is defined as the consciousness of a perfect or unimpeded energy, whether of the functions of sense or of intellect.[1]

This way of viewing pleasure has been followed by most modern thinkers. Thus, according to Descartes,[2] pleasure is nothing but the consciousness of some one or other of our perfections, a definition which reappears in the doctrine of Leibniz.[3]

2a. The tendency of this view has been to intellectualise feeling by resolving pleasure and pain into a *quasi*-presentative phenomenon, a mode of intellectual consciousness. This tendency, which seems to be involved in Locke's treatment of pleasure and pain as simple " ideas,"[4] shows itself plainly enough in the doctrine of Wolff that pleasure is the intuitive *cognition* of any perfection whatever (whether true or apparent).[5] The same feature is prominent in the teaching of Sulzer, who, setting out with the assumption that the natural activity of the soul is the production of ideas, connects pain with the presence of an impediment to this activity, pleasure with a full or ' lively ' excitation of this activity.[6] A somewhat similar view is given by Hamilton, who says : " Pleasure is the reflex of the spontaneous and unimpeded exertion of a power of whose energy we are conscious, pain a reflex of the overstrained or repressed exertion of such a power ".[7]

The restriction of pleasure and pain to intellectual activity is very apparent in the theory of Herbart and his school.[8] Here both modes of feeling are viewed as incidents in the mechanical interaction of presentations. Thus, according to Herbart, pleasure is the consciousness of the support which presentations yield one another when this is more than sufficient to effect a rise of a presentation above the threshold of consciousness ; pain, on the other hand, is the sense of conflict which arises when one and the same presentation is simultaneously acted upon by other presentations, some of which tend to suppress, others to support it. A similar theory will be found in the work of Volkmann already referred to.[9]

---

[1] *Nicomachean Ethics*, bks. vii. and x.

[2] *Epistles*, first part, lect. vi.

[3] *Nouveaux Essais*, lib. ii. chap. xxi. § 41.

[4] *Essay on the Human Understanding*, bk. ii. chap. xx.

[5] *Psychol. Empirica*, § 511.

[6] *Ueber den Ursprung der angenehm. und unangenehm. Empfindungen.*

[7] *Lectures on Metaphysics*, ii. xlii. and xliii.

[8] See Stout's exposition of the Herbartian psychology, *Mind*, xiii. p. 487 ff.

[9] See his *Lehrbuch der Psychologie*, vol. ii. § 127. A view somewhat similar to that of the Herbartians, though emphasising the factor of attention as the essential constituent of intellectual activity, is given by Ward, *op. cit.*, p. 71, col. a.

2*b*. The Aristotelian conception that pleasure is positive and connected with activity has been modified and enlarged so as to take on the appearance of a psycho-physical or biological truth. That is to say, pleasure is regarded as the concomitant of a vital motion or a nervous action. The germ of this theory may be found in Hobbes, who views pleasure and pain as the appearance (*i.e.*, the subjective side) of appetite, which, *in reality*, is motion. He adds that the 'emotion' in the case of pleasure seems to be a *corroboration* of vital motion, the contrary in the case of molestation.[1] This reference to organic processes (which is faintly indicated in Kant's conception of pleasure and pain as correlated with a furtherance and a hind-rance of *life*) grows distinct in the doctrine of Lotze that pain arises when the nerve is injured, pleasure when excitation exceeds the usual level of nervous activity without involving loss of functional efficiency.[2]

2*c*. The reference to nervous conditions becomes more distinctly a reference to the organism as a whole, so that pleasure is viewed as the concomitant of that which benefits, pain, of that which injures, the sum of life-functions. Here we see a tendency to bring in the idea of a normal equilibrium, departure from which, whether by way of excess or of defect, causes pain, whereas action within the limit of excess is pleasurable. This idea, already adumbrated in the doctrines of Plato and Kant, becomes definite in most modern theories which take account of nervous conditions. Thus it is plainly set forth in the system of Bain, where states of plea-sure are said to be connected with an increase, states of pain with an abatement of some or all of the vital functions. The writer seeks to reach this generalisation by a careful inductive examination both of the agents which cause pleasure and pain and of the manifestations or effects of these states.[3] Herbert Spencer, as pointed out above carries the idea of the organic solidarity or consensus still further. Accord-ing to him pleasure is the concomitant of a medium activity of an organ, and as such is beneficial to the organism as a whole, whereas pain, being the concomitant either of excessive or of defective action, may be supposed to be injurious to the organism.[4]

The latest form of this theory is that which connects pleasure and pain with quantitative variations of the sum of vital energies. Thus in the ingenious doc-trine of L. Dumont pleasure and pain are viewed as correlated with particular ratios of accumulation to expenditure of energy. The former arises as the concomitant either of storage (pleasure of repose) or of diminished friction in expenditure.[5] This idea is developed with an appearance of scientific precision by Zöllner in his *Kometenbuch*. He regards pleasure as the psychical aspect of the process of trans-

---

[1] *Leviathan*, pt. i. chap. vi.

[2] *Med. Psychologie*, pp. 285, 286 ; *cf. Allgem. Pathologie*, i. p. 539.

[3] *The Senses and the Intellect*, p. 283.

[4] *Principles of Psychology*, vol. i. pt. ii. chap. ix.; *cf.* Grant Allen, *Physiological Æsthetics*, chap. ii.

[5] *Théorie Scientifique de la Sensibilité*, p. 67.

formation of potential energy (Spannkraft) into living force, pain of the transformation of kinetic into potential energy.[1]

It may be doubted whether any one of the above theories is adequate to the
variety of the phenomena. Thus, as Fechner points out in his criticism of Zöllner,
pleasure does not seem to be uniformly concomitant with the raising of the life-
process, for excessive enjoyment consumes life-force just as pain does. That is to
say, expenditure, even up to the point of injury, is itself pleasurable. Hence Bain
has to supplement his reference to the vital functions by a principle of Stimulation.
It may be added that the pleasures of repose do not easily adapt themselves to
Zöllner's definition of pleasure-conditions. There must, it would seem, be a
reference at once to the excitation evoked in the particular organ, and to the
secondary effects of this in the *ensemble* of organs. It may be added that, according
to Fechner, the first reference must include not only the quantity but also the form
of the stimulation.[2]

---

[1] Modifications of this view may be found in Delboeuf, *Élements de la Psycho-
physique*, p. 182, and *Théorie Générale de la Sensibilité*, p. 44, and in Horwicz,
Psychol. Analysen, ii. 2, p. 40.

[2] For Fechner's view, see *Vorschule der Æsthetik*, ii. p. 263 ff. On the different historical views, see Hamilton, *Lectures on Metaphysics*, vol. ii. xliii. ; Wundt,
op. cit., ii. cap. x. 4 ; L. Dumont, *op. cit.*, 1ère pt. chap. ii. ; Volkmann, *Lehrbuch
der Psychologie*, ii. p. 291 ff. ; and an article on " The Physical Basis of Pleasure
and Pain," by H. R. Marshall, in *Mind*, xvi. p. 327 ff.

357

APPENDIX J.

CLASSIFICATIONS OF THE EMOTIONS.

As hinted above, the classification of the emotions is a problem beset with special difficulties, which arise in part out of the peculiar complexity and interpenetration of the phenomena themselves, and in part out of the exceptional obstacles to a calm introspective fixation and analysis of these turbulent mental states. These difficulties are clearly illustrated in the utter want of agreement among those who have attempted the task.

Although, as has been observed, but little thought was given to the feelings in the earlier stages of psychological history, we find that some attention was paid to the division and orderly arrangement of the more important varieties. Thus Aristotle, though, as Grote remarks, he did not carry out a systematic classification, went some way in this direction by distinguishing bodily from mental pleasures, and further by dividing emotions (πάθη) into those which have a preponderance of pleasure, as love, and those which have a preponderance of pain, as anger, fear, and so forth, thus plainly showing that his basis of classification was the distinction of the agreeable and the disagreeable.[1]

Coming now to modern philosophy we find the problem receiving attention at the hands of its founder, Descartes, who, in his treatise, *Les Passions de l'Âme*, divides the feelings (like ideas) into two groups, theoretical and practical. He recognises six primary affections, *viz.*, wonder, love, hate, desire, sorrow, joy. The only valuable feature of this confused arrangement is the unique position given to wonder, a position not unlike that assigned it by a recent psychologist, Dr. Bain. In the case of this passion the movements of the vital spirits are supposed to be confined to the brain, in that of the others, to be extended to the heart and felt there.[2]

A more serious attempt at a scientific classification is to be met with in Spinoza's *Ethica*. This thinker, who connects the genesis of the passions with the fundamental impulse towards self-maintenance against hindrance, distinguishes, as fundamental divisions, sorrow and joy, according to the prevalence of obstacle and of effort. From these, together with desire (cupiditas), also a primary affection, he seeks to derive the others.[3]

---

[1] See Siebeck, *Geschichte der Psychologie*, 1er theil, 2e abtheilung, p. 87 and p. 90 ff.

[2] See Erdmann, *History of Philosophy*, ii. p. 28 f. Malebranche reduced the passions to two *passions mères*, love and aversion, *ibid.* p. 50.

[3] See Erdmann, *op. cit.*, p. 78 f.; Ueberweg, *Geschichte der Phil.* ii. p. 81; and Pollock, *Spinoza*, p. 224 and following.

If, now, we examine more recent attempts at classification, we find that, although more elaborate, they do not satisfy the conditions of a rigorous scientific arrangement. In the psychology of Reid and Stewart there is hardly a pretence at a classification. Brown's arrangement of the emotions, according to their relation to time, into (1) immediate, as wonder, (2) retrospective, as anger and gratitude, and (3) prospective, *i.e.*, the desires, ambition, and the like, is too artificial to repay examination.[1] Hamilton gives us a more satisfactory scheme. He groups the feelings into the two genera, sensations (*i.e.*, sense-feelings), and internal or mental feelings or sentiments. These last he sub-divides into contemplative and practical, and carries his scheme further by distinguishing the contemplative feelings according to the intellectual faculty concerned, and the practical according to the active impulse engaged, *e.g.*, self- and race-preservation. It is evident, however, that this scheme, good as it undoubtedly is, is not drawn up on any one clear principle.[2]

Dr. Bain, one of the most recent and vigorous workers in the department of emotional psychology, has, with a full recognition of the difficulties of the task, essayed an exhaustive classification of the emotions. He proceeds by first marking off certain feelings, as novelty, mainly determined by change or transition ("emotions of relativity"), and then setting forth certain well-recognised "genera," more particularly, fear, anger, and love or tender emotion, passing from these as primary and original to the secondary and more composite emotions. Here, it is evident, we have a kind of compromise, due to a combination of the two principles of division, simplicity, and originality or instinctiveness.[3]

A bolder attempt to exhibit a scale or serial order of emotion is to be found in Mr. Herbert Spencer's classification. This arranges the feelings according to degree of representativeness (or indirectness of presentation) into the following stadia corresponding to those of cognition : (1) presentative feelings, *i.e.*, actual sense-feelings ; (2) presentative-representative feelings, actual and revived sense-feelings ; (3) representative feelings, revived sense-feelings : and (4) re-representative feelings, involving a more abstract or indirect mode of representation, as the sentiment of property.[4] This arrangement, though valuable as indicating complexity and height in the developmental scale, is, it is obvious, open to the objection that it does not correspond to recognised qualitative differences among our emotive states. Thus anger, fear, and so forth, would (as Mr. Spencer expressly tells us) find a place under (2), and (as ideally excited) under (3) also. It may be urged, too, that the attempt to trace back all feeling to sense-feeling ignores the co-operation of certain *central* conditions, as reaction of attention, change, and harmony.

The mode of classification current in Germany is based on the distinction : (*a*) formal feelings, those not connected with any particular quality of presentation,

---

[1] Brown's arrangement is indicated in his *Lectures on the Philosophy of the Human Mind*, lii.

[2] Hamilton's classification is given in his *Lectures on Metaphysics*, ii. lect. xlv.

[3] *The Emotions and the Will*, chap. iii.

[4] *Principles of Psychology*, ii. pt. viii. chap. ii. (*cf.* i. pt. iv. chap. viii.).

but with the mode of interaction of presentations (including effects of relief and harmony); and (b) qualitative feelings, those which depend on special qualities in presentation, and are sub-divisible into sense-feelings and certain higher feelings (intellectual, æsthetic, etc.).   In addition to these feelings proper, other mixed states involving an effect on conation, as fear, rage, etc., are recognised.[1]

In France one of the most ingenious attempts to give a truly scientific arrangement to our emotions is that of the late Léon Dumont, who, following the hint given by Aristotle, aims at developing all varieties of feeling from the fundamental distinction of pleasure and pain.   By conceiving of pleasure as connected with a balance of accumulation of force over expenditure, pain with an excess of expenditure above accumulation, Dumont seeks to classify all our feelings into four groups, viz., positive pleasures (arising from increase of nervous excitation), negative pleasures (arising from diminution of expenditure), positive pains (which arise from too great expenditure), and negative pains (arising from insufficient accumulation).   This classification, however, though scientifically conceived, fails to exhibit the psychological affinities among our emotive states.   Thus it groups together feelings so dissimilar as the pains of lesion and the emotion of doubt and of fear.[2]

Lastly, reference may be made to more than one recent attempt to conceive of emotion biologically as an organic reaction, and to arrange the several emotive states with reference to the teleological function which each subserves, and the place of this function in the scale of importance.   A painstaking and elaborate attempt in this direction has recently been made by Dr. Mercier.   Valuable as this mode of arrangement is here shown to be, the attempt illustrates the difference in the psychological and the biological point of view, and the practical impossibility of getting an arrangement that at once corresponds to teleological distinctions and to psychical differences inherent in the feelings themselves.[3]

---

[1] The Herbartian mode of classification is well illustrated by Nahlowsky, *Das Gefühlsleben.* (*Cf.* Horwicz, *Psychol. Analysen*, 2er theil, 2e hälfte, § 6.) Bain gives a full account of the common German scheme in his interesting bibliographical note, *op. cit.*, Appendix B.

[2] See his work, *Théorie Scientifique de la Sensibilité*, pt. ii. chap. i.

[3] See Mercier's book, *The Nervous System and the Mind*, chap. xii. and following. For a fuller account of the various ways of arranging the emotions, see Bain, *The Emotions and the Will*, Appendix B ; and Volkmann, *op. cit.*, ii. p. 341 f.

APPENDIX K.

THE PSYCHOLOGY AND PHILOSOPHY OF BEAUTY.

As in the case of pure cognition so in that of the appreciation of the beautiful we have to distinguish a psychological from a philosophical problem. The former is concerned with the mental process in the contemplation and enjoyment of beautiful things, with an analysis and a genetic account of the peculiar psychosis which we call æsthetic delight. The latter has to do with the objective nature of beauty, what it is in objects that constitutes the real attribute as distinguished from its false semblance. This latter problem belongs, strictly speaking, not to pure psychology but to that branch of philosophy which we call æsthetics, and which has for its practical function to supply us with certain ideal conceptions, or a standard by which the validity of any particular judgment of taste may be estimated. This philosophical inquiry into the objective nature of beauty may proceed metaphysically and deductively, as when Plato constructed a theory of beauty on the basis of his system of objective ideas. Here, it is evident, the connexion of æsthetics with psychology is only very remote. In most attempts, however, to deal with the problem some attention is paid to the empirical facts, the particular objects which men of taste agree (approximately at least) in pronouncing beautiful. Such an analytical inductive inquiry into the perceptible constituents of objects called beautiful has always formed the main part of æsthetical speculation, in this country at least. Here, it is equally evident, the philosophical and the psychological domain touch and overlap. For to refer the beautiful to particular ingredients or associative concomitants of objects is plainly, by implication, to point out certain initial factors in the whole psychical process of grasping and enjoying beauty. Thus, if beauty be wholly resolved into an indirect or associated effect, or into that of a certain order or arrangement among the parts of the contemplated object, as unity in variety, this is only another way of saying that particular modes of association or perception excite, by means of known psychological laws, the distinctive feeling or psychosis.[1]

Hence we need not wonder that the several well-known æsthetic theories, especially those developed in this country, have been worked out by help of, and as applications of, known or supposed psychological truths. A glance at one or two of the more famous theories will make this plain.

The earliest, and on the whole the commonest, attempt to reduce all forms of beauty to one simple principle is that which connects it with orderly arrangement of parts, variously described as unity in variety, relations of proportion, and so forth. We have the germ of this view in Aristotle's fragmentary account of beauty. It has

---

[1] Unless, indeed, as apparently with Hutcheson, and still more plainly with E. Gurney, the effect of beauty is referred to a unique psychical faculty or sense.

been the common doctrine among French and German writers, and is represented in this country by writers otherwise so far apart as Hutcheson, Hume, and Hamilton. According to this view the enjoyment of beauty is dependent on that mode of intellective consciousness which we call the perception of relations.[1]

As a second, strongly contrasting theory, we may take that of Alison, who sought to eliminate the 'direct factor,' as it has since been called, altogether, and to reduce the whole effect of beauty to association or suggestion. This he did in the interests of imagination as against the curious doctrine of Burke that beauty is throughout a sensuous phenomenon, *viz.*, a peculiar sense-feeling, as we should now call it, produced by a relaxation of the fibres, an organic effect specially brought about by *smoothness*, to which sweetness in the region of taste and in that of smell is analogous. Alison contended that beauty, so far from residing in any sensuous quality or effect, is the correlative of an emotion which is excited through a peculiar use of the imagination. Here, again, we see æsthetical theory betaking itself to a psychological consideration of the mental process involved.

Probably enough has been said, without going further into the history of æsthetical theory, to show how the attempt to formulate a principle respecting the essential conditions of beauty leads back to psychological theory. How much this is the case is illustrated in the fact that attempts to frame theories of the beautiful have not infrequently been elaborated as a part, or a special application, of a general psychological theory of pleasure ; and, further, in the marked tendency among later theorists to recognise a plurality of constituent factors in the effect of the beautiful, a tendency which is manifestly the result of a more complete psychological investigation of the mental processes involved.

Further points of contact may be just alluded to. It is evident from our treatment of the subject that an attempt to define the distinguishing marks of æsthetic pleasure is as much a psychological as an æsthetic problem. Not only so, the more manifestly philosophical problems, as the determination of the relation of beauty to utility, a point discussed from the time of Plato downwards, also the inquiry how far a uniform standard of taste is possible, all carry us on into the psychological domain, *i.e.*, the theory of the underlying mental processes. We may say, then, that of all the branches of philosophical discipline æsthetics is most directly and most completely based on psychology.[2]

---

[1] This is well brought out by Hume, *Concerning the Principles of Morals*, Appendix I., who says in all decisions as to external (in contradistinction to moral) beauty the relations of the parts are "obvious to the eye". The insufficiency of any such principle for explaining all the various effects of beautiful objects is well argued by E. Gurney, *The Power of Sound*, chap. ix.

[2] For a fuller account of the problems of æsthetics and their relations to psychology, see my article " Æsthetics," *Encyclop. Britann. ; cf.* the paper " Art and Psychology " in *Mind*, i. p. 467.

*

APPENDIX L.

PSYCHOLOGY AND ETHICS.

In the case of the moral or ethical sentiment, as in that of the æsthetic, we have, in addition to a psychological treatment of the mental processes, a philosophical examination into the objective validity of the ideas involved, and a practical or regulative determination of a standard of common reference. Thus we have, corresponding to the problems of æsthetics, the problems of ethics, such as the definition of the fundamental conceptions of the science, *viz.*, 'the good,' and the correlative ideas 'right ' and ' duty ' or ' obligation,' the laying down of a first principle of morality which may serve as a canon by which to judge of the rightness and wrongness of particular acts, and (in the more practical branch of ethics) the construction of a theory of motives or inducements by which men may be led to pursue the ethical end.

Here, again, we must, in the first place, carefully distinguish the psychological, or, as it may somewhat loosely be called, the subjective, from the ethical or objective aspect. It is one thing to trace out the conditions and mode of development of the ethical consciousness or psychosis, another to determine its objective import, and the normal standard to which it should conform. In the case of this philosophical aspect we have, as in that of the æsthetic faculty, to do with an *ideal* conception, that is, of something which is recognised as good and desirable rather than as actually existent. It is obvious that there is a considerable step from the psychological determination of men's actual moral approvals and disapprovals to the ethical determination of the proper object of these feelings.

Nevertheless, though the demarcation of the problems of what is and what ought to be, of genesis and validity, is, in the case of morality, particularly clear, it has commonly been recognised by moralists that psychology forms a necessary auxiliary to their science. Thus, however severely in the interests of a science of the ideal the moralist seeks to keep clear the true or valid moral judgment from the actual judgments of everyday life, he must, as a practical man who aims at guiding and improving actual moral ideas, give vitality and possible reality to his ideal end by bringing it into a general agreement with the prevalent moral conceptions of mankind. And here we see that ethics approaches the psychological boundary : for it seems impossible to establish in a complete scientific manner what men customarily approve and disapprove without taking account of the nature and origin of the sentiment of approval itself.

Accordingly, we find from the history of ethics in ancient and modern times that psychological discussion has formed a large feature in the science. Thus, even in Greek thought about the subject, which was concerned with the individual problem of the highest and best living, we meet with ample psychological reference, as in the elaborate theories of happiness and of volition which may be gathered from the *Ethics* of Aristotle. It is, however, in modern times, in which the direction of ethical inquiry has shifted from the question of the highest (individual) good to that of the nature and grounds of duty, that we see this connexion with psychology most plainly illustrated. In our own country more particularly the treatment of this modern question very early took the form of an inquiry into the nature and *origin* of conscience. On the one hand, we have had intuitionalists (or, as they might in this connexion be better called, naturalists or instinctivists) like Butler and Hutcheson, who asserted that conscience is a simple, unique and instinctive faculty supplying us, independently of external influences, such as social custom and education, with natural intuitions of what is right and wrong, and the correlative feelings ; and, on the other hand, the derivativists or associationists, represented by Hume, and more distinctly by Hartley and his followers, who contend that the moral faculty is a compound product, involving the external agencies of education, especially authority together with punishment and reward. The explanation of this concentration of ethical thought on the problem of origin is similar to that which we found in the case of epistemological inquiry. Butler, Hutcheson, and the rest emphasised the original and instinctive character of our moral judgments because they thought that by so doing they specially secured for them a *divine* authority.[1] Since, however, it has been seen that moral ideas and sentiments may retain all their authority even after it has been shown that they are derivative, the psychological question has fallen more into the background of ethical discussion.

Nevertheless, psychology still occupies a firm footing, if a smaller space, in the ethical domain. Thus in the current sociological, as distinguished from the older and more individualistic treatment of morality, in which the moral feelings and judgments are viewed as attributes of the community and a product of social evolution, reference has to be made to those psychological principles by help of which sociological phenomena are to be explained.[2] Not only so, in the more practical development of ethics we find that the psychological point of view has again and again to be introduced. Thus in dealing with the conception of moral accountability or responsibility with the correlative of idea of desert reference has to be made to the all-important psychological question of free-will. So, too, in attempting to ascertain by what forces men may be got to act rightly, the modern moralist has to go somewhat fully into the psychology of motive and of the volitional process as a whole. Witness the attention still given to the Greek question how far morality is a product of intelligence, and to the problem raised by Hobbes and his opponents, Shaftesbury and Butler, *viz.*, the relation of benevolence or disinterested sentiment to the self-regarding or egoistic feelings.[3]

[1] On the explanation of the prominence of the psychological factor in English ethics, see Sidgwick, *History of Ethics*, Introduction iii. and chap. iv.

[2] *Cf.* above, I. p. 34 f. The sociological or anthropological factor in ethical inquiry is made prominent in L. Stephen's *Science of Ethics* and in Wundt's *Ethik*.

[3] The connexions between psychology and ethics are fully brought out by Dr. Bain in his *Ethics*, pt. i. ; *cf. Mind*, xiii. p. 534 ff. A recent example of a system of morality based on psychology is Fowler's *Principles of Morals*. The demarcation of the two regions has been carefully carried out by H. Sidgwick. (See his *Methods of Ethics*, especially chap. i. Introduction ; also his discussion of the relation of evolution to ethics in *Mind*, i. p. 52, and v. p. 216.)

APPENDIX M.

PSYCHOLOGY AND FREE-WILL.

In the above account of free-will we have considered the phenomenon from a strictly psychological or psycho-physical point of view, taking the peculiar psychosis, consciousness of freedom, and showing what psychical conditions go to produce it. But there is another aspect of the subject. The doctrine of man's freedom has constituted a prominent feature in the history of theology. Here the human will is looked on in its relation to the divine, and the question becomes whether man has been predestined to act as he does act, and so to sin (predestination, and in a slightly altered form, fatalism), or has been left free to determine whether he shall or shall not sin by an exercise of individual choice.

In this theological discussion we have suggested by implication in the theory of divine fore-knowledge and predestination the idea of volition as a causally determined result. Yet it was only after the question passed out of the theological domain that the real issue between determinism and indeterminism became clear. By this is meant the question whether our actions are fully explained as the effect of antecedent conditions, which are commonly summed up as ' motives,' though, as pointed out above, they must be taken to include the semi-conscious organic factor that, as we have seen, co-operates in the production of voluntary action. The determinists assert that this is so : the upholders of (philosophical) free-will maintain that in the higher processes of choice we have the interposition of a new active principle, a transcendental *ego*, which independently of all motive decides on a particular line of conduct.

The philosophical doctrine that our actions are "free" has commonly been based on the consciousness of freedom, more particularly in its moral aspect, *viz.*, the sense of accountability or responsibility.[1] Hence it is necessary in order to test the validity of the belief to look at consciousness on another, *viz.*, the objective, side, that is, in its relation to objective reality. Here, again, then, we encounter the distinction between the psychological and the philosophical treatment of mental phenomena. Psychology, which is concerned only with unfolding and genetically explaining the phenomena of volition, cannot, it is manifest, directly settle the point whether the consciousness of freedom, which undoubtedly exists as a psychical fact, guarantees the objective existence of any such thing as a self-determining principle. So far the psychological and the philosophical problems are distinct.

At the same time, this distinction is by no means an absolute one. Though genesis and history are different from validity, a study of the first may, as we saw

---

[1] It is for this reason that the free-will question has filled a large place in ethical discussion.

in the case of cognition, put us on the track for discovering the other. An examination of the psychical phenomenon or psychosis which we popularly describe as sense of freedom is a necessary preliminary to the philosophical inquiry. Such a psychological analysis of the conviction on its subjective side will, at least, help us to see what the *consciousness* of freedom amounts to, and may suggest that, properly considered, this particular mode of consciousness does not involve any such idea as that of a transcendental force or active principle ; and, should it succeed in doing this, it would manifestly cut away the psychological ground of the common form of the doctrine of liberty.

Psychology stands in another connexion with the philosophical problem of free-will. As pointed out above, as a connected scientific account of mental phenomena, it presupposes the applicability of the law of causation throughout. So far, then, as it succeeds in establishing itself as a systematic and verified science, it will have something to say on the question of freedom. It may rightly demand that any philosophic conception of a transcendental self-determining will must be consistent with the psychological generalisation that phenomenal volitions, which are what our experience immediately tells us of, are empirically determined by their phenomenal conditions. How far Kant and others, who, while allowing the empirical fact that actions follow uniformly from the presence of certain groups of conditions, postulate the existence of a transcendental or noumenal will behind, have succeeded in thus harmonising metaphysical and psychological conceptions, this is not the place to consider.[1]

---

[1] The student who desires to know more of the history of the theologico-philosophic problem of liberty and necessity may consult Dr. Bain's sketch in his *Mental Science*, bk. iv. chap. xi. The present state of the question may be ascertained by reference to J. S. Mill, *Examination of Sir W. Hamilton's Philosophy*, chap. xxvi. ; Bain, *The Emotions and the Will*, pt. ii. chap. xi. ; and W. James, *op. cit.*, p. 569 ff. ; also the discussion of the question in *Mind*, by W. G. Ward, Bain, and Shadworth Hodgson, vol. v. (*cf.* vol. x. p. 532). The German reader may further consult the historical *résumés* given by Volkmann, *op. cit.*, § 151, and Wundt, *op. cit.*, ii. cap. xx. 2. The ethical aspect of free-will, made prominent in Kant's system, is dealt with, among others, by Sidgwick, *Methods of Ethics*, bk. i. chap. v. ; L. Stephen, *The Science of Ethics*, chap. vii. ; and F. H. Bradley, *Ethical Studies*, essay i.

APPENDIX N.

PHILOSOPHICAL VIEW OF MIND AND BODY.

As pointed out, empirical psychology does not inquire into the ultimate nature of mind as an entity or "substance," or into the closely connected question of the metaphysical interpretation of the connexion of mind with a seemingly heterogeneous substance, *viz.*, the bodily organism. At the same time, the study of the phenomena of mind naturally leads on to these metaphysical questions. Moreover, the bearing of these questions on the highly interesting problem of the immortality of the soul has led psychologists in general to trespass so far into the domain of rational or metaphysical psychology as to attempt to define their attitude in relation to these problems.

The bearings of empirical psychology on these problems may be briefly indicated as follows: (1) What view does a consideration of the phenomena of mind lead us to entertain respecting the inmost nature and ultimate sources of mental activity? More particularly, does it lead us to the hypothesis of a spiritual substance or soul, distinct from, and independent of, material things? (2) What does a thorough-going study of the physiological concomitants of mental phenomena lead us to regard as the real relation between mind and body? And how is this relation to be interpreted from a philosophical point of view?

These different lines of inquiry have been pursued together. The discussion as to what mind is in itself necessarily embraces that of its relation to its foreign companion, a material organism. Hence in tracing out the metaphysical bearings of our psychological study on the problem of soul we may most conveniently deal with the interpretation of the known relation between mind and body.

Here it is at once evident that we have to do with a problem partly similar, partly dissimilar to that already touched on in connexion with the philosophical treatment of cognition. In dealing with mind and body in their temporal juxtaposition we are not primarily concerned with this last as an object of cognition. At the same time, since this inquiry is a metaphysical one, *viz.*, into the nature of body and of mind *as substances*, it inevitably goes back upon the philosophical or epistemological question. Thus, in conceiving of the body as a being or substance, we are led on to ask in what sense this material thing can have any detached or absolute existence apart from the mind which knows it.[1]

If now we consider the conclusions reached by an examination of the concomitant phenomenal processes, psychical and neural, we find, in the first place, that it in no way justifies the reduction of the former to terms of the latter, *viz.*, modes of movement of material particles. The phenomena of consciousness are

---

[1] On the distinction between metaphysic (or ontology) as theory of being, and philosophy as theory of knowledge, see Prof. Seth's article "Philosophy" in the *Encyclop. Britannica.*

sui generis. Thus, if we take the psychical elements, sensations, we discover no affinity between a taste or a sound and the accompanying series of molecular movements. The same holds good *a fortiori* of the higher psychoses, as ideas, the emotion of love, and so forth. We have thus in the evolution of consciousness a succession of phenomena altogether disparate from material or physical phenomena.

On the other hand, we have found that this unique series of events is correlated with another and disparate series, *viz.,* the molecular movements in particular portions of the nervous system. This correlation is before all else a concomitance in time or a synchronising of two series of events. The fact of this concomitance has been established with some degree of exactness in the case of the more elementary psychical phenomena, and it is presumable that it holds good throughout.

Again, these two series, the psychical and the physical, not only run on together in time but exhibit certain correspondences in their respective variations. Thus, as we have seen, differences in intensity, extent, and complexity in our psychical states answer to similar differences in the nervous processes.[1] On the other hand, qualitative psychical differences, though, as we have seen, probably correlated with differences in the neural concomitants,[2] are not correlated with homologous differences. Thus though the peculiarities of the several classes of sensations, as colours, tones, and so forth, are supposed to implicate differences in the mode of molecular movement, these last are as utterly dissimilar to the corresponding sensational differences as the particular series of molecular movements in the case of a sensation red is dissimilar to this last. A like remark applies, as we have seen, to the processes by which psychical elements are organically combined and transformed into higher products. Discrimination, assimilation, and association are with good ground supposed to have each its distinct neural base: yet it seems impossible to conceive of these neural concomitants as akin to the psychical functions otherwise than in respect of those common features, magnitude, complexity, and temporal flow, already referred to.

Coming now to the exact nature of the temporal concomitance, we find that the conscious series runs parallel with an intermediate central portion of the neural series. In the process of peripheral stimulation and the propagation of its effect to the centre, and in the transmission of a motor impulse from the brain to the muscles, we have a physiological process without any conscious concomitant. How are we to conceive of this partial parallelism? Does it point to any true causal relation between the psychical and the neural factor, or does it rather suggest a parallelism of two disconnected processes, as of two rivers flowing side by side?

These questions have not yet been satisfactorily answered by scientific methods. According to a common view, more especially among physiologists, we have to think of the chain of nervous events as complete and self-sufficient throughout. It follows from this supposition that there can be no causal action of consciousness upon the neural series, just because, in accordance with the principles of modern physical science, more particularly the law of the conservation of energy, every phase of a series of movements is fully accounted for by a knowledge of the

---

[1] See above, I. p. 52 and p. 87 ff.   [2] See above, I. p. 53 and p. 91.

preceding phases. This view looks upon the appearance of consciousness at a certain point in the physical succession as something collateral and apparently accidental. This doctrine is known as that of human automatism, the doctrine that we are essentially nervous machines with a useless appendage of consciousness somehow added. It may be objected, however, that such a view cannot, strictly speaking, consistently regard the psychical phenomenon as conditioned by the neural process; for according to it the effect of the physical action is exhausted by pointing out the subsequent stages of the neural process, *so that there ought to be no second series of conscious phenomena at all.*

Opposed to this view, we have another which regards the psychical processes as at least as real as the physiological, having a reality which cannot, without setting at nought fundamental distinctions, be subsumed under, or even made subordinate to, physical action. This view is naturally the one to which students of psychology have on the whole inclined. It concedes that mental events are conditioned by nervous processes in the sense explained above, *i.e.*, as varying in certain ways when these last are made to vary; but it declines to regard the one as in any sense the outcome of the other. Some psychologists go further, and argue that a scientific examination of the facts goes to support the idea that the conscious chain in its turn stands in a causal relation to the nervous, since its intervention serves to modify the form of the succeeding organic reaction.[1]

This slight sketch of the different scientific conceptions of the psycho-physical relation may suffice to show that we have here to do with a unique fact of our experience. We know of no analogous correlation by the help of which science is able to elucidate it.[2]

The difficulties of the problem are further illustrated in the attempts of metaphysic to find in a theory of the ultimate nature of mind and of body an explanation of the facts. The first and crudest of these theories is the dualism of Descartes, according to which mind and body are distinct independent beings having no real interaction, the appearance of such interaction being due to special interpositions of divine power. This theory offers, it is plain, no scientific explanation, but only a theological or supernatural solution of the fact of concomitance. Hence it has not satisfied the thought of the modern world, which has endeavoured to transcend the duality of mind and body by help of a single uniting principle.

Such a unification may be attempted, first of all, in one of the two directions pointed out above, *viz.*, by resolving either of the two factors into the other. In this way we obtain the metaphysical conceptions known as Materialism and Spiritualism. The former attempts to reduce all substance to matter, and to view conscious mind as a product of this. This way of dealing with the connexion is

[1] *Cf.* above, I. p. 79, footnote [1].

[2] For a full setting forth of the psycho-physical parallelism, see Münsterberg, *Die Willenshandlung*, iii.; and *Beiträge zur exper. Psychologie*, heft i. einleitung ii. iii.; *cf.* Wundt, *Physiol. Psychol.* cap. xxiv. The doctrine of human automatism is criticised by Lewes, *Physical Basis of Mind*, prob. iii., especially chap. vii.; and James, *Principles of Psychology*, chap. v. The causal interaction of mind and body is vigorously maintained by Ladd, *Elements of Physiol. Psychology*, pt. iii.

open to the objections already pointed out, *viz.*, that consciousness is a reality wholly disparate from material processes, and cannot, therefore, be resolved into these. It is further open to the criticism that it makes that which is immediately known (our mental states) subordinate to that which is only indirectly or inferentially known (external things); also to the philosophic objection of idealism, that a material entity existing *per se* out of relation to a cogitant mind is an absurdity.

The other doctrine, spiritualism, escapes the objections just urged. By conceiving of mind as the one real substance, and viewing the material body as in its inmost nature spiritual also, it simplifies the metaphysical question. In its turn, however, it has to encounter the difficulty of the juxtaposition of the two in the case of at least certain of the higher organisms. The history of spiritualistic doctrines shows that it is not fitted to furnish a simple and satisfactory theory of the concomitance.

This leads us to the latest and most popular attempt to grapple with the problem, *viz.*, that first defined by Spinoza, and now known as Monism. According to this theory, both the mental and the material are real, or self-existent, but are not independent realities. Consciousness and the fundamental property of material things, extension, are conjoint attributes of one and the same substance. Thus the ultimate reality or fundamental substance is neither spiritual alone, nor material alone, but both. The parallelism of the two chains of events in the case of certain organisms points to the conclusion that the two attributes are inseparable, and only different aspects of the same reality, like the convex and the concave side of a curve. This doctrine plainly gets rid of the difficulties which beset the apparent interaction of mind and body. There is no interaction as the dualist would conceive of this, but merely a parallelism due to the uniform co-manifestation of the two co-inherent attributes. It may be added that it escapes the difficulty lurking in dualism and materialism, if not also in spiritualism, *viz.*, that of the apparent limitation of consciousness to particular modes of material action, *viz.*, functional movements of central nervous organs. According to monism, the parallelism runs through all things, inert as well as living matter, so that we have to conceive of every particle of matter as having its *quasi*-mental aspect. In this extension of the realm of mind, however, to material things generally, monism separates itself widely from the point of view of our common psychology, which conceives of mind and consciousness as co-extensive, and correlates this exclusively with particular collocations and formations of material particles.[1]

[1] For an account of the several theories of body and mind in their connexion with psycho-physical facts, see Bain, *Mind and Body*, chap. vii.; Volkmann, *op. cit.*, §§ 18-22; Wundt, *op. cit.*, cap. xxiii.; Höffding, *Psychology*, ii.; and Ziehen, *Leitfaden der physiol. Psychologie*, p. 173 ff. It is to be noted that these writers are by no means agreed as to the right way of classifying the theories. Thus, while Volkmann classifies them as above, Wundt recognises three main types, materialism and spiritualism, each of which has a dualistic and a monistic form, and animism, or the theory that the material universe is an animated organism. Other classifications are given by Bain and Höffding. The relation of the psycho-physical problem to the philosophical or metaphysical is also touched on by James, *op. cit.*, i. p. 176 and by Münsterberg, *Ueber Aufgabe und Methoden der Psychologie*, ii.

INDEX.

INDEX.

    ✱

*

24*

I.

*

J.

K.

Kammler, on degrees of pressure, 104.

Kant, on negative characteristics of mind, 7; on inadequacy of introspection as psychological method, 17, 21; on time-intuition, 329, 344; on feeling, 14; the sublime characterised by, 146, 148, 197 *note;* on moral autonomy, 157; classification of mental functions by, 327; on space, 332; his philosophy of perception, 337; on pleasure and pain, 353, 355; free-will doctrine of, 365.

Knowing, intellection; as fundamental mental function, 60; stages of, 61; relation of, to feeling and willing, 67, 296; pleasures of, 127.

Knowledge, distinguished from knowing, 4; imagination involved in acquisition and discovery of, 372; as systematised belief, 493; as social product, 496; philosophical theory of, 380.

Körner, on priority of feeling, 70.

Kries, Von, on aural sense of direction, 266 *note.*

L.

Ladd, G. T., on taste, 99 *note;* on nervous elements in colour-sensation, 117, 119 *note.*

Lange, C., passions and emotions distinguished by, 56 *note.*

Lange, N., on motor concomitant in ideation, 150; in perception, 151 *note.*

Language, as medium of reproducing knowledge, 310; as instrument of thought, 411; psychology of, 427; physiology of, 427; as social phenomenon, 428; origin of, 348; growth of, in race and individual, 348.

Laughter, excitants of, 148; theories concerning, 153.

Laws of mind, 27.

Lehmann, A., on nervous process in attention, 152 *note,* 316 *note;* on recognition, 184 *note,* 189, 196, 286, 407 *note;* on duration of after-image, 282 *note;* on comparison, 407; on double sensation, 312 *note;* on physiology of hypnotic state, 316.

Leibniz, G. W., theory of retention as affected by metaphysics of, 187; feeling according to, 328; place of intellect in mind according to, 328; theory of persisting presentations of, 340; pleasure defined by, 354.

Lewes, G. H., on animal consciousness, 21; on triple process of mind, 70 *note;* on faculty and function, 72; physiological and psychological process correlated by, 74.

Lewis, Dr. Bevan, on melancholia, 321; on mania, 321 *note;* on cerebral dissolution, 322 *note;* on impairment of object-consciousness, 323 *note.*

Light, sensations of, 116; pleasures of, 92.

Likeness, relation of, 179; Herbartian view of, 180; discovery of, in comparison, 397; judgments respecting, 439. (See Assimilation.)

Local discrimination, of sensation, 94; in touch, 106; in sight, 121.

Localisation of mental functions. (See Phrenology.)

Localisation, of organic sensations, 208, 262; of skin-sensations, 221, 262; of simultaneous touch-sensations, 223, 262; of bodily sensations aided by sight, 263; faulty, 263.

Locke, John, treatment of knowledge by, 4, 29; reflexion according to, 15; on defects of memory, 358 *note;* on clearness and distinctness of ideas, 433 *note;* on individual character, 307; on visual perception, 331; on association of ideas, 341; on time, 343; on pleasure and pain, 354.

Logic, how related to psychology, 12, 350; treatment of thought by, 410, 430, 434; of judgment by, 436; of reasoning by, 474.

Logical sentiment, characteristics of the, 124; forms of, 127.

Lombroso, C., on genius, 309 *note.*

Longinus, on the sublime, 147.

N.

*

R.

Richet, Dr. Ch., on temporary image, 279 *note*.
Right and wrong, perception of, **155, 168**; criterion of, **362.**
Rivalry of impulses, **250.**
Robertson, Prof. Croom, on Hobbes' doctrine of sense-perception, **330** *note.*
Romanes, G. J., on animal mind, 21; on generic images of lower animals, 415 *note ;* on instinct, **187.**
Rote, learning by, 315.
Roughness, perception of, 232.
Routine, as exemplification of habit, **228.**

S.

Schiff, M., on anæsthesia, **10**; on paths of sensation and pain, **14.**
Schneider, G. H., germ of intellection in lowest types of consciousness asserted by, 69 *note,* **329** *note ;* on impulse as basis of voluntary action, **195** *note.*
Schopenhauer, A., on pleasure, **18, 353**; on self-esteem, **102**; on art-contemplation, **135** *note ;* on will, **329.**
Schumann, F., on weight, 231.
Science, function of imagination in, 374.
Secondary qualities of bodies, 215.
Self, differentiation of, from not-self, 235, **223**; bodily organism as, 264, 476; development of idea of, 475, **224**; inner or mental, 477; idea of, as enduring, 479; feeling of, **97.**
Self-consciousness, how related to consciousness, 8, 77; relation of, to pleasure and pain, **13**; relation of self-feeling to, **99**; as element in higher volition, **291, 293**; disturbance of, **323.**
Self-control, process of, **263**; particular forms of, **266, 268, 271**; limits of, **278.**
Self-feeling, nature of, **97**; development of, **98**; relation of, to self-consciousness, **99**; teleology of, **102**; as element in moral sentiment, **167.**
Self-esteem, feeling of, **101.**
Semi-circular canals, function of, 114, 266.
Sensation, defined, 81; presentative and affective element in, 82, **9**; general or common, 83; intensity of, 86; quality of, 90; extensity or local distinctness of, 94; duration of, 97; organic variations of, 98; of taste, 99; of smell, 101; of touch, 103; thermal, 107; auditory, 109; musical, 111; visual, 115; colour, 116; muscular, 122; of movement, 127; of resistance, 130; concomitant, 136; secondary, 136, **312**; relation of attention to, 143; elaboration of, 169; differentiation of, 171; discrimination of, 174; assimilation of, 179; recognition of, 181; integration of, 185; associative revival of, 189; relation of perception to, 206; localisation of, 208; comparative revivability of, 355.
Sense-feeling, nature of, 83; definition of, **47**; varieties of, **48.**
Sense-organ, definition of, 85.
Senses, the special, 85; the series of, 99.
Sensibility, defined, 82; absolute, 87; discriminative, 89, 403.
Sensuous element in beauty, **138.**
• Sentiments, characteristics of abstract, **123.**
Sergi, Prof. G., on physiology of perception, 213 *note.*
Sight, sense of, 115; pleasures and pains of, **82.** (See Visual Perception.)
Similarity, as law of reproduction, 330; relation of contiguous suggestion to that by, 334; historical view of place of, among laws of association, **340.** (See Likeness and Assimilation.)
Single vision, 243.
Singular judgments, 436.

*